Methods in Cell Biology

VOLUME 73

Cumulative Index

Volume 53–71

Series Editors

Leslie Wilson
Department of Biological Sciences
University of California
Santa Barbara, California

Paul Matsudaira
Whitehead Institute for Biomedical Research
Department of Biology
Massachusetts Institute of Technology
Cambridge, Massachusetts

Methods in Cell Biology

Prepared under the auspices of the American Society for Cell Biology

VOLUME 73

Cumulative Index

Volumes 53–71

ELSEVIER
ACADEMIC PRESS

AMSTERDAM • BOSTON • HEIDELBERG • LONDON
NEW YORK • OXFORD • PARIS • SAN DIEGO
SAN FRANCISCO • SINGAPORE • SYDNEY • TOKYO
Academic Press is an imprint of Elsevier

Academic Press
An Elsevier Imprint.
525 B Street, Suite 1900, San Diego, California 92101-4495, USA
http://www.academicpress.com

Academic Press
84 Theobald's Road, London WC1X 8RR, UK
http://www.academicpress.com

International Standard Book Number: 0-12-411218-8

PRINTED IN THE UNITED STATES OF AMERICA
Transferred to Digital Printing, 2011

Cumulative Subject Index 1
Contributor Index 203

CONTENTS OF VOLUMES 53–71

VOLUME 53

Nuclear Structure and Function

Miguel Berrios

PART I Cell Fractionation Preparation and Characterization of Probes

1. Isolation and Characterization of Karyoskeletal Protein-Enriched Fractions from Vertebrate Livers
Miguel Berrios 3

2. Preparation of Karyoskeletal Protein-Enriched Fractions from *Drosophila melanogaster* Cells and Tissues *Saccharomyces cerevisiae*
Paul A. Fisher 23

3. Isolation of Nuclei and Nucleoli from the Yeast
J. E. Dove, J. S. Brockenbrough, and J. P. Aris 33

PART II Structural Analysis of the Interphase Nucleus

4. Determining Nuclear Structure Using the Fluorescence Light Microscope
Michael R. Paddy 49

5. Localization of Single Nuclear Pore Complexes by Confocal Laser Scanning Microscopy and Analysis of Their Distribution
Ulrich Kubitscheck and Reiner Peters 79

6. Nuclear Ultrastructure: Transmission Electron Microscopy and Image Analysis
Andrew S. Belmont 99

7. Three-Dimensional Surface Structure Analysis of the Nucleus
T. D. Allen, S. A. Rutherford, G. R. Bennion, C. Wiese, S. Riepert, E. Kiseleva, and M. W. Goldberg 125

8. Scanning Transmission Electron Microscopy of Nuclear Structures
Joseph S. Wall, James F. Hainfield, and Martha N. Simon 139

PART III Chromatin and Subnuclear Structures

9. Electron Microscopic Imaging of Chromatin with Nucleosome Resolution
C. L. Woodcock and R. A. Horowitz 167

10. Mapping Three-Dimensional Chromosome Architecture in Situ
Abby F. Dernburg and John W. Sedat 187

11. Structural Analysis of Meiotic Chromosomes and Synaptonemal Complexes in Higher Vertebrates
Alberto J. Solari 235

12. Genetic and Morphological Approaches for the Analysis of Meiotic Chromosomes in Yeast
Josef Loidl, Franz Klein, and JoAnne Engebrecht 257

13. Mapping Proteins to Nuclear Pore Complexes by Immunogold Electron Microscopy
Monika Grote and Reiner Peters 287

14. Methods Used to Study Structure and Function of the Nucleolus
Robert L. Ochs 303

15. Identification of Base-Unpairing Region-Binding Proteins and Characterization of Their *in Vivo* Binding Sequences
T. Kohwi-Shigematsu, I. deBelle, L. A. Dickinson, S. Galande, and Y. Kohwi 323

PART IV Nuclear Assembly/Disassembly in Cell-Free Systems

16. Cell-Free Nuclear Reassembly in Mammalian Mitotic Homogenates
Brian Burke 357

17. Analysis of Nuclear Envelope Assembly Using Extracts of *Xenopus* Eggs
M. J. Lohka 367

18. Cell-Free Nuclear Assembly and Disassembly in *Drosophila*
Paul A. Fisher and Miguel Berrios 397

19. Methods for Studying *in Vitro* Assembly of Male Pronuclei Using Oocyte Extracts from Marine Invertebrates: Sea Urchins and Surf Clams
Philippe Collas and Dominic Poccia 417

PART V Localization of Nuclear Processes

20. Mapping of DNA Replication Sites *in Situ* by Fluorescence Microscopy
Roel van Driel, Erik M. M. Manders, Luitzen de Jong, Jan Stap, and Jacob A. Aten 455

21. EM Visualization of Transcriptionally Active Genes after Injection into *Xenopus* Oocyte Nuclei
Yvonne N. Osheim and Ann L. Beyer 471

22. Cell-Free Systems to Study Chromatin Remodeling
Gregory H. Leno 497

23. In Vitro Systems for the Reconstitution of snRNP and Protein Nuclear Import
Colin Dingwall and Isabel Palacios 517

24. *In Vivo* Nuclear Transport Kinetics in *Saccharomyces cerevisiae*
Paul M. Roberts and David S. Goldfarb 545

25. Nuclear Transport of RNAs in Microinjected *Xenopus* Oocytes
Michael P. Terns and David S. Goldfarb 559

26. *In Vivo* Systems to Study the Dynamics of Nuclear Lamins
Ricardo Benavente and Georg Krohne 591

VOLUME 54
Cumulative Subject Index, Volumes 31–52
Leslie Wilson and Paul Matrudaira

VOLUME 55
Laster Tweezers in Cell Biology
Michael Sheetz

1. Forces of a Single-Beam Gradient Laser Trap on a Dielectric Sphere in the Ray Optics Regime
A. Ashkin 1

2. Basic Laser Tweezers
Ronald E. Sterba and Michael P. Sheetz 29

3. A Simple Assay for Local Heating by Optical Tweezers
Scot C. Kuo 43

4. Reflections of a Lucid Dreamer: Optical Trap Design Considerations
Amit D. Mehta, Jeffrey T. Finer, and James A. Spudich 47

5. Laser Scissors and Tweezers
Michael W. Berns, Yona Tadir, Hong Liang, and Bruce Tromberg 71

6. Optical Force Microscopy
Andrea L. Stout and Watt W. Webb 99

7. Single Molecule Imaging and Nanomanipulation of Biomolecules
Yoshie Harada, Takashi Funatsu, Makio Tokunaga, Kiwamu Saito, Hideo Higuchi, Yoshiharu Ishii, and Toshio Yanagida 117

8. Signals and Noise in Micromechanical Measurements
Frederick Gittes and Christoph F. Schmidt 129

9. Cell Membrane Mechanics
Jianwu Dai and Michael P. Sheetz 157

10. Application of Laser Tweezers to Studies of the Fences and Tethers of the Membrane Skeleton that Regulate the Movements of Plasma Membrane Proteins
Akihiro Kusumi, Yasushi Sako, Takahiro Fujiwara, and Michio Tomishige 173

11. *In Vivo* Manipulation of Internal Cell Organelles
Harald Felgner, Franz Grolig, Otto Müller, and Manfred Schliwa 195

12. Optical Chopsticks: Digital Synthesis of Multiple Optical Traps
Justin E. Molloy 205

VOLUME 56
Video Microscopy
Greenfield Sluder and David E. Wolf

1. Video Basics: Use of Camera and Monitor Adjustments
Greenfield Sluder and Edward H. Hinchcliffe 1

2. Electronic Cameras for Low-Light Microscopy
Keith Berland, Ken Jacobson, and Todd French 19

3. Cooled CCD Versus Intensified Cameras for Low-Light Video—Applications and Relative Advantages
Masafumi Oshiro 45

4. Techniques for Optimizing Microscopy and Analysis through Digital Image Processing
Ted Inoué and Neal Gliksman 63

5. Introduction to Image Processing
Richard A. Cardullo and Eric J. Alm 91

6. Quantitative Video Microscopy
David E. Wolf 117

7. Proper Alignment of the Microscope
H. Ernst Keller 135

8. Mating Cameras To Microscopes
Jan Hinsch 147

9. High-Resolution Video-Enhanced Differential Interference Contrast (VE-DIC) Light Microscopy
E. D. Salmon and Phong Tran 153

10. A High-Resolution Multimode Digital Microscope System
E. D. Salmon, Sidney L. Shaw, Jennifer Waters, Clare M. Waterman-Storer, Paul S. Maddox, Elaine Yeh, and Kerry Bloom 185

11. Ratio Imaging Instrumentation
Kenneth Dunn and Frederick R. Maxfield 217

12. Ratio Imaging: Practical Considerations for Measuring Intracellular Calcium and pH in Living Tissue
Randi B. Silver 237

13. Perfusion Chambers for High-Resolution Video Light Microscopic Studies of Vertebrate Cell Monolayers: Some Considerations and a Design
Conly L. Rieder and Richard W. Cole 253

14. Fluorescence Lifetime Imaging Techniques for Microscopy
Todd French, Peter T. C. So, Chen Y. Dong, Keith M. Berland, and Enrico Gratton 278

15. Digital Deconvolution of Fluorescence Images for Biologists
Yu-li Wang 305

VOLUME 57
Animal Cell Culture Methods
Jennie Mather and David Barens

SECTION I Principles of Cell Culture

1. Animal Cell Culture Equipment and Techniques
Angela Helmrich and David Barnes 3

2. Making Informed Choices: Medium, Serum, and Serum-Free Medium How to Choose the Appropriate Medium and Culture System for the Model You Wish to Create
Jennie P. Mather 19

3. Cell Line Availability: Where to Get the Cell Lines You Need
Robert J. Hay 31

4. Cell Culture Contamination: Sources, Consequences, Prevention, and Elimination
Carolyn Kay Lincoln and Michael G. Gabridge 49

SECTION II Establishing Cell Lines

5. Immortalization by Gene Transfection
Yoshinori Katakura, Shahabuddin Alam, and Sanetaka Shirahata 69

6. Establishment of Mammalian Testicular Cell Lines
Marie-Claude C. Hoffmann and Jose Luis Millan 93

7. Cell Hybridization, Hybridomas, and Human Hybridomas
Sanetaka Shirahata, Yoshinori Katakura, and Kiichiro Teruya 111

8. Establishing Human Glioma-Derived Cell Lines
Manfred Westphal and Hildegard Meissner 147

9. Culture Methods for Selective Growth of Normal Rat and Human Schwann Cells
Ronghao Li 167

10. Invertebrate Cell Culture Considerations: Insects, Ticks, Shellfish, and Worms
Christopher J. Bayne 187

11. Cell Line Characterization and Authentication
Joseph Kaplan and Bharati Hukku 203

SECTION III Specialized Culture Techniques

12. Laboratory Scale-up of Cell Cultures (0.5–50 Liters)
Jennie P. Mather 219

13. Cell Synchronization
Gary F. Merrill 229

14. Measurement of Cell Death
Deryk T. Loo and Jill R. Rillema 251

15. Simultaneous Measurement of Cell Cycle and Apoptotic Cell Death
Alison Moore, Christopher J. Donahue, Kenneth D. Bauer, and Jennie P. Mather 265

16. Embryonic Stem Cells, Creating Transgenic Animals
Melinda Pirity, Anna-Katerina Hadjantonakis, and András Nagy 279

SECTION IV Microscopy and Morphology

17. Electron Microscopy: Use of Transmission and Scanning Electron Microscopy to Study Cells in Culture
David M. Phillips 297

18. Indirect Immunofluorescence Microscopy in Cultured Cells
Sally P. Wheatly and Yu-li Wang 313

19. Cellular Localization of mRNA and Protein: *In Situ* Hybridization Histochemistry and *in Situ* Ligand Binding
Teresa K. Woodruff 333

VOLUME 58
Green Fluorescent Proteins
Kevin Sullivan and Steve Kay

1. Biophysics of the Green Fluorescent Protein
F. G. Prendergast 1

2. Understanding Structure–Function Relationships in the *Aequorea victoria* Green Fluorescent Protein
Andrew B. Cubitt, Leslie A. Woollenweber, and Roger Heim 19

3. Quantitative Imaging of the Green Fluorescent Protein (GFP)
David W. Piston, George H. Patterson, and Susan M. Knobel 31

4. Single-Molecule Fluorescence Detection of Green Fluorescence Protein and Application to Single-Protein Dynamics
Daniel W. Pierce and Ronald D. Vale 49

5. Targeting GFP to Organelles
Francesca De Giorgi, Zimran Ahmed, Carlo Bastianutto, Marisa Brini, Laurence Sophie Jouaville, Robert Marsault, Marta Murgia, Paolo Pinton, Tullio Pozzan, and Rosario Rizzuto 75

6. Cytoskeletal Dynamics in Yeast
Janet L. Carminati and Tim Stearns 87

7. Analysis of Nuclear Transport *in Vivo*
Paul Ferrigno and Pamela A. Silver 107

8. GFP Fusion Proteins as Probes for Cytology in Fission Yeast
Kenneth E. Sawin 123

9. GFP Variants for Multispectral Imaging of Living Cells
James Haseloff 139

10. GFP Fusions to a Microtubule Motor Protein to Visualize Meiotic and Mitotic Spindle Dynamics in *Drosophila*
Sharyn A. Endow 153

11. GFP as a Cell and Developmental Marker in the *Drosophila* Nervous System
Andrea Brand 165

12. Using Time-Lapse Confocal Microscopy for Analysis of Centromere Dynamics in Human Cells
Kevin F. Sullivan and Richard D. Shelby 183

13. Visualization of Large-Scale Chromatin Structure and Dynamics Using the *lac* Operator/*lac* Repressor Reporter System
Andrew S. Belmont, Gang Li, Gail Sudlow, and Carmen Robinett 203

14. Centrosome Dynamics in Living Cells
Aaron Young, Richard Tuft, Walter Carrington, and Stephen J. Doxsey 223

15. Transfections of Primary Muscle Cell Cultures with Plasmids Coding for GFP Linked to Full-Length and Truncated Muscle Proteins
Guissou A. Dabiri, Kenan K. Turnacioglu, Joseph C. Ayoob, Jean M. Sanger, and Joseph W. Sanger 239

16. Monitoring the Dynamics and Mobility of Membrane Proteins Tagged with Green Fluorescent Protein
J. Lippincott-Schwartz, J. F. Presley, K. J. M. Zaal, K. Hirschberg, C. D. Miller, and J. Ellenberg 261

17. Synchronous Real-Time Reporting of Multiple Cellular Events
Jeffrey D. Plautz and Steve A. Kay 283

18. Visualizing Protein Interactions in Living Cells Using Digitized GFP Imaging and FRET Microscopy
Ammasi Periasamy and Richard N. Day 293

19. Flow Cytometric Analysis and FACS Sorting of Cells Based on GFP Accumulation
David W. Galbraith, Michael T. Anderson, and Leonard A. Herzenberg 315

20. GFP Biofluorescence: Imaging Gene Expression and Protein Dynamics in Living Cells
Paul C. Goodwin 343

VOLUME 59
The Zebrafish: Biology
H. William Detrich III, Monte Westerfield, and Leonard I. Zon

PART I Introduction

1. Overview of the Zebrafish System
H. William Detrich, III, Monte Westerfield, and Leonard I. Zon 3

2. Cell Cycles and Development in the Embryonic Zebrafish
Donald A. Kane 11

PART II Cell Culture and General Methods

3. Zebrafish Embryonal Cell Culture
Angela Helmrich and David Barnes 29

4. Primary Fibroblast Cell Culture
Barry H. Paw and Leonard I. Zon 39

5. Production of Haploid and Diploid Androgenetic Zebrafish (Including Methodology for Delayed *in Vitro* Fertilization)
Graham E. Corley-Smith, Bruce P. Brandhorst, Charline Walker, and John H. Postlethwait 45

PART III Gene Expression and Function in Development

6. Analysis of Protein and Gene Expression
Trevor Jowett 63

7. Strategies to Perturb Zebrafish Development
Matthias Hammerschmidt, Patrick Blader, and Uwe Strähle 87

8. Vectors and Techniques for Ectopic Gene Expression in Zebrafish
Tana M. Hyatt and Stephen C. Ekker 117

9. Analysis of Zebrafish Development Using Explant Culture Assays
Yevgenya Grinblat, Mary Ellen Lane, Charles Sagerstrom, and Hazel Sive 127

PART IV Early Embryonic Development

10. Embryonic Axis Formation in the Zebrafish
Mary C. Mullins 159

11. Confocal Microscopic Analysis of Morphogenetic Movements
Mark S. Cooper, Leonard A. D'Amico, and Clarissa A. Henry 179

12. Cytoskeletal Dynamics of the Zebrafish Embryo
Jacek Topczewski and Lilianna Solnica-Krezel 205

13. Kinesin-like Microtubule Motors in Early Development
Ming-Chyuan Chen and H. William Detrich, III 227

PART V Organogenesis

14. Techniques in Neural Development
Cecilia B. Moens and Andreas Fritz 253

15. Development of the Retina
Jarema Malicki 273

16. Growth Control in the Ontogenetic and Regenerating Zebrafish Fin
Stephen L. Johnson and Paul Bennett 301

17. Vascular and Blood Gene Expression
Leon H. Parker, Leonard I. Zon, and Didier Y. R. Stainier 313

18. Analysis of Hemostasis in the Zebrafish
Pudur Jagadeeswaran, Yuan C. Liu, and John P. Sheehan 337

19. Cell Lineage Tracing in Heart Development
Fabrizio C. Serluca and Mark C. Fishman 359

20. Neurogenesis in Zebrafish Embryos
Ajay B. Chitnis and Igor B. Dawid 367

VOLUME 60
The Zebrafish: Genetics and Genomics
H. Detrich, III, Leonard Zon, and Monte Westerfield

1. A Gynogenesis-Based Screen for Maternal-Effect Genes in the Zebrafish, *Danio Rerio*
Francisco Pelegri and Stefan Schulte-Merker 1

2. Developmental Mutant Screens in the Zebrafish
Fredericus J. M. van Eeden, Michael Granato, Jorg Odenthal, and Pascal Haffter 21

3. Haploid Screens and Gamma-Ray Mutagenesis
Charline Walker 43

4. Early Pressure Screens
Christine E. Beattie, David W. Raible, Paul D. Henion, and Judith S. Eisen 71

5. Retrovirus-Mediated Insertional Mutagenesis in Zebrafish
Adam Amsterdam and Nancy Hopkins 87

6. Genetic Applications of Transposons and Other Repetitive Elements in Zebrafish
Zoltán Ivics, Zsuzsanna Izsvák, and Perry B. Hackett 99

7. Transgenesis
Anming Meng, Jason R. Jessen, and Shuo Lin 133

8. The Zebrafish Genome
John Postlethwait, Angel Amores, Allan Force, and Yi-Lin Yan 149

9. Using Random Amplified Polymorphic DNAs in Zebrafish Genomic Analysis
John H. Postlethwait, Yi-Lin Yan, and Michael A. Gates 165

10. Simple Sequence-Length Polymorphism Analysis
Eric C. Liao and Leonard I. Zon 181

11. Gene Mapping in Zebrafish Using Single-Strand Conformation Polymorphism Analysis
Dorothee Foernzler and David R. Beier 185

12. Mapping Zebrafish Mutations by AFLP
David G. Ransom and Leonard I. Zon 195

13. Zebrafish Expressed Sequence Tags and Their Applications
Zhiyuan Gong 213

14. Zebrafish YAC, BAC, and PAC Genomic Libraries
Chris T. Amemiya, Tao P. Zhong, Gary A. Silverman, Mark C. Fishman, and Leonard I. Zon 235

15. Positional Cloning of Mutated Zebrafish Genes
William S. Talbot and Alexander F. Schier 259

16. Construction and Characterization of Zebrafish Whole Genome Radiation Hybrids
Cheni Kwok, Ricky Critcher, and Karin Schmitt 287

17. Zebrafish/Mouse Somatic Cell Hybrids for the Characterization of the Zebrafish Genome
Marc Ekker, Fengchun Ye, Lucille Joly, Patricia Tellis, and Mario Chevrette 303

18. Banded Chromosomes and the Zebrafish Karyotype
Angel Amores and John H. Postlethwait 323

19. Zebrafish Informatics and the ZFIN Database
Monte Westerfield, Eckehard Doerry, Arthur E. Kirkpatrick, and Sarah A. Douglas 339

Appendix 1. Genetic Backgrounds and Some Standard Stocks and Strains Used in Zebrafish Developmental Biology and Genetics
Stephen L. Johnson and Leonard I. Zon 357

Appendix 2. Centromeric Markers in the Zebrafish
Don Kane, Leonard I. Zon, and H. William Detrich, III 363

Appendix 3. Collection, Storage, and Use of Zebrafish Sperm
David G. Ransom and Leonard I. Zon 365

Appendix 4. Zebrafish Web Site Listings
Pat Edwards 373

VOLUME 61
Mitosis and Meosis
Conly Rieder

1. Isolation of Centrosomes from *Drosophila* Embryos
Michelle Moritz and Bruce M. Alberts 1

2. Studying the Composition and Function of Centrosomes in Vertebrates
Michel Bornens and Mohammed Moudjou 13

3. Isolation of Centrosomes from *Spisula solidissima* Oocytes
Robert E. Palazzo and Jacalyn M. Vogel 35

4. Methods for *in Situ* Localization of Proteins and DNA in the Centromere–Kinetochore Complex
A. Van Hooser and B. R. Brinkley 57

5. Three-Dimensional Transmission Electron Microscopy and Its Application to Mitosis Research
Bruce F. McEwen and Michael Marko 81

6. Enlightening Mitosis: Construction and Expression of Green Fluorescent Protein Fusion Proteins
Kevin F. Sullivan 113

7. Recombinant p50/Dynamitin as a Tool to Examine the Role of Dynactin in Intracellular Processes
Torsten Wittmann and Tony Hyman 137

8. *In Vitro* Assays for Studying *Saccharomyces cerevisiae* Kinetochore Activity
Fedor Severin, Ken Kaplan, Peter Sorger, and Tony Hyman 145

9. Fluorescent Speckle Microscopy of Spindle Microtubule Assembly and Motility in Living Cells
Clare Waterman-Storer, Arshad Desai, and E. D. Salmon 155

10. Polarized Light Microscopy of Spindles
Rudolf Oldenbourg 175

11. Micromanipulation of Chromosomes and Spindles in Insect Spermatocytes
Dahong Zhang and R. Bruce Nicklas 209

12. Microinjection of Mitotic Cells
Patricia Wadsworth 219

13. Obtaining Antibodies to Spindle Components
Ryoko Kuriyama and Kathy Ensrud 233

14. Using Antisense Technology to Study Mitosis
Linda Wordeman and Mike Wagenbach 245

15. The Use and Action of Drugs in Analyzing Mitosis
Mary Ann Jordan and Leslie Wilson 267

16. Correlative Light and Electron Microscopy of Mitotic Cells in Monolayer Cultures
Conly L. Rieder and Grisel Cassels 297

17. Identification and Characterization of Mitotic Mutations in *Drosophila*
William E. Theurkauf and Margarete M. S. Heck 317

18. Methods for Isolating and Analyzing Mitotic Mutants in *Aspergillus nidulans*
Berl R. Oakley 347

19. Using Green Fluorescent Protein Fusion Proteins to Quantitate Microtubule and Spindle Dynamics in Budding Yeast
Kerry Bloom, Dale L. Beach, Paul Maddox, Sidney L. Shaw, Elaine Yeh, and E. D. Salmon 369

20. The Use of *Xenopus* Egg Extracts to Study Mitotic Spindle Assembly and Function *in Vitro*
Arshad Desai, Andrew Murray, Timothy J. Mitchison, and Claire E. Walczak 385

21. Methods for Studying Cell Division in Higher Plants
Jan W. Vos, Aline H. Valster, and Peter K. Hepler 413

22. Using Sea Urchin Gametes for the Study of Mitosis
Greenfield Sluder, Frederick J. Miller, and Edward H. Hinchcliffe 439

VOLUME 62
Tetrahymena Thermophila
David Asai and James Forney

PART I Overview

1. Laboratory and Evolutionary History of *Tetrahymena thermophila*
David L. Nanney and Ellen M. Simon 3

2. Cell Biology of *Tetrahymena thermophila*
Joseph Frankel 27

3. *Tetrahymena* Genetics: Two Nuclei Are Better Than One
Kathleen M. Karrer 127

PART II Methods

4. *Tetrahymena* as a Laboratory Organism: Useful Strains, Cell Culture, and Cell Line Maintenance
Eduardo Orias, Eileen P. Hamilton, and Judith D. Orias 189

5. Long-Term Storage
Peter J. Bruns, Helen R. Smith, and Donna Cassidy-Hanley 213

6. Genetic Crosses: Setting Up Crosses, Testing Progeny, and Isolating Phenotypic Assortants
Eileen P. Hamilton and Eduardo Orias 219

7. Methods for Genetic Analysis
Peter J. Bruns and Donna Cassidy-Hanley 229

8. Isolation of Micronuclear and Macronuclear DNA
Sally Lyman Allen 241

9. Genetically Sorting a Collection of *Tetrahymena* Mutants
Eduardo Orias and Eileen P. Hamilton 253

10. Genetically Mapping New Mutants and Cloned Genes
Eileen P. Hamilton and Eduardo Orias 265

11. Selection of Motility Mutants
David G. Pennock 281

12. Nuclear and Cytoskeletal Fluorescence Microscopy Techniques
K. R. Stuart and E. S. Cole 291

13. Nuclear and Cortical Histology for Brightfield Microscopy
E. S. Cole and K. R. Stuart 313

14. Fixation of *Tetrahymena* Cells for Electron Microscopy
William Dentler 323

15. Immunoelectron Microscopy of *Tetrahymena*
R. H. Gavin, John G. Hoey, and Jorge A. Garcés 333

PART III Cell Biology

16. Regulated Protein Secretion in *Tetrahymena thermophila*
Aaron P. Turkewitz, N. Doane Chilcoat, Alex Haddad, and John W. Verbsky 347

17. Electrophysiology of *Tetrahymena*
Todd M. Hennessey and Heather G. Kuruvilla 363

18. Isolation and Characterization of in Vivo Modified Histones and an Activity Gel Assay for Identification of Histone Acetyltransferases
Emily A. Wiley, Craig A. Mizzen, and C. David Allis 379

19. *Tetrahymena* Telomerase Activity, Purification, and Reconstitution
Chantal Autexier 395

20. Studying the Telomerase RNA in *Tetrahymena*
E. Blackburn, D. Gilley, T. Ware, A. Bhattacharyya, K. Kirk, and H. Wang 417

21. Isolation and Characterization of 22S Outer Arm Dynein from *Tetrahymena* Cilia
Tracie M. Gibson and David J. Asai 433

22. Preparation of Cytoskeletal Fractions from *Tetrahymena thermophila*
Norman E. Williams 441

23. Immunoprecipitation Procedures
Norman E. Williams 449

24. *Tetrahymena* Calcium-Binding Proteins, TCBP-25 and TCBP-23
Osamu Numata, Kazuko Hanyu, Tetsuya Takeda, and Yoshio Watanabe 455

PART IV Manipulating Genes

25. Microinjection of *Tetrahymena thermophila*
Douglas L. Chalker, John G. Ward, Caterina Randolph, and Meng-Chao Yao 469

26. Transient and Stable DNA Transformation of *Tetrahymena thermophila* by Electroporation
Jacek Gaertig and Geoffrey Kapler 485

27. Biolistic Transformation of Macro- and Micronuclei
Peter J. Bruns and Donna Cassidy-Hanley 501

28. Knockout Heterokaryons Enable Facile Mutagenic Analysis of Essential Genes in *Tetrahymena*
Bing Hai, Jacek Gaertig, and Martin A. Gorovsky 513

29. Creation and Use of Antisense Ribosomes in *Tetrahymena thermophila*
Qichang Fan, Rosemary Sweeney, and Meng-Chao Yao 533

30. Protein Tagging in *Tetrahymena*
Lanlan Yu and Martin A. Gorovsky 549

Appendix I. Genetic Nomenclature Rules for *Tetrahymena thermophilo* 561

Appendix II. Codon Usage in *Tetrahymena thermophila* 565

VOLUME 63
Cytometry, Part A
Zbigniew Darzynkiewicz, John Robinson, and Harry Crissman

PART I Principles of Cytometry and General Methods

1. A Brief History of Flow Cytometry and Sorting
Myron R. Melamed 3

2. Principles of Flow Cytometry: An Overview
Alice L. Givan 19

3. Laser Scanning Cytometry
Louis A. Kamentsky 51

4. Principles of Confocal Microscopy
J. Paul Robinson 89

5. Optical Measurements in Cytometry: Light Scattering, Extinction, Absorption, and Fluorescence
Howard M. Shapiro 107

6. Flow Cytometric Fluorescence Lifetime Measurements
Harry A. Crissman and John A. Steinkamp 131

7. Principles of Data Acquisition and Display
Howard M. Shapiro 149

8. Time as a Flow Cytometric Parameter
Larry Seamer and Larry A. Sklar 169

9. Protein Labeling with Fluorescent Probes
Kevin L. Holmes and Larry M. Lantz 185

PART II Cell Preparation

10. Preparation of Cells from Blood
J. Philip McCoy, Jr. 207

11. Cell Preparation for the Identification of Leukocytes
Carleton C. Stewart and Sigrid J. Stewart 217

12. Strategies for Cell Permeabilization and Fixation in Detecting Surface and Intracellular Antigens
Steven K. Koester and Wade E. Bolton 253

PART III Standardization, Quality Assurance

13. Stoichiometry of Immunocytochemical Staining Reactions
James W. Jacobberger 271

14. Standardization and Quantitation in Flow Cytometry
Robert A. Hoffman 299

PART IV Cell Proliferation

15. Methods to Identify Mitotic Cells by Flow Cytometry
Gloria Juan, Frank Traganos, and Zbigniew Darzynkiewicz 343

16. Cell Cycle Kinetics Estimated by Analysis of Bromodeoxyuridine Incorporation
Nicholas H. A. Terry and R. Allen White 355

17. Flow Cytometric Analysis of Cell Division History Using Dilution of Carboxyfluorescein Diacetate Succinimidyl Ester, a Stably Integrated Fluorescent Probe
A. Bruce Lyons, Jhagvaral Hasbold, and Philip D. Hodgkin 375

18. Antibodies against the Ki-67 Protein: Assessment of the Growth Fraction and Tools for Cell Cycle Analysis
Elmar Endl, Christiane Hollmann, and Johannes Gerdes 399

19. Detection of Proliferating Cell Nuclear Antigen
Jørgen K. Larsen, Goran Landberg, and Goran Roos 419

20. Lymphocyte Activation Associated Antigens
Andrea Fattorossi, Alessandra Battaglia, and Cristiano Ferlini 433

PART V Cell Death/Apoptosis

21. Analysis of Mitochondria during Cell Death
Andrea Cossarizza and Stefano Salvioli 467

22. Cytometry of Caspases
Steven K. Koester and Wade E. Bolton 487

23. Analysis of Apoptosis in Plant Cells
Iona E. Weir 505

24. Difficulties and Pitfalls in Analysis of Apoptosis
Zbigniew Darzynkiewicz, Elżbieta Bedner, and Frank Traganos 527

PART VI Cell–Cell, Cell–Environment Interactions

25. Analysis of Cell Migration
Nicole Dodge Zantek and Michael S. Kinch 549

26. Three-Dimensional Extracellular Matrix Substrates for Cell Culture
Sherry L. Voytik-Harbin 561

27. Three-Dimensional Imaging of Extracellular Matrix and Extracellular Matrix–Cell Interactions
Sherry L. Voytik-Harbin, Bartlomiej Rajwa, and J. Paul Robinson 583

28. Cytometric Analysis of Cell Contact and Adhesion
Michael S. Kinch 599

29. Invadopodia: Unique Methods for Measurement of Extracellular Matrix Degradation *in Vitro*
Emma T. Bowden, Peter J. Coopman, and Susette C. Mueller 613

VOLUME 64
Cytometry, Part B

Zibgniew Darzynkiewicz, John Robinson, and Harry Crissman

PART VII Cytogenetics and Molecular Genetics

30. Sorting of Plant Chromosomes
Jaroslav Doležel, Martin A. Lysák, Marie Kubaláková, Hana Šimková, Jií Macas, and Sergio Lucretti 3

31. Quantitative DNA Fiber Mapping
Heinz-Ulli G. Weier 33

32. Primed *in Situ* Labeling
Johnny Hindkjaer, Lars Bolund, and Steen Kølvraa 55

33. Measurements of Telomere Length on Individual Chromosomes by Image Cytometry
Steven S. S. Poon and Peter M. Lansdorp 69

34. Detection of Chromosome Translocation Products in Single Interphase Cell Nuclei
Jingly Fung, Santiago Munné, and Heinz-Ulli G. Weier 97

PART VIII Cell Function and Differentiation

35. Analysis of Mitochondria by Flow Cytometry
Martin Poot and Robert H. Pierce 117

36. Analysis of RNA Synthesis by Cytometry
Peter Østrup Jensen, Jacob Larsen, and Jørgen K. Larsen 129

37. Flow Cytometry of Erythropoiesis in Culture: Bivariate Profiles of Fetal and Adult Hemoglobin
Ralph M. Bohmer 139

38. Flow Cytometric Analysis of Human Hemopoietic Progenitor Differentiation by Assessing Cell Division Rate and Phenotypic Profile
Luca Pierelli, Giovanni Scambia, and Andrea Fattorossi 153

PART IX Experimental Oncology

39. Cytometry of Antitumor Drug–Intracellular Target Interactions
Paul J. Smith and Marie Wiltshire 173

40. Monitoring of Cellular Resistance to Cancer Chemotherapy: Drug Retention and Efflux
Awtar Krishan 193

41. Resistance of Tumor Cells to Chemo- and Radiotherapy Modulated by the Three-Dimensional Architecture of Solid Tumors and Spheroids
Ralph E. Durand and Peggy L. Olive 211

42. Analysis of DNA Damage in Individual Cells
Peggy L. Olive, Ralph E. Durand, Judit P. Banáth, and Peter J. Johnston 235

43. Cytometric Methods to Analyze Ionizing-Radiation Effects
William D. Wright, Isabelle Lagroye, Peng Zhang, Robert S. Malyapa, and Joseph L. Roti Roti 251

44. Cytometric Methods to Analyze Thermal Effects
Robert P. VanderWaal, Ryuji Higashikubo, Mai Xu, Douglas R. Spitz, William D. Wright, and Joseph L. Roti Roti 269

PART X Clinical Oncology

45. Multiparameter Data Acquisition and Analysis of Leukocytes by Flow Cytometry
Carleton C. Stewart and Sigrid J. Stewart 289

46. Immunophenotyping of Hematological Malignancies by Laser Scanning Cytometry
Richard J. Clatch 313

47. Immunophenotyping of Acute Leukemia: Utility of CD45 344 for Blast Cell Identification
J-P. Vial and F. Lacombe 343

48. Cell Proliferation Markers in Human Solid Tumors: Assessing Their Impact in Clinical Oncology
Maria Grazia Daidone, Aurora Costa, and Rosella Silvestrini 359

49. Detection of Minimal Residual Disease
Andrzej Deptala and Sharon P. Mayer 385

50. Analysis of Human Tumors by Laser Scanning Cytometry
Wojciech Gorczyca, Andrzej Deptala, Elzbieta Bedner, Xun Li, Myron R. Melamed, and Zbigniew Darzynkiewicz 421

51. Laser Cytometry of Human Tissues and Tumors: Proliferation and Therapeutic Applications
David A. Rew 445

52. Prediction and Precise Diagnosis of Diseases by Data Pattern Analysis in Multiparameter Flow Cytometry: Melanoma, Juvenile Asthma, and Human Immunodeficiency Virus Infection
Günter Valet, Hanna Kahle, Friedrich Otto, Edeltraut Bräutigam, and Luc Kestens 487

PART XI Microorganisms and Infectious Diseases

53. Flow Cytometric Analysis of Microorganisms
S. A. Sincock and J. Paul Robinson 511

54. Staining and Measurement of DNA in Bacteria
Harald B. Steen 539

55. Flow Cytometric Monitoring of Bacterial Susceptibility to Antibiotics
Mette Walberg and Harald B. Steen 553

56. Flow Cytometry for Evaluation and Investigation of Human Immunodeficiency Virus Infection
Thomas W. Mc Closkey 567

VOLUME 65
Mitochondria

Liza Pon and Eric Schon

1. Isolation and Subfractionation of Mitochondria from Animal Cells and Tissue Culture Lines
Francesco Pallotti and Giorgio Lenaz 1

2. Isolation and Subfractionation of Mitochondria from the Yeast *Saccharomyces cerevisiae*
Kerstin Diekert, Anton I. P. M. de Kroon, Gyula Kispal, and Roland Lill 37

3. Isolation and Subfractionation of Mitochondria from Plants
A. H. Millar, A. Liddell, and C. J. Leaver 53

4. Assessing Functional Integrity of Mitochondria *in Vitro* and *in Vivo*
Mauro Degli Esposti 75

5. Assaying Mitochondrial Respiratory Complex Activity in Mitochondria Isolated from Human Cells and Tissues
Mark A. Birch-Machin and Douglass M. Turnbull 97

6. *In Vivo* Measurements of Respiration Control by Cytochrome c Oxidase and *in Situ* Analysis of Oxidative Phosphorylation
Gaetano Villani and Giuseppe Attardi 119

7. Assay of Mitochondrial ATP Synthesis in Animal Cells
Giovanni Manfredi, Antonella Spinazzola, Nicoletta Checcarelli, and Ali Naini 133

8. Measurement of Membrane Permeability and Permeability Transition of Mitochondria
Naoufal Zamzami, Carine Maisse, Didier Métivier, and Guido Kroemer 147

9. Assaying Actin-Binding Activity of Mitochondria in Yeast
Istvan R. Boldogh and Liza A. Pon 159

10. Analysis and Prediction of Mitochondrial Targeting Peptides
Olof Emanuelsson, Gunnar von Heijne, and Gisbert Schneider 175

11. Assaying Protein Import into Mitochondria
Michael T. Ryan, Wolfgang Voos, and Nikolaus Pfanner 189

12. Analysis of Protein–Protein Interactions in Mitochondria by Coimmunoprecipitation and Chemical Cross-Linking
Johannes M. Herrmann, Benedikt Westermann, and Walter Neupert 217

13. Blue-Native Gels to Isolate Protein Complexes from Mitochondria
Hermann Schägger 231

14. Application of Electron Tomography to Mitochondrial Research
Carmen A. Mannella 245

15. Epitope Tagging and Visualization of Nuclear-Encoded Mitochondrial Proteins in Yeast
Dan W. Nowakowski, Theresa C. Swayne, and Liza A. Pon 257

16. Targeting of Green Fluorescent Protein to Mitochondria
Koji Okamoto, Philip S. Perlman, and Ronald A. Butow 277

17. Assessment of Mitochondrial Membrane Potential *in Situ* Using Single Potentiometric Dyes and a Novel Fluorescence Resonance Energy Transfer Technique
James A. Dykens and Amy K. Stout 285

18. Optical Imaging Techniques (Histochemical, Immunohistochemical, and *in Situ* Hybridization Staining Methods) to Visualize Mitochondria
Kurenai Tanji and Eduardo Bonilla 311

19. Visualization of Mitochondrial Movement in Yeast
Hyeong-Cheol Yang, Viviana Simon, Theresa Swayne, and Liza Pon 333

20. Targeting of Reporter Molecules to Mitochondria to Measure Calcium, ATP, and pH
Anna M. Porcelli, Paolo Pinton, Edward K. Ainscow, Anna Chiesa, Michela Rugolo, Guy A. Rutter, and Rosario Rizzuto 353

21. Genetic Transformation of *Saccharomyces cerevisiae* Mitochondria
Nathalie Bonnefoy and Thomas D. Fox 381

22. Transmitochondrial Technology in Animal Cells
Carlos T. Moraes, Runu Dey, and Antoni Barrientos 397

23. Diagnostic Assays for Defects in Mitochondrial DNA Replication and Transcription in Yeast and Human Cells
Bonnie L. Seidel-Rogol and Gerald S. Shadel 413

24. Analysis of Mitochondrial Translation Products *in Vivo* and *in Organello* in Yeast
Benedikt Westermann, Johannes M. Herrmann, and Walter Neupert 429

25. Numerical Methods for Handling Uncertainty in Microarray Data: An Example Analyzing Perturbed Mitochondrial Function in Yeast
Charles B. Epstein, Walker Hale IV, and Ronald A. Butow 439

VOLUME 66
Apotosis

Lawrence Schwartz and Jonathan Ashwell

1. Isolation and Assay of Caspases
Srinivasa M. Srinivasula, Ayman Saleh, Manzoor Ahmad, Teresa Fernandes-Alnemri, and Emad S. Alnemri 1

2. Cloning and Analysis of Bcl-2 Family Genes
Enrique Cepero, Bryan W. Johnson, and Lawrence H. Boise 29

3. Flow Cytometric Analysis of Cell Shrinkage and Monovalent Ions during Apoptosis
Carl D. Bortner and John A. Cidlowski 49

4. Use of Flow and Laser-Scanning Cytometry in Analysis of Cell Death
Zbigniew Darzynkiewicz, Xun Li, and Elzbieta Bedner 69

5. Analysis of Protein Transglutamylation in Apoptosis
Zoltán Nemes, András Mádi, Lyuben N. Marekov, Mauro Piacentini, Peter M. Steinert, and Lászió Fésüs 111

6. Analysis of Sphingomyelin and Ceramide Levels and the Enzymes Regulating their Metabolism in Response to Cell Stress
Rick T. Dobrowsky and Richard N. Kolesnick 135

7. Cell-Free Systems to Study Apoptosis
Howard O. Fearnhead 167

8. Role of c-Jun N-terminal Kinase in Apoptosis
Zheng-gang Liu, Joseph Lewis, Tzu-Hao Wang, and Amy Cook 187

9. Methods for Studying Pro- and Antiapoptotic Genes in Nonimmortal Cells
Mila E. McCurrach and Scott W. Lowe 197

10. Calcium Flux Measurements in Apoptosis
David J. McConkey and Leta K. Nutt 229

11. Proteinase Families and Their Inhibitors
Donald L. Mykles 247

12. Identification and Analysis of Caspase Substrates: Proteolytic Cleavage of Poly(ADP-ribose)polymerase and DNA Fragmentation Factor 45
Claudia Boucher, Stéphane Gobeil, Kumiko Samejima, William C. Earnshaw, and Guy G. Poirier 289

13. Analysis of Reactive Oxygen Species in Cell Death
Ivan Stamenkovic 307

14. Methods for Studying Apoptosis and Phagocytosis of Apoptotic Cells in *Drosophila* Tissues and Cell Lines
Kristin White, Simonetta Lisi, Phani Kurada, Nathalie Franc, and Peter Bangs 321

15. Phosphatidylserine Exposure and Phagocytosis of Apoptotic Cells
Patrick Williamson, Stefan van den Eijnde, and Robert A. Schlegel 339

16. The (Holey) Study of Mitochrondria in Apoptosis
Nigel J. Waterhouse, Joshua C. Goldstein, Ruth M. Kluck, Don D. Newmeyer, and Douglas R. Green 365

17. *In Situ* Detection of Dying Cells in Normal and Pathological Tissues
Christos Valavanis, Stephen Naber, and Lawrence M. Schwartz 393

18. Model Cell Lines for the Study of Apoptosis *in Vitro*
Christos Valavanis, Yanhui Hu, Yili Yang, Barbara A. Osborne, Salem Chouaib, Lloyd Greene, Jonathan D. Ashwell, and Lawrence M. Schwartz 417

19. Programmed Cell Death Assays for Plants
Alan M. Jones, Silvia Coimbra, Angelika Fath, Mariana Sottomayor, and Howard Thomas 437

20. Studies of Apoptosis Proteins in Yeast
Hong Zhang and John C. Reed 453

21. Methods to Study Cell Death in *Dictyostelium discoideum*
Jean-Pierre Levraud, Myriam Adam, Sophie Cornillon, and Pierre Golstein 469

22. Methods of Study of Tumor Necrosis Factor-Related Ligands in Apoptosis
Isabelle A. Rooney, Chris A. Benedict, Paula S. Norris, and Carl F. Ware 499

VOLUME 67
Centrosomes and Spindle Pole Bodies
Robert Palazzo and Trisha Davis

1. Purification and Reconstitution of *Drosophila* γ-Tubulin Complexes
Ruwanthi N. Gunawardane, Yixian Zheng, Karen Oegema, and Christiane Wiese 1

2. Using Rapid Freeze and Freeze-Substitution for the Preparation of Yeast Cells for Electron Microscopy and Three-Dimensional Analysis
Thomas H. Giddings, Jr., Eileen T. O'Toole, Mary Morphew, David N. Mastronarde, J. Richard McIntosh, and Mark Winey 27

3. Digital Fluorescence Microscopy of Cell Cytoplasts with and without the Centrosome
Vladimir Rodionov, Elena Nadezhdina, John Peloquin, and Gary Borisy 43

4. Methods for the Study of Pericentrin in Centrosome Assembly and Function
Aruna Purohit, German A. Pihan, and Stephen J. Doxsey 53

5. Molecular Dissection of Yeast Spindle Pole Bodies by Two Hybrid, *in Vitro* Binding, and Co-purification
C. Schramm, C. Janke, and E. Schiebel 71

6. Genetic Analysis of Yeast Spindle Pole Bodies
Trisha N. Davis 95

7. Methods for the Study of Centrosomes in *Drosophila* during Embryogenesis
Uyen Tram, Blake Riggs, Carol Koyama, Alain Debec, and William Sullivan 113

8. Methods for Identification of Centrosome-Associated Proteins
Ryoko Kuriyama, Toshiro Ohta, Jacalyn Vogel, and Gang Peng 125

9. Reconstitution of Centrosome Microtubule Nucleation in *Drosophila*
Michelle Moritz, Michael B. Braunfeld, Bruce M. Alberts, and David A. Agard 141

10. Reconstitution of Centrosome Microtubule Nucleation in *Spisula*
Bradley J. Schnackenberg and Robert E. Palazzo 149

11. *In Vitro* Approaches for the Study of Microtubule Nucleation at the Fission Yeast Spindle Pole Body
Hirohisa Masuda, Saeko Takada, Takehiko Shibata, W. Zacheus Cande, and Yasushi Hiraoka 167

12. Gamma Tubulin and Microtubule Nucleation in Mammalian Cells
Harish C. Joshi and Jun Zhou 179

13. Gamma Tubulin in Plant Cells
Yulia Ovechkina and Berl R. Oakley 195

14. Centrosomes and Parthenogenesis
Frédéric Tournier and Michel Bornens 213

15. *In Vitro* Approaches for the Study of Molecular Motors in Aster Formation
Duane A. Compton 225

16. Methods for the Study of Centrosome-Independent Spindle Assembly in *Xenopus* Extracts
Sarah M. Wignall and Rebecca Heald 241

17. Methods for the Study of Centrosome Reproduction in Mammalian Cells
Ron Balczon 257

18. Centrosome Reproduction in *Xenopus* Lysates
Edward H. Hinchcliffe and Greenfield Sluder 269

19. Centrosome Reproduction *in Vitro*: Mammalian Centrosomes in *Xenopus* Lysates
Matthieu Piel and Michel Bornens 289

20. Identification of Centrosome Kinases
Andrew M. Fry and Alison J. Faragher 305

21. Methods for the Analysis of Centrosome Reproduction in Cancer Cells
Wilma L. Lingle and Jeffrey L. Salisbury 325

22. Isolation of Centrosomes from *Dictyostelium*
Ralph Gräf 337

VOLUME 68
Atomic Force Microscopy in Cell Biology
Bhanu Jena and J. K. Horber

1. Local Probe Techniques
J. K. Heinrich Horber 1

2. The Atomic Force Microscope in the Study of Membrane Fusion and Exocytosis
Bhanu P. Jena and Sang-Joon Cho 33

3. Atomic Force Microscope Imaging of Cells and Membranes
Eric Lesniewska, Pierre Emmanuel Milhiet, Marie-Cécile Giocondi, and Christian Le Grimellec 51

4. Measuring the Elastic Properties of Living Cells by the Atomic Force Microscope
Manfred Radmacher 67

5. Cell Adhesion Measured by Force Spectroscopy on Living Cells
Martin Benoit 91

6. Molecular Recognition Studies Using the Atomic Force Microscope
Peter Hinterdorfer 115

7. The Biophysics of Sensory Cells of the Inner Ear Examined by Atomic Force Microscopy and Patch Clamp
Matthias G. Langer and Assen Koitschev 141

8. Biotechnological Applications of Atomic Force Microscopy
Guillaume Charras, Petri Lehenkari, and Mike Horton 171

9. Cellular Membranes Studied by Photonic Force Microscopy
Arnd Pralle and Ernst-Ludwig Florin 193

10. Methods for Biological Probe Microscopy in Aqueous Fluids
Johannes H. Kindt, John C. Sitko, Lia I. Pietrasanta, Emin Oroudjev, Nathan Becker, Mario B. Viani, and Helen G. Hansma 213

11. Supported Lipid Bilayers as Effective Substrates for Atomic Force Microscopy
Daniel M. Czajkowsky and Zhifeng Shao 231

12. Cryo-Atomic Force Microscopy
Sitong Sheng and Zhifeng Shao 243

13. Conformations, Flexibility, and Interactions Observed on Individual Membrane Proteins by Atomic Force Microscopy
Daniel J. Müller and Andreas Engel 257

14. Single-Molecule Force Measurements
Aileen Chen and Vincent T. Moy 301

15. Forced Unfolding of Single Proteins
S. M. Altmann and P.-F. Lenne 311

16. Developments in Dynamic Force Microscopy and Spectroscopy
A. D. L. Humphris and M. J. Miles 337

17. Scanning Force Microscopy Studies on the Structure and Dynamics of Single DNA Molecules
Giampaolo Zuccheri and Bruno Samorì 357

VOLUME 69
Methods in Cell–Matrix Adhesion
Josephine Adams

PART I Preface and Perspectives on Cell–Matrix Adhesion

Preface
Josephine C. Adams 3

Cell–Matrix Interactions: The View from the Outside
Paul Bornstein 7

Matrix and Meaning
Martin A. Schwartz 13

Cell–Matrix Adhesion Research and the Development of Biotherapeutics
Roy R. Lobb 17

PART II Matrix Methodologies

1. Detection and Purification of Instructive Extracellular Matrix Components with Monoclonal Antibody Technologies
Susana G. Gil, Randy O. Sigle, and William G. Carter 27

2. Isolation and Purification of Proteoglycans
John M. Whitelock and Renato V. Iozzo 53

3. Expression of Recombinant Matrix Components Using Baculoviruses
Deane F. Mosher, Kristin G. Huwiler, Tina M. Misenheimer, and Douglas S. Annis 69

4. Heparan Sulfate–Growth Factor Interactions
Alan C. Rapraeger 83

5. Analysis of Basement Membrane Self-Assembly and Cellular Interactions with Native and Recombinant Glycoproteins
Peter D. Yurchenco, Sergei Smirnov, and Todd Mathus 111

6. Preparation and Analysis of Synthetic Multicomponent Extracellular Matrix
Kim S. Midwood, Iwona Wierzbicka-Patynowski, and Jean E. Schwarzbauer 145

7. Analysis of Matrix Dynamics by Atomic Force Microscopy
Helen G. Hansma, Dennis O. Clegg, Efrosini Kokkoli, Emin Oroudjev, and Matthew Tirrell 163

8. Analysis of Matrix Degradation
H. G. Munshi and M. Sharon Stack 193

PART III Adhesion Receptor Methodologies

9. Expression Cloning Strategies for the Identification of Adhesion Molecules
Joe W. Ramos and Mark Ginsberg 209

10. Purification of Integrins and Characterization of Integrin-Associated Proteins
Johannes A. Eble and Fedor Berditchevski 223

11. Methods for Analysis of the Integrin Ligand Binding Event
Jeffrey W. Smith 247

12. Intracellular Coupling of Adhesion Receptors: Molecular Proximity easurements
Maddy Parsons and Tony Ng 261

PART IV Functional Applications of Cell–Matrix Adhesion in Molecular Cell Biology

13. Functional Analysis of Cell Adhesion: Quantitation of Cell–Matrix Attachment
Steven K. Akiyama 281

14. Measurements of Glycosaminoglycan-Based Cell Interactions
J. Kevin Langford and Ralph D. Sanderson 297

15. Applications of Adhesion Molecule Gene Knockout Cell Lines
Jordan A. Kreidberg 309

16. Flexible Polyacrylamide Substrata for the Analysis of Mechanical Interactions at Cell–Substratum Adhesions
Karen A. Beningo, Chun-Min Lo, and Yu-Li Wang 325

17. Cell Migration in Slice Cultures
Donna J. Webb, Hannelore Asmussen, Shin-ichi Murase, and Alan F. Horwitz 341

18. Application of Cell Adhesion to Study Signaling Networks
Cindy K. Miranti 359

19. Use of Micropatterned Adhesive Surfaces for Control of Cell Behavior
Philip LeDuc, Emanuele Ostuni, George Whitesides, and Donald Ingber 385

20. Adenoviral–Mediated Gene Transfer in Two-Dimensional and Three-Dimensional Cultures of Mammary Epithelial Cells
Harriet Watkin and Charles H. Streuli 403

PART V General Information

Appendix A: List of Suppliers 427

Appendix B: Relevant Microarray Dataset Experiments 431

Appendix C: Web Site Resources of Interest 435

VOLUME 70
Cell Biological Applications of Confocal Microscopy
Brian Matsumoto

1. Introduction to Confocal Microscopy
Shirley J. Wright and David J. Wright 1

2. Direct-View High-Speed Confocal Scanner: The CSU-10
Shinya Inoué and Ted Inoué 87

3. Introduction to Multiphoton Excitation Imaging for the Biological Sciences
Victoria E. Centonze 129

4. Confocal Microscopy: Important Considerations for Accurate Imaging
Lars Majlof and Per-Ola Forsgren 149

5. Multicolor Laser Scanning Confocal Immunofluorescence Microscopy: Practical Application and Limitations
T. Clark Brelje, Martin W. Wessendorf, and Robert L. Sorenson 165

6. Practical Aspects of Objective Lens Selection for Confocal and Multiphoton Digital Imaging Techniques
Gerald S. Benham 247

7. Resolution of Subcellular Detail in Thick Tissue Sections: Immunohistochemical Preparation and Fluorescence Confocal Microscopy
Irene L. Hale and Brian Matsumoto 301

8. Confocal Fluorescence Microscopy for *Tetrahymena thermophila*
E. S. Cole, K. R. Stuart, T. C. Marsh, K. Aufderheide, and W. Ringlien 337

9. Confocal Imaging of *Drosophila* Embryos
Stephen W. Paddock 361

10. Confocal Fluorescence Microscopy of the Cytoskeleton of Amphibian Oocytes and Embryos
David L. Gard 379

11. Confocal Fluorescence Microscopy Measurements of pH and Calcium in Living Cells
Kay-Pong Yip and Ira Kurtz 417

12. Confocal and Nonlinear Optical Imaging of Potentiometric Dyes
Leslie M. Loew, Paul Campagnola, Aaron Lewis, and Joseph P. Wuskell 429

13. Measurement of Intracellular Ca^{2+} Concentration
Nicolas Demaurex, Serge Arnaudeau, and Michal Opas 453

14. Running and Setting Up a Confocal Microscope Core Facility
Susan DeMaggio 475

VOLUME 71
Neurons: Methods and Applications for the Cell Biologist
Peter J. Hollenbeck and James R. Bamburg

1. Comparing the Properties of Neuronal Culture Systems: A Shopping Guide for the Cell Biologist
Peter J. Hollenbeck and James R. Bamburg 1

2. Growing and Working with Peripheral Neurons
Yan He and Peter W. Baas 17

3. Dissection and Culturing of Chick Ciliary Ganglion Neurons: A System Well Suited to Synaptic Study
Barbara W. Bernstein 37

4. The Culture of Chick Forebrain Neurons
Steven R. Heidemann, Matthew Reynolds, Kha Ngo, and Phillip Lamoureux 51

5. Growing and Working with Spinal Motor Neurons
Thomas B. Kuhn 67

6. Avian Purkinje Neuronal Cultures: Extrinsic Control of Morphology by Cell Type and Glutamate
Peter L. Jeffrey, Vladimir J. Balcar, Ornella Tolhurst, Ron P. Weinberger, and Jenny A. Meany 89

7. Culturing Hippocampal and Cortical Neurons
Peter J. Meberg and Matthew W. Miller 111

8. Working with *Xenopus* Spinal Neurons in Live Cell Culture
Timothy M. Gómez, Dan Harrigan, John Henley, and Estuardo Robles 129

9. Culturing Neurons from the Snail *Helisoma*
Christohper S. Cohan, James L. Karnes, and Feng-Quan Zhou 157

10. The Tibial-1 Pioneer Pathway: An *in Vivo* Model for Neuronal Outgrowth and Guidance
Jennifer Bonner, Kimberly A. Gerrow, and Timothy P. O'Connor 171

11. Techniques to Dissect Cellular and Subcellular Function in the *Drosophila* Nervous System
Heinrich J. G. Matthies and Kendal Broadie 195

12. PC12 Cells as a Model for Studies of Regulated Secretion in Neuronal and Endocrine Cells
T. F. J. Martin and R. N. Grishanin 267

13. B35 Neuroblastoma Cells: An Easily Transfected, Cultured Cell Model of Central Nervous System Neurons
Carol A. Otey, Malika Boukhelifa, and Patricia Maness 287

14. Live-Cell Imaging of Slow Axonal Transport in Cultured Neurons
Anthony Brown 305

15. Making Proteins into Drugs: Assisted Delivery of Proteins and Peptides into Living Neurons
Gianluca Gallo 325

16. Tranfection of Primary Central and Peripheral Nervous System Neurons by Electroporation
Cecilia Y. Martinez and Peter J. Hollenbeck 339

17. Biolistic Transfection
Paul C. Bridgman, Michael E. Brown, and Irina Balan 353

18. Expression of Transgenes in Primary Neurons from Chick Peripheral and Central Nervous Systems by Retroviral Infection of Early Embryo
Peter J. Hollenbeck and D. M. Fekete 369

19. Production and Use of Replication-Deficient Adenovirus for Transgene Expression in Neurons
L. S. Minamide, A. E. Shaw, P. D. Sarmiere, O. Wiggan, M. T. Maloney, Barbara W. Bernstein, J. M. Sneider, J. A. Gonzalez, and James R. Bamburg 387

SUBJECT INDEX

Boldface numerals indicate volume number.

A

7AAD, **66,** 331
7A6 antigen
  detection, **66,** 409
AB strain
  development, **60,** 357
  haploid screen, **60,** 52, 62
AB-derived strain
  characteristics, **60,** 357–358
  haploid screen, **60,** 52–53
ABC method
  mitochondrial proteins
    immunolocalization, **65,** 327–328
Abf2p-green fluorescent protein, **65,** 279
  fluorescence patterns, **65,** 280
Aborting, of drop charging: purity protected by, **63,** 47–48
Absorption signals, flow cytometric, **63,** 112
AC mode, SFM of DNA, **68,** 367
Acetic acid, dissolving rat collagen in, **71,** 22
Acetone, fixation of cell suspensions by, **63,** 405
N-Acetylcysteine, maintenance of $\Delta\Psi$, **63,** 479
Achromat, confocal digital imaging, **70,** 256
Acid ceramidase, **66,** 159–160
Acid hydrolysis
  ceramide, **66,** 147–148
Acid sphingomyelinase, **66,** 149–152
  materials, **66,** 151
  molecular probes, **66,** 152
  principle, **66,** 151
  procedure, **66,** 151–152
  signal regulation, **66,** 149–150
ACMA (9-amino-6-chloro-2-methoxy acridine), **65,** 9
Acousto-optic deflectors, multiple trap synthesis, laser tweezers, **55,** 208–209
Acousto-optic tunable filter
  confocal laser scanning microscopes, **70,** 18–19, 27
  Drosophila embryo confocal imaging, **70,** 369–370
Acousto-optical deflector, video-rate CLSM
  Bio-Rad laboratories, **70,** 43
  NORAN instruments, **70,** 39–42
  overview, **70,** 39
Acridine orange (AO), **66,** 322, 323
  Drosophila
    embryos, **66,** 325
    imaginal discs, **66,** 325–326
Acridine orange, as lysosomal probe, **63,** 10
Actin, **62,** 53, 57–58, 63–64, 441
Actin
  functions in yeast, **58,** 87–88, 97
  green fluorescent protein fusion studies in yeast, **58,** 101–102
  sedimentation assay, **65,** 167
Actin-binding assay
  yeast actin purification, **65,** 160–166
    DE_{52}, **65,** 165
    DNase I affinity chromatography, **65,** 163
    DNase I column, **65,** 160–161
    polymerization and depolymerization, **65,** 165–166
    sedimentation assay, **65,** 166–170
    solutions, **65,** 160–161
    yeast cell extract preparation, **65,** 162–163
  yeast mitochondria, **65,** 150–172
Actin filaments, AFM studies, **68,** 78–80
Actin, *Xenopus* oocyte, **70,** 412
α-Actinin-green fluorescent protein fusion protein
  antibody competition studies **58,** 257
  construction
    primer design, **58,** 242–243
    terminus selection for protein fusion, **58,** 241–242
  fibroblast expression, **58,** 251–252
  fluorescence microscopy
    assembly in cleavage furrow of former metaphase plate, **58,** 250–251
    charge-coupled device camera, **58,** 256
    culture chambers, **58 255–256**
    myofibrillogenesis studies, **58,** 249

α-Actinin-green fluorescent protein
fusion protein (*cont.*)
time-lapse acquisition and image
processing, **58,** 256
microinjection, **58,** 247
photobleaching, **58,** 258
rhodamine phalloidin competition
study, **58,** 257
subcellular localization by
immunofluorescence, **58,** 247–248
transfection of cells
calcium phosphate, **58,** 246–247
efficiency, **58,** 246
embryonic avian cardiomyocyte
preparation, **58,** 243–244
embryonic avian myoblast
preparation, **58,** 243, 245–246
lipofection, **58,** 246
troubleshooting, **58,** 257–258
Actinomycin D
inhibition of nucleolar function, **53,** 316
segregation of nucleolus, **53,** 314–315
Activated caspase-3
detection, **66,** 408
Activated caspase-7
detection, **66,** 408
Activated caspase-9
detection, **66,** 408
Activation antigens
inclusion of Bcl-2, **63,** 435
subset-specific differences, **63,** 445
uptake, **63,** 445, 447
ActiveMotif Inc., **71,** 328
Active Q control
overview, **68,** 340
practical implementation, **68,** 340–344
Active Q-control atomic force microscopy
dynamic force spectroscopy
applications, **68,** 350–351
overview, **68,** 347–349
imaging, **68,** 344–345
Active transport, **62,** 192
Activity Western Blot
PARP-1, **66,** 294–296
electrophoresis, **66,** 294
extract preparation, **66,** 294
renaturation, **66,** 294–296
vs. Western Blot, **66,** 295
Acute lymphoblastic leukemia
B-lineage, **64,** 392–393
IgH gene rearrangements, **64,** 406
molecular analysis, **64,** 412
T-cell, **64,** 392
Acute myeloid leukemia
FAB classification, **64,** 345, 348–351
false negativity and false positivity, **64,** 352
intermediate cell populations in, **64,** 353–354
MRD detection in, **64,** 393–395
showing features of CD45 expression,
64, 354
ADC, *see* Analog-to-digital converters
AdEasy system, **71,** 390
advantages of, **71,** 391
Adenocarcinomas, clinical studies of
proliferation, **64,** 462–463
Adenoviral expression of proteins
advantages of, **71,** 407
disadvantage of, **71,** 392
Adenoviral luciferase vectors
generating, **65,** 371
Adenoviral titers, with two-day
immunoassay, **71,** 403–405
Adenoviral vectors
advantages of, **71,** 388
amplification of, **71,** 400
materials for making, **71,** 391–392
preparing, **71,** 391–397
producing in HEK293 cells, **71,** 397–400
safety issues regarding, **71,** 405–407
selecting, **71,** 386
titer assays for, **71,** 402–405
transfection of HEK293 cells with, **71,** 367
Adenovirus
tumor necrosis factor-related
ligands, **66,** 520–523
Adenovirus, immortalization of
cells, **57,** 72–73
Adenovirus, mammary epithelial cell
generation in *Escherichia coli*, **69,** 410–413
HEK293 E1 cells, **69,** 415–416
high-titer viral stocks, **69,** 418–420
infection, **69,** 420–422
overview, **69,** 409–410
screening, **69,** 413–415
virus titration, **69,** 417–418
Adenovirus mediation, **71,** 310–314
in cortical and hippocampal neuronal
cultures, **71,** 113
replication-deficient, **71,** 388
Adenovirus stocks, purification and
storage of, **71,** 400–401
Adenylate cyclase, **62,** 38
Adenylate kinase (ADK), **66,** 366
mitochondria released, **66,** 378

Adherent cells
cell-free systems, **66,** 170–171
transmitochondrial cybrid generation, **65,** 402–404
Adhesion, slow *vs.* fast in *Heliosoma* cultures, **71,** 169
Adhesion assays
cell, *see* Cell adhesion assay
cell–matrix, *see* Cell–matrix adhesion assay
deposited ECM, **69,** 35–36
instructive ECM, **69,** 37–40, 45
integrin-mediated, *see* Integrin-mediated adhesion assay
Adhesion force
cell–cell
research results, **68,** 105–109
single cell immobilization, **68,** 104–105
cell layers
monolayer immobilization, **68,** 99–100
research results, **68,** 100–104
cell–surface
research results, **68,** 95–99
sphere immobilization, **68,** 94
cells
determination techniques, **68,** 91–92
spectroscopy, *see* Cell adhesion force spectroscopy
$\alpha_v\beta_3$ integrin on osteoclasts, **68,** 182–184
molecular, spectroscopy, **68,** 14–15
Adhesion receptor
integrin–null ES cells, **69,** 320
$\alpha3\beta1$ integrin–null cells, **69,** 319–320
$\alpha5$ integrin–null ES cells, **69,** 320–321
$\beta1$ integrin–null ES cells, **69,** 317–318
$\alpha4$ and $\alpha5$ integrin–null fibroblasts, **69,** 320
$\beta1$ integrin–null somatic cell lines, **69,** 318–319
Adhesive islands, ECM-coated, living cell adherence, **69,** 388–390
Adhesive substrate, preparation, **69,** 283–284
ADK, **66,** 366
mitochondria released, **66,** 378
Adobe Photoshop
Action feature in task automation, **58,** 198
colorizing fluorescence images, **58,** 196–197
colorizing stacks, **58,** 198–199
cropping stacks, **58,** 199–200
grayscale images, **58,** 197–198
importing images, **58,** 195–196
merging fluorescence onto a continuous tone image, **58,** 198
text layer resolution, **58,** 362
ADP, **65,** 137
ADP, PFM studies, **68,** 26
Adrenal medullary tumor, **71,** 268
Adriamycin, **66,** 211
Adriamycin, *see* Doxorubicin
Adsorption
DNA
charged molecules in solution, **68,** 364
electrochemical adsorption, **68,** 359
electrostatic adsorption, **68,** 361–363
surface treatments, **68,** 365
protein, force spectroscopy, **68,** 14–15
Aequorea victoria, **65,** 277, 358
GFP, **65,** 373
green fluorescent protein comparison to *Renilla reniformis*, **58,** 1–2, 8
lumisomes, **58,** 140
Aequorin, **61,** 114, **65,** 354, 357–365
calcium imaging, **66,** 239–240
expression, **65,** 361
low buffering capacity, **65,** 360
low-light emission, **65,** 360–361
mitochondrial calcium measurements, **65,** 358
procedure, **65,** 361–362
signal-to-noise ratio, **65,** 360
targeted probes, **65,** 358–360
Aequorin, green fluorescent protein interactions
dimerization and isoform effects, **58,** 19–20
resonance energy transfer, **58,** 15–16, 19
Affi-Gel 10
Apaf-1 CARD, **66,** 24
Affinity blotting, **65,** 218–219
Affinity chromatography, **65,** 219
Apaf-1 purification, **66,** 21–22
basement membrane self-assembly, **69,** 124–125
DNase I
actin-binding assay, **65,** 163
integrin-associated proteins, **69,** 237–239
integrin, lectin column, **69,** 234
proteoglycans, **69,** 61–63
solubilized integrins, **69,** 228–233
Affinity coelectrophoresis, basement membrane self-assembly, **69,** 125–127
Affinity cross-linking, integrin ligand binding pocket, **69,** 248–250
Affinity isolation
TGase glutamyl donor substrate proteins, **66,** 122–123
Affinity labeled
TGase glutamyl donor substrate proteins identification, **66,** 124–125
Affinity labeling assay
caspase assays, **66,** 176

Affinity labeling caspases
caspase assays, **66,** 176–177, 178
Affinity precipitation, SPB components
antibody couplings, **67,** 84
cell lysis, **67,** 85–86
controls, **67,** 89
gene fusion construction, **67,** 82–84
protein A fusion proteins, **67,** 87–88
Affinity purification
antibodies, **65,** 212–213
Affinity purification, γ-tubulin antibodies, **67,** 203–205
Affinity retardation chromatography, basement membrane self-assembly, **69,** 125
AFLPs, *see* Amplified fragment length polymorphisms
AFM, *see* Atomic force microscopy
Agar dish protocol, **66,** 479–480
Agar, incubation on, for colony formation, **64,** 553–554
Agarose drop assay, cell migration assay, **69,** 291–292
Agarose embedding, thick tissue sections, **70,** 330
Aggregation assays, measurement of cell–cell adhesion, **63,** 600–603
Aging, **62,** 128
Agonistic antibodies
tumor necrosis factor-related ligands, **66,** 519
Agricultural guidelines and practices, **71,** 57
Agriculture schools, ease of obtaining species through, **71,** 13, 57
AIF
mitochondria released, **66,** 377–378
staurosporin, **65,** 153
Air, AFM probe microscopy
ECM macromolecules, **69,** 183–185
fibrillar collagens, **69,** 187–188
FN–heparin–gold complexes, **69,** 185–187
overview, **69,** 182–183
Air bubbles, AFM cantilever, **68,** 216
Air-cooled argon ion laser, multicolor LSCM
detector–double-labeled specimen cross-talk, **70,** 213–217
fluorophore visualization specificity, **70,** 209–212
three-color LSCM, **70,** 217–221
two-color LSCM, **70,** 212–213
Air-cooled krypton–argon ion laser, multicolor LSCM
double-labeled specimen imaging, **70,** 225–228
examples, **70,** 228–231
fluorophore visualization specificity, **70,** 221–225
two-color LSCM, **70,** 225
Air resistance, as factor in biolistics, **71,** 357–359
Airy disks, confocal microscopy, **70,** 9–10
Ajuba cytoplasmic protein, **71,** 92
Albumen, removing in preparation for retroviral infection, **71,** 379–380
Alcian Blue, **62,** 354–357
Alcohols, cell fixation, **63,** 256, 277
Algorithm
built into LSC software, **64,** 337
for multiparameter data analysis, **64,** 488–492
quantification, IFI, **64,** 88–91
segmentation, IFI, **64,** 80–82
TFL-TELO, **64,** 77
Algorithms
decision, **63,** 83
LSCM software, **63,** 60–61
Alignment
fixed, checking for, **64,** 570–571
instrument for flow karyotyping, **64,** 15–16
Alkaline ceramidase, **66,** 159–160
Alkaline comet assay
developed by Olive, **64,** 256–257
measurement of DNA damage, **64,** 253–254
Alkaline comet method, **64,** 239
Alkaline lysis solutions, in QDFM, **64,** 36, 41–42
Alkaline methanolysis
metabolic labeling, **66,** 139
Alkaline phosphatase, in Purkinje neuronal cultures, **71,** 95
Alkanethiolate SAMs
micropatterned culture surface, **69,** 395–397
synthesis, **69,** 393
ALL, *see* Acute lymphoblastic leukemia
Allophycocyanin, as fluorescent label, **63,** 195
Allophycocyanin, emission distribution, **64,** 296–297
Alpha-glucosidase, **62,** 45
α-Amanitin, inhibition of nucleolar function, **53,** 317
α-Hemolysin
AFM studies, **68,** 270
staphylococcal, AFM studies, **68,** 270
α-Tubulin, **67,** 181–182
$\alpha V\beta_3$ Integrin
cell-matrix adhesion, **69,** 17–18
RGD-containing ECM proteins, **69,** 17
$\alpha_v\beta_3$ Integrin, adhesion force measurement, **68,** 182–184
$\alpha 3\beta 1$ Integrin-null cells, **69,** 319–320
$\alpha 5$ Integrin-null ES cells, **69,** 320–321

α4 Integrin-null fibroblasts, **69,** 320
α5 Integrin-null fibroblasts, **69,** 320
α3 Integrin-null mouse, **69,** 311–313
$\alpha_5\beta_1$ Protein, immobilization, **69,** 179–181
α-Tubulin–green fluorescent protein, fusion protein studies in yeast, **58,** 97–99
Alphoid DNA, **61,** 66
Alveolar membrane, **62,** 34–35, 38–40
Alzheimer's disease
  associated with Purkinje neurons, **71,** 91
  cortical and hippocampal involvement in, **71,** 112
  toxicologic effects of lead on, **71,** 290
American Type Culture Collection
  authentication of cell lines, **57,** 41–43
  categorization of cell lines, **57,** 34–35
  development of cell lines, **57,** 38–40
  fees, **57,** 35
  functions, **57,** 32
  patent culture repository, **57,** 36, 38
  representative cell lines, **57,** 33, 35–37
Amersham Australia, **71,** 95
Amgel, preparation from amniotic membranes, **63,** 570
Amicon Ultrafree-MC centrifugal concentrator, **71,** 313
Amicronucleate cells, **62,** 7, 21, 75, 148–149
9-amino-6-chloro-2-methoxy acridine, **65,** 9
Amino acid sequences
  Bcl-2 family, **66,** 31
Amino acids
  tryptic involucrin peptides, **66,** 129
7-amino actinomycin D (7AAD), **66,** 331
7-Amino-actinomycin, DNA-specific, **64,** 423–425
3-Aminopropyltriethoxy silane, slide preparation, **64,** 38
Amiprophos-methyl solutions, **64,** 8
AML, *see* Acute myeloid leukemia
Amniotic membranes
  Amgel preparation from, **63,** 570
  intact, preparation, **63,** 565, 567
Amoeboid cells, **66,** 478
L-Aminopeptidase, kinetic activity measurement, **63,** 77–78
Amplification
  in situ, **63,** 84
  linear and logarithmic, **63,** 38–40
  logarithmic
    and fluorescence compensation, **63,** 124–127
    fluorescence signal, **63,** 534
Amplified fragment length polymorphisms
  advantages and disadvantages as polymorphic markers, **60,** 170–171, 209–210
  applications, **60,** 195–196
  gene mapping in zebrafish
    band recovery from gels, **60,** 207–208
    electrophoresis of products and analysis, **60,** 197, 206–207
    genomic DNA preparation
      cell lysis, **60,** 198
      digestion, **60,** 196, 199
      extraction, **60,** 198–199
    ligation of adapters, **60,** 196, 199–200
    map construction, **60,** 210
    overview of steps, **60,** 196–197
    polymerase chain reaction
      preamplification, **60,** 196–197, 201–202
      selective amplification, **60,** 197, 203, 206
    primer labeling, **60,** 202
  kits for analysis, **60,** 196
  positional cloning project initiation, **60,** 264, 266–267, 270
  sequence-tagged site construction, **60,** 271
  single strand conformation polymorphism conversion, **60,** 208–209
Amplifiers, instrumental noise, **55,** 59–60, 151–152
Amyotrophic lateral sclerosis, GluT in spines of patients with, **71,** 92
Anaglyphs, CSU-10 digital signal processing, **70,** 116
Analog processing
  definition, **56,** 70
  digital processing compared, **56,** 70–73
  signal input, analog-to-digital converter, **56,** 74
  video-enhanced differential interference contrast light microscopy
    components, **56,** 176–177
    contrast enhancement, **56,** 162–165
Analog-to-digital conversion
  circuits with comparators, **63,** 155
  flow cytometric data, **63,** 109
  in confocal microscopy, **63,** 92
  in signal processing, **63,** 40
Analog-to-digital converters
  flash and hybrid, **63,** 157–158
  resolution, **63,** 158–159
  single slope, **63,** 156
  successive approximation, **63,** 156–157
Anaphase, multimode imaging, *in vitro*, **56,** 210–211
Anaplastic astrocytoma, *see* Glioma
Aneuploidy, for pericentrin level elevation, **67,** 66

Angel, *see* Miniature inverted-repeat transposable elements
Angiogenesis
cell–matrix adhesion studies, **69,** 19
synthetic provisional matrix, **69,** 156
Anguilla anguilla, **65,** 28
Animal cell culture, *see also* Glioma, Schwann cell, Testicular cell
cloning, **57,** 16
contamination, *see* Contamination, Cell culture
equipment
autoclave, **57,** 8, 53–54
cell counter, **57,** 8–9
centrifuges, **57,** 9
freezer, **57,** 9
hoods, **57,** 4–6
incubators, **57,** 6–7, 52
liquid nitrogen storage tank, **57,** 9
medium filtration, **57,** 8
microscopes, **57,** 7–8
water baths, **57,** 9
water purification, **57,** 8, 53
freezing, **57,** 9, 16, 40–41
glassware, **57,** 13, 54–55
immortalization, *see* Immortalization, Cell lines
laboratory design, **57,** 10–11, 52–53
large-scale culture, *see* Scaleup, Cell culture
passaging of cultures, **57,** 15
plasticware, **57,** 13, 54–55
primary culture establishment, **57,** 14–15
reagents, medium, and serum, **57,** 11–13
sterile technique, **57,** 13–14, 51–52
Animal cells
mitochondrial ATP synthesis, **65,** 133–143
Animal-holding facilities, **71,** 10
Animal-use issues, **71,** 14, 57
Anion-exchange chromatography, proteoglycans
experimental protocols, **69,** 58–59
overview, **69,** 57–58
soluble proteoglycans, **69,** 59
Anlagen, **62,** 131, 142, 234
Annexin assays
lipid asymmetry, **66,** 342
Annexin-liposome binding, **66,** 345
Annexin staining
flow cytometry, **66,** 346–349
Annexin V, **66,** 409
apoptosis assays, **66,** 211–213
apoptotic cells, **66,** 84–87
data analysis, **66,** 86–87
flow cytometry, **66,** 85
LSC, **66,** 85
materials, **66,** 85
in differentiation of apoptosis and necrosis, **63,** 535–536
purification, **66,** 342–346
staining for phosphatidylserine exposure, **63,** 493
Annexin V purification
bacterial spheroblasts, **66,** 344
liposomes, **66,** 344–345
protein storage, **66,** 345
reagents and materials, **66,** 345
Annexin V, apoptosis/cell cycle analysis with flow cytometry
filters, **57,** 273
performance of assay, **57,** 273–274
preparation of cells, **57,** 272
principle, **57,** 267–268
staining of cells, **57,** 272–273
Annulin, **71,** 182
Antagonistic antibodies
antagonistic, **66,** 519
tumor necrosis factor-related ligands, **66,** 519
Antennapedia PTD, **71,** 327
peptides conjugated by, **71,** 334
Anterior–posterior pattern, organizers in neural plate, **59,** 253–259
Anthracyclines
fluorescence quenching, **64,** 199
transport and retention monitoring, **64,** 195
Anthraquinones, cytotoxicity, **64,** 177
Anti-zKLP antibodies, production, **59,** 233–242
Antiapoptotic Bcl-2 family, **66,** 30
schematic diagrams, **66,** 32
Antibiotic-antimycotic, **62,** 193, 524
Antibiotics
effects on target organisms, **64,** 526–527
inhibition of bacteria, **64,** 558
time parameters for effects of, **64,** 561
Antibodies
acid sphingomyelinase, **66,** 152
affinity purification, **65,** 212–213
affinity purified, **71,** 213
against bacterial fusion protein, **67,** 131–132
against centrosomes, **67,** 126–127
against γ-tubulin
incubation *in vitro*, **67,** 181–183
microinjection, **67,** 184–185
preparation, **67,** 180
against plant tubulin, **67,** 200–202
affinity purification, **67,** 203–205
nonspecific reactivity, **67,** 205–206
agonistic, **66,** 519

anti-PCNA, **63,** 422
anticentrosomal, **67,** 126–127
anticlam centrosomal, **67,** 127
antidinitrophenyl, **66,** 131–132
antiphospho-JNK, **66,** 192
antitubulin, viability staining with, **63,** 491–492
binding kinetics, **63,** 220–224
blocking, in Tibia-1 pathway studies, **71,** 191
bound per cell, **63,** 332–333
calibrated measurements applied to, **63,** 330–331
ceramide synthase, **66,** 157
conjugated, brightness, **64,** 298, 300
coupling to protein beads, **67,** 84
cross-linking
protein-protein interactions, **65,** 222
directly conjugated, **63,** 232
Drosophila embryo confocal imaging, **70,** 372–373
exhaustion, fluorescence saturation due to, **64,** 150
FLIM, staining, **69,** 269
for MT nucleation, **67,** 328
histone H_3 plus, as marker of mitotic cells, **63,** 344–345
immunofluorescence staining techniques, **70,** 181–184
in chromosome translocation study, **64,** 99
in detection of cyclins, **63,** 347
in *Drosophila* cultures, **71,** 208–212
in QDFM, **64,** 35
labeling, FLIM, **69,** 266–267
MIB-1, staining pattern, **63,** 410–414
monoclonal, *see* Monoclonal antibodies
monoclonal, against *Dictyostelium* centrosomes, **67,** 347–351
monoclonal, bacterium-specific, **64,** 519–520
neutral sphingomyelinase, **66,** 154–155
one and two, staining of cells with, **63,** 239–240
phospho-specific, adhesion assays, **69,** 374–375
polyclonal, in SPB immunoprecipitation, **67,** 87
primary and secondary, in immunophenotyping, **63,** 228–229
protein detection
Bcl-2 family, **66,** 37
reliable, for immunophenotyping, **64,** 314
resolving discrete and nondiscrete populations, **63,** 245–247
screening in PC12 cell lines, **71,** 272
selection, **63,** 259
single antibody histogram analysis, **64,** 289–290
specific binding, verification of, **63,** 247–248
and specific species, **71,** 9
specificity, **63,** 287–292
sphingomyelin synthase, **66,** 158–159
staining of tissue culture cells, **67,** 307–308
staining protocols, **71,** 217
structure, **63,** 219–220
suppliers for confocal microscopy, **70,** 76–77
thick tissue labeling with fluorochrome-conjugated probes, **70,** 315–319
three and four, staining of cells with, **63,** 241–242
titering, **63,** 244–248
to antidynein, **67,** 61–63
Xenopus oocytes
keratin filament staining, **70,** 412
microtubule staining, **70,** 411
overview, **70,** 412–413
whole-mount confocal microscopy, **70,** 391–392
yeast cell walls, **65,** 270–271
Antibody
humanization, **57,** 112–113, 127
immunofluorescence microscopy
double labeling, **57,** 321
handling, **57,** 319
incubation protocol, **57,** 329–330
primary antibodies, **57,** 319–320
secondary antibodies, **57,** 320–321
titration, **57,** 319
monoclonal antibody, *see* Hybridoma
Antibody column, integrin affinity chromatography, **69,** 231–232
Antibody–antigen reactions
kinetics
in permeabilized cells, **63,** 282–287
in solution, **63,** 281–282
technical problems, **63,** 271–272
Antibody–antigen recognition, force microscopy, **68,** 134–137
Antibody labeling, membrane protein AFM imaging, **68,** 264–265
Anticancer agents, cytotoxic
disruption of DNA integrity, **64,** 174–175
DNA topoisomerase inhibitors, **64,** 175–177
drug–DNA interactions, **64,** 183–188
flow cytometric analysis, quality control, **64,** 178–182
inhibition of
DNA metabolism, **64,** 175
mitotic spindle function, **64,** 175
Anticentrosome antibodies, **67,** 126–127
Anticlam centrosomal antibodies, **67,** 127

Anticytoskeletal drugs, applying to peripheral neurons, **71,** 18
Antidinitrophenyl antibodies
apoptosis
hydrazine-linked beads, **66,** 131–132
protein transglutamylation, **66,** 131–132
Antidynein, antibodies against, **67,** 61–63
Antifade reagents, **70,** 76–77
Antifade reagents, in confocal microscopy, **63,** 99
Anti-Fas (CD95/Apo-1) induced cytotoxicity assay
Bcl-2 family, **66,** 40
Antigen quantitation, **64,** 582
Antigen receptor, transgenic models, **63,** 395–396
Antigenicity, COCO system proteins, **69,** 79–80
Antigens
alterations during HIV infection, **64,** 580–582
major lymphocyte subpopulations distinguished by, **64,** 568–569
multiple, simultaneous assessment, **64,** 314
proliferation-associated, **64,** 364
LSCM analysis, **64,** 430–434
suitable for monitoring AML, **64,** 394–395
surface, blast cell populations, **64,** 351–352
Antigens, for *Dictyostelium* centrosomes, **67,** 350
Antigens, *see also* Activation antigens
cell surface: simultaneous staining with
APO2.7, **63,** 263
Ki-67, **63,** 407–408
PCNA, cytokeratin, DNA, and tubulin, **63,** 264–266
T cell receptor z, **63,** 263–264
expression, drug effects, **63,** 227
immunolocalization, **63,** 292
intracellular, immunofluorescence, **63,** 12
quantitation per cell, **63,** 333–334
target, properties, **63,** 261–262
Antigens, Mab identification, **69,** 41–42
Antimicrobial agents, flow cytometric analyses, **64,** 526–529
Antimitotic agents, reducing nonneuronal contamination with, **71,** 31
Antiphospho-JNK antibody
Western Blotting
JNK activation, **66,** 192
Antiproliferative agents, **71,** 10–11
Antisense ribosomes, **62,** 167–168, 533–546
Antisense RNA
and specific species, **71,** 9
in B35 cells, **71,** 295
in *Xenopus* spinal neuron cultures, **71,** 131
Antisense RNA, integrin-mediated signaling pathways, **69,** 381
Antisense technology
antisense definition, **61,** 245–246
antisense reagent choice
antisense transcript length considerations, **61,** 251
control oligo categories, **61,** 250
cotransfection of ribozyme constructs, **61,** 253
cryptic sense strand transcription, **61,** 251
DNA *vs.* RNA, **61,** 248–249
hammerhead ribozyme constructs, **61,** 251–252
nonsequence-specific inhibition, **61,** 250
nucleotide length and specificity, **61,** 249
phosphorothioate oligo toxicity, **61,** 250–251
phosphorothioate substitution modifications, **61,** 249
ribozyme constructs, **61,** 252–253
sequence structure and composition affects, **61,** 249–250
tandem arrays, **61,** 251
target protein depletion scoring, **61,** 254
types, **61,** 248
antisense reagents in cell division studies
mitotic kinesin-related proteins
end capping, **61,** 255
interspecies use, **61,** 256
mitotic midzone motor nonmitotic function, **61,** 256
redundant function identification, **61,** 255
Xklp1 antisense oligo effects, **61,** 255
uniquely mitotic problems
abortive cytokinesis, **61,** 257
gamma-tubulin study, **61,** 256–257
mitotic centromere-associated kinesin (MCAK) difficulties, **61,** 257
rapid cell growth disadvantages, **61,** 256
scrape loading, **61,** 257–258
traction-mediated cytofission, **61,** 257
transfection efficiency, **61,** 257
utility and limitations
accumulation of multinucleate and aneuploid cells, **61,** 260
auxiliary studies, **61,** 260
checkpoint arrest, **61,** 260–261
double and triple knockouts, **61,** 261
one-step endogenous protein replacement, **61,** 261
assaying target protein levels
Northern blots, **61,** 254
quantitative immunofluorescence with light microscopy, **61,** 254

transcription/translation system
in vitro, **61,** 254–255
Western immunoblots, **61,** 254
mechanism of action
cytoplasmic events
assays, **61,** 248
AUG start codon, **61,** 248
cis-acting autoregulatory process, **61,** 248
inhibition of translation initiation complex, **61,** 248
peptide nucleic acids (PNC), **61,** 248
RNase H hydrolyzing triggers, **61,** 248
steric inhibition of ribosome binding, **61,** 248
nuclear events
accumulation, **61,** 247
intronless genes, **61,** 247
RNase sensitivity, **61,** 247
transfected antisense constructs, **61,** 247
uses
protein depletion, **61,** 246
regulation of eukaryotic gene expression, **61,** 246
Antitumor agents, delivery
clinical implications, **64,** 224–225
physiological considerations, **64,** 223–224
slow drug penetration, **64,** 224
AO, **66,** 322, 323
Drosophila
embryos, **66,** 325
imaginal discs, **66,** 325–326
AOD, *see* Acousto-optical deflector
AOTF, *see* Acousto-optic tunable filter
Apaf-1, **66,** 182, 454
procaspase 9, **66,** 464
Apaf-1 apoptosome
cytosolic S-100 extracts, **66,** 23
pure Apaf-1, **66,** 23
Apaf-1 apoptosome assay
caspase-9, **66,** 18–20
procaspase 3, **66,** 20
procaspase 9, **66,** 19
Apaf-1 caspase-associated recruitment domain (CARD)
Affi-Gel 10, **66,** 24
Apaf-1 function
oligomerization, **66,** 23–25
procaspase 9, **66,** 23
Apaf-1 purification
affinity chromatography, **66,** 21–22
caspase assay
baculovirus expression systems, **66,** 21–25
gel-filtration chromatography
Superose 12 Column, **66,** 22–23
ion-exchange chromatography
Mono Q Column, **66,** 22
Aphasia, **71,** 291
Aphidicolin, synchronization of cells by, **64,** 276–277
Apical filament ring, **62,** 57
Aplysia, **71,** 3
organelle transport in, **71,** 8
APO2.7
and cell surface antigens, simultaneous staining, **63,** 263
expression, staining cells for, **63,** 490–491
Apoptosis
application of LSCM cell position feature, **63,** 74
c-Jun N-terminal kinase, **66,** 187–193
calcium flux, **66,** 229–242
caspase 3
flow cytometry, **66,** 63
cell size
propidium iodide, **66,** 54–57
cellular proteins
protein substrate assay, **66,** 16–17
c-Jun N-terminal kinase, **66,** 187–193
commercial detection kits, **63,** 538–539
detection, **66,** 398
DNA laddering analysis, **57,** 259–261
duration, **63,** 530
enzyme-linked immunosorbent assay of histone-associated DNA fragments, **57,** 261
flow cytometric quantification, **63,** 11
flow cytometry, **66,** 50
caspases, **66,** 50
cell size, **66,** 52–57
cell sorting, **66,** 64–67
flow cytometry assays
annexin V/cell cycle analysis
filters, **57,** 273
performance of assay, **57,** 273–274
preparation of cells, **57,** 272
principle, **57,** 267–268
staining of cells, **57,** 272–273
propidium iodide/surface staining, **57,** 256–257
TUNEL/cell cycle analysis
filters, **57,** 269
fixation of cells, **57,** 268
labeled nucleotides, **57,** 275–276

Apoptosis (*cont.*)
performance of assay, **57,** 269–272
principle, **57,** 259, 267
staining, **57,** 268–269, 271–272
TUNEL/surface staining, **57,** 257–259
frequency, estimation, **63,** 531–532
in plants, **63,** 506–509
flow cytometric analysis:
problems, **63,** 509–512
in vitro
assays, **66,** 171–181
cell line models, **66,** 417–432
initiation, **66,** 51–52
ISEL assay, **57,** 255–256
JC-1
LSC, **66,** 82
JNK
evaluation, **66,** 193
lactate dehydrogenase assay, **57,** 261
laser light scattering, **66,** 79–80
light microscopy, **66,** 396
LSC
vs. flow cytometry, **66,** 100–103
uniqueness, **66,** 100–102
LSCM analysis, **63,** 541–543
mechanism, **57,** 266
microscopy, **66,** 104
mitochondrial role during, **63,** 470–472
mitochondrial transmembrane
potential, **66,** 80–84
flow cytometry, **66,** 81–82
LSC, **66,** 82
monovalent intracellular ions
flow cytometry, **66,** 57–63
morphological features, **66,** 395
morphology of cells
characteristics, **57,** 251–252, 266
electron microscopy, **57,** 253
light microscopy, **57,** 253
MTT assay, **57,** 261–262
nuclear staining assays, **57,** 254–255
oncogenic mutations, **66,** 198
plant cells, physiological changes
during, **63,** 520–523
poly(ADP-ribose) polymerase, **66,** 97–100
protein transglutamylation, **66,** 111–132
ε-(γ-glutamyl)-lysine cross link
density, **66,** 115–119
antidinitrophenyl antibodies, **66,** 131–132
steps, **66,** 290
thymus cell subpopulations, **63,** 143
triggering mechanisms, **63,** 468
tumor necrosis factor-related
ligands, **66,** 499–524
TUNEL assay, principle and
applications, **57,** 255–256
vital dye staining assays, **57,** 253–254
vs. necrosis and *vs.* necrotic stage of
apoptosis, **63,** 535–536
vs. necrosis
flow cytometry, **66,** 71
XTT assay, **57,** 262
Apoptosis, **61,** 280, 282
Apoptosis assays
annexin V, **66,** 211–213
DAPI staining, **66,** 213
FACS analysis, **66,** 213–214
nonimmortal cells, **66,** 211–214
propidium iodide staining, **66,** 211–213
Apoptosis-inducing factor (AIF), **66,** 366
mitochondria released, **66,** 377–378
Apoptosis induction
tumor necrosis factor-related
ligands, **66,** 517–520
Apoptosis, LSCM analysis, **64,**
425–430
Apoptosis protein expression
yeast, **66,** 458–462
Apoptosis proteins
yeast, **66,** 453–467
plasmids, **66,** 456
Apoptosomes, **66,** 366, 454
fractionation, **66,** 24
Apoptotic bodies
engulfed by live cells, **63,** 538
misclassification as single apoptotic
cells, **63,** 533–534
Apoptotic cells
annexin V, **66,** 84–87
flow cytometry, **66,** 85
DNA strand breaks, **63,** 352
identification, **66,** 397
based on enzyme activity, **63,** 532
by flow cytometry, **63,** 528
cell morphology for, **63,** 539, 541
immunoprecipitation
TGase-interacting proteins, **66,** 119
in situ identification
TGase glutamyl donor substrate
proteins, **66,** 119–124
LSC
annexin V, **66,** 85
red maximal pixel, **66,** 78, 79
morphological changes, **63,** 506

phagocytosis, **66,** 358–362
selective loss during sample preparation, **63,** 537
single, misclassification of apoptotic bodies as, **63,** 533–534
subpopulations, ethidium bromide bound to, **63,** 143
and viable cells, distinguished, **63,** 492–493
Apoptotic index
cell death incidence, **66,** 104–105
Apoptotic index, correlation with incidence of cell death, **63,** 529–531
Application markings, objective lens, **70,** 260
Aqueous fluids, AFM imaging
bubble removal from cantilever, **68,** 216
cantilevers, **68,** 218
different solutions, **68,** 218–219
modes, **68,** 217
parameters, **68,** 217–218
when not to image, **68,** 221–223
when to image, **68,** 221
with O ring, **68,** 216
without O ring, **68,** 215–216
Arabidopsis mitochondrial protein
two-dimensional gel electrophoresis, **65,** 69
Arabidopsis thaliana, **65,** 60–61
confocal imaging
FM 1-43 counterstaining of plasma membranes, **58,** 146
growth in sterile culture, **58,** 144–145
instrumentation, **58,** 143–145
multichannel microscopy
cyan and yellow-shifted mutants, **58,** 148
double labeling with green fluorescent protein variants, **58,** 149
KNAT3–green fluorescent protein studies, **58,** 148
rationale, **58,** 147–148
principle, **58,** 143–144
propidium iodide counterstaining of cell walls, **58,** 145–146
development and intercellular interactions, **58,** 139–140
green fluorescent protein expression
cell type marking, **58,** 146–147
cryptic intron removal, **58,** 141
subcellular targeting, **58,** 141
thermotolerant mutants, **58,** 141–142
mitochondrial genome size, **65,** 420
protoplast
flow cytometry, **58,** 326, 331–333, 337
preparation, **58,** 322–323
recombinant green fluorescent protein construct preparation, **58,** 323
transfection, **58,** 323–324
AraC, in long-term hippocampal cultures, **71,** 121–122
Arc epi-illumination, for fluorescence image analysis, **63,** 52
Arc light, mercury, **63,** 6
Area feature, LSCM list mode, **63,** 67–71
Argon ion gas laser, **71,** 151
Argon ion laser
in combination with other lasers, **63,** 22
wavelengths, **63,** 21
Argon ion lasers
detector–double-labeled specimen cross-talk, **70,** 213–217
fluorophore visualization specificity, **70,** 209–212
multicolor LSCM, **70,** 187–188
three-color LSCM, **70,** 217–221
two-color LSCM, **70,** 193–195, 201–204, 212–213
Aristotle, **71,** 68
Arithmetic precision, CSU-10 digital signal processing, **70,** 111–112
ARP, autonomously replicating piece, **62,** 138, 263
Arterioles, intracellular calcium, **70,** 423–425
Artificial neural networks, in detection of bacteria, **64,** 520–521
Aspartate-specific serine proteases, **66,** 4–5
Aspartic proteinase inhibitors, **66,** 251
Aspergillus nidulans mitotic mutant isolation and analysis
characteristics of organism
advantages, **61,** 348–349
cell cycle
chromosomal condensation, **61,** 353
mitotic apparatus, **61,** 350, 352–353
spindle pole bodies (SPB), **61,** 350
time requirements, **61,** 350
description, **61,** 348
life cycle
ascospore, **61,** 350
ascus, **61,** 350
asexual reproduction, **61,** 349–350
cleistothecium formation, **61,** 350
conidia formation, **61,** 350
croziers, **61,** 350
sexual reproduction, **61,** 350
conidial harvest and suspension preparation
harvest
glass bead, **61,** 354
glass rod, **61,** 354

Aspergillus nidulans mitotic mutant isolation and analysis (*cont.*)
hemacytometry concentration determination, **61,** 355
hyphae and conidiophore removal, **61,** 354
storage, **61,** 355
suspension procedure, **61,** 354
genetic analysis
complementation tests, **61,** 364
dominant *vs.* recessive determination
diploid formation, **61,** 363
heterokaryon formation, **61,** 363
non-disjunction and aneuploids, **61,** 363
gene mapping
to chromosomal locus, **61,** 365
to linkage groups, **61,** 364–365
rationale, **61,** 363
isolation
conditionally lethal mutations
advantages and findings, **61,** 356
chromosome mitotic indices (CMI), **61,** 358
definition, **61,** 355–356
extent of mutagenesis, **61,** 357
heat shock, **61,** 358
immunofluorescence microscopy, **61,** 358
irrelevant mutation elimination, **61,** 358
procedure, **61,** 357
rapid initial screen, **61,** 357–358
rationale, **61,** 356
replica plating, **61,** 357
spindle mitotic index (SMI), **61,** 358
temperature-sensitive types, **61,** 356
temperature shift experiment screen, **61,** 358
mutations as revertants of existing mutations
back mutation, **61,** 360
conditional lethality, **61,** 360
genetic analysis, **61,** 361
intergenic suppressor, **61,** 360
intragenic suppressor, **61,** 360
tubulin gene studies, **61,** 360
mutations resistant to anti-microtubule agents
beta-tubulin mutations, **61,** 359
drug resistant *vs.* random mutations, **61,** 359
selection, **61,** 359
temperature sensitivity screening, **61,** 359
molecular genetic methods
cloning of genes involved in mitosis
cosmids, **61,** 366
DNA circularization, **61,** 365–366
increasing transformation frequencies, **61,** 366
plasmid recovery, **61,** 365, 366
sib-selection approach, **61,** 366
Southern hybridization, **61,** 366
transformation, **61,** 365
creating mutant alleles of cloned genes
gene replacement procedures, **61,** 367
heterokaryon gene disruptions, **61,** 366–367
morphological analysis
electron microscopy
fixation, **61,** 362
freeze substitution, **61,** 363
immunofluorescence microscopy
antibody specificity, **61,** 362
cell wall digestion, **61,** 362
DM1A, **61,** 362
secondary antibodies, **61,** 362
YOL1/34, **61,** 362
light microscopy
acetoorcein staining, **61,** 361
DAPI staining, **61,** 361
Feulgen staining, **61,** 361
mitotic index, **61,** 362
mutagenesis
mutagenesis indicators, **61,** 355
4–nitroquinoline 1–oxide, **61,** 355
kill curve, **61,** 355
UV light procedure, **61,** 355
strains and media
Fungal Genetics Stock Center, **61,** 353–354
MAG medium, **61,** 354
simple defined media, **61,** 354
YAG medium, **61,** 354
YG medium, **61,** 354
Assays
duplication, *see* Duplication assays
for Bacmid DNA transfections, **67,** 15
for centrosome proteins, **67,** 317–318
for β-galactosidase, **67,** 81–82
for SPB proteins, **67,** 91–92
for γTuSC and γTuRC, **67,** 19–21
functional, *see* Functional assays
immobilized centrosome recovery assay, **67,** 153–155
MT in extract, **67,** 241–243
MT nucleation, *see* Microtubule nucleation assays
MT reassembly *in vitro*, **67,** 182
Aster assembly
assay for centrosome kinases, **67,** 316

centrosomal and noncentrosomal protocols, **67,** 243
in centrosome reproduction studies, **67,** 281–283
MTs
induction, **67,** 228–229
mitotic extract preparation, **67,** 226–228
motor function, **67,** 229–233
noncentrosomal, **67,** 243
Asthma
data pattern classification, **64,** 494, 497–503
immunophenotype classifications, **64,** 507
multiparameter flow cytometric analysis, **64,** 490
Astrocytes
contaminating cortical cultures, **71,** 120
effects on Purkinje neurons, **71,** 101
GFAP as marker of, **71,** 119
in Purkinje neuronal cultures, **71,** 93
ATA, **66,** 264
Ataxia telangiectasia (AT), associated with Purkinje neuron degeneration, **71,** 91
ATCC, *see* American Type Culture Collection
Atebrine, **65,** 9
NADH oxidation, **65,** 10
Atmosphere, and choice of neuronal culture, **71,** 12
Atomic force microscope
applications
intermolecular force measurements, **55,** 101–104
optical force microscopy, **55,** 100–101
overview, **55,** 99–100
Atomic force microscopy
active Q-control
dynamic force spectroscopy, **68,** 347–351
imaging, **68,** 344–345
advanced fluid handling
DNA digestion by DNase, **68,** 227
future outlook, **68,** 227
operation principles, **68,** 226–227
overview, **68,** 225–226
advantages, **68,** 337–338
apical plasma membrane structures, **68,** 39–40
aqueous fluids
cantilever, bubble removal, **68,** 216
cantilevers, **68,** 218
different solutions, **68,** 218–219
modes, **68,** 217
parameters, **68,** 217–218
when not to image, **68,** 221–223
when to image, **68,** 221
with O ring, **68,** 216
without O ring, **68,** 215–216
bacterial pore, **68,** 287–289
basic setup, **68,** 151–153
basic technique, **68,** 142
bilayers
charged lipids, **68,** 238–240
ligand-linked lipids, **68,** 237–238
binding force measurement on intact cells, **68,** 178–179
binding map analysis, **68,** 179–180
biomechanics, **68,** 184
cantilever
AFM studies, **68,** 218
bubble removal, **68,** 216
JAR cell culture, **68,** 109–110
materials, **68,** 303–304
oscillation, **68,** 339
preparation, **68,** 303–305
quality factor, **68,** 339
selection, **68,** 303
tip calibration, **68,** 304–305
tip functionalization, **68,** 303
cell adhesion receptors, **68,** 307–308
cell applications
actin filaments, **68,** 78–80
contact mode, **68,** 57–58
cytoskeleton elasticity, **68,** 76–77
elasticity, **68,** 74–76
homogeneity, **68,** 83–84
imaging consequences, **68,** 80–82
limitations, **68,** 61–63, 82
oscillating mode, **68,** 58–61
sample thickness, **68,** 82–83
cell immobilization, **68,** 53–54
cell viability, **68,** 56
cellular dynamics, **68,** 84–86
complementary structural information, **68,** 291
confocal imaging, live cells, **68,** 184
confocal microscopy, **68,** 176–177
conventional physiological methods, **68,** 142
cryo-, *see* Cryo-atomic force microscopy
depressions
fusion pores, **68,** 40, 43
transient fusion, **68,** 43–47
dynamic, *see* Dynamic force microscopy
equipment, **68,** 52–53
features, **68,** 67–68
finite element modeling, **68,** 182
frame grabbing, **68,** 176–177
hair bundle mechanics, lateral link effects, **68,** 155–159

Atomic force microscopy (*cont.*)
hair cell measurement, **68,** 148–150
high-resolution, *see* High-resolution atomic force microscopy
important symbols, **68,** 86–87
improvement areas, **68,** 187–188
individual protein conformation, **68,** 291
induced strain calculations, **68,** 181–182
instrumentation and function, **68,** 2–3
live cell studies, **68,** 37–38, 47–48
live osteoblasts, **68,** 184
macromolecular structures, **68,** 290
material property analysis, **68,** 180–181
matrix dynamics, **69,** 164–165
membrane imaging, **68,** 63
membrane proteins
antibody labeling, **68,** 264–265
assembly, **68,** 285, 291–292
force-induced conformational changes, **68,** 272–274
oligomerization, **68,** 270–272
polypeptide end removal, **68,** 267–270
polypeptide loop removal, **68,** 267
polypeptide loop replacement, **68,** 265–267
methodological strategies, **68,** 142
mica, **69,** 166
microscope–AFM interface, **68,** 174–176
modes, **68,** 338
native membrane protein
HPI layer changes, **68,** 280–282
OmpF porin conformation, **68,** 282–285
surface structure changes, **68,** 278–280
OHC transduction channel gating, **68,** 161–165
operation, **68,** 34–35
operation modes, **68,** 35–36, 53
optical and patch–clamp technique combinations, **68,** 19
optical microscopy combination
infected monkey kidney cells, **68,** 8–9
living cell imaging, **68,** 7–8
pipetting concepts, **68,** 10–11
osteoclasts
$\alpha_v\beta_3$ integrin adhesion force, **68,** 182–184
receptor mapping, **68,** 184
other techniques, **68,** 143
overview, **68,** 172–174, 214
patch–clamp technique
AFM/patch–clamp setup, **68,** 150–151
instrumentation, **68,** 11–12
membrane patch, **68,** 12
setup, **68,** 153
voltage-sensitive ion channels, **68,** 12–13
pharmaceutical applications, **68,** 188–190
probe microscopy, **69,** 167–168
probe microscopy, air
ECM macromolecules, **69,** 183–185
fibrillar collagens, **69,** 187–188
FN–heparin–gold complexes, **69,** 185–187
overview, **69,** 182–183
probe microscopy, fluid
bioartificial membranes, **69,** 177–179
collagen degradation, **69,** 171–174
collagen III assemblies, **69,** 174
force measurements, **69,** 181–182
$\alpha_5\beta_1$ immobilization, **69,** 179–181
laminin-1 dynamic motion, **69,** 168–171
overview, **69,** 174–175
unfolding tenascin, **69,** 175–177
probes, **68,** 36
protein adsorption, **68,** 15
protein engineering
overview, **68,** 314–315
sample preparation, **68,** 315–316
protein folding, **68,** 15–19
protein modular structure, **68,** 15
protein–protein interactions, **68,** 292
protein unfolding
biological implications, **68,** 328
calibration, **68,** 332
data acquistion, **68,** 331–332
data fitting procedures, **68,** 323
double-sensor-stabilized AFM, **68,** 328–331
force curves, **68,** 317–318
force–clamp experiments, **68,** 318–319
force spectroscopy mode, **68,** 317
general scheme, **68,** 312–313
Monte Carlo simulations, **68,** 327, 334
multiple pickup, **68,** 322–323
pathway questions, **68,** 324–325
refolding, **68,** 319–321
single-parameter model, **68,** 325–327
sorting data, **68,** 321
range of analysis, **68,** 70–74
recognition imaging
dynamic recognition force microscopy, **68,** 134–137
lateral force mapping, **68,** 133–134
resonant frequency, **68,** 339
sample, **68,** 37
sample preparation, **69,** 166; **68,** 305–306
sample qualities, **68,** 231–232
scanning forces, **68,** 54–56
schematics, **68,** 302–303

single-molecule imaging, **68,** 292
soft sample force curves, **68,** 70
spectrin, **68,** 313
streptavidin–biotin unbinding, **68,** 306–307
substrate choice, **68,** 232
substrates, **68,** 215
supported bilayer substrates
Langmuir trough, **68,** 234–235
Langmuir trough and small Teflon wells, **68,** 235–236
small vesicle fusion, **68,** 232–234
surfaces, **68,** 215
tapping, *see* Tapping mode atomic force microscopy
technique improvements, **68,** 289–290
technology overview, **68,** 147–148
tips
cleaning, **68,** 117–118
crosslinker binding, **68,** 119
esterification, **68,** 118
ligand binding, **68,** 119
ligand density and funtionality, **68,** 119–120
modifications, **68,** 177–178
probe surfaces, **68,** 120–121
shape, **68,** 82
topographs
bacteriorhodopsin surface, **68,** 276–278
OmpF porin, **68,** 275
structural information, **68,** 290–291
transverse dynamic
application, **68,** 353–354
overview, **68,** 351–353
ultralow force measurement, **68,** 115
unbinding force
distribution, **68,** 123
force–distance cycle, **68,** 121–123
very small force application, **68,** 143
z-range scanner, **68,** 177
Atomic force microscopy (AFM), chromatin and nucleosomes, **53,** 182
Atomic model, AFM topograph comparison
bacteriorhodopsin surface, **68,** 276–278
OmpF porin, **68,** 275
ATP, **65,** 365–371
glucose-induced changes, **65,** 370
measurement in vivo, **65,** 367–368
measuring, **65,** 365–366
mitochondria, **66,** 382
standard assay, **65,** 142
synthesis
mitochondrial disease cellular models, **65,** 141–143
ATP synthases, AFM studies
III_x oligomer, **68,** 271–272
overview, **68,** 270–271
ATP synthesis
animal cells, **65,** 133–143
ATP, turnover observation, low-background total internal reflection fluorescence microscopy
kinesin molecule, **55,** 125–127
myosin molecule, **55,** 123–125
ATPase oligomers
coimmunoprecipitation, **65,** 223–225
Atractyloside, **65,** 137
CMXRos, **65,** 155
PT pore opening, **65,** 155
AtT-20α5 cells, cell-assembled ECM, **69,** 152–153
Aufderheide rotocompressor device
assembly, adjustment, repair, **70,** 351–352
background information, **70,** 347–349
lubrication and maintenance, **70,** 353
parts description, **70,** 349–351
use, **70,** 352–353
Aurintricarboxylic acid (ATA), **66,** 264
Authentication, cell lines
cell bank protocol, **57,** 41–43
chromosome analysis, **57,** 209–211
DNA fingerprinting, **57,** 211–214
history, **57,** 204–206
importance in various applications, **57,** 204–205
invertebrate cell culture, **57,** 198
isoenzyme phenotyping, **57,** 207–209
species-specific immunofluorescence, **57,** 206–207
Autoclave, animal cell culture, **57,** 8, 53–54
Autocorrelation function, spatial arrangement, **53,** 61–63
Autofluorescence
cancerous tissue imaging *in vivo*, **70,** 141
Drosophila embryo confocal imaging, **70,** 368–369
during immunophenotyping, **63,** 438
in studying ECM–cell interactions, **63,** 591–593
plant cells: problems with, **63,** 511
reduction, **63,** 410
Autofluorescence, *Drosophila*, **58,** 285, 288
Automated amino acid analysis
ε-(γ-glutamyl)-lysine cross link density, **66,** 118–119
Autonomously replicating vectors, **62,** 495
Autosomal recessive disorders, ataxia telangiectasia (AT), **71,** 91

Avalanche photodiode, **63,** 121–122
Average multiplicity index, **64,** 492
Averaging, confocal microscopy, **70,** 54
Averaging filters
  mechanics, **56,** 81
  spatial noise reduction, **56,** 100–101
  temporal noise reduction, **56,** 95–96
Avian leukemia virus vectors, **71,** 370
Avian Purkinje neuronal cultures, **71,** 89–105
Avian retroviral vector RCAS, **71,** 369
Avidin, **65,** 319
Avidin/biotin, protein labeling, **63,** 191–193
Avidin, deglycosylated form, **63,** 192
Axenic strains
  Dictyostelium, **66,** 473–476
Axial resolution, wide-field confocal microscopy, **70,** 250
Axon guidance
  B35 cell lines in study of, **71,** 287, 291–292
  in chick embryos, **71,** 67
  in *Heliosoma* cultures, **71,** 158
  in vivo study of, **71,** 154
  in *Xenopus* cultures, **71,** 130–131
  lipofection for studies of, **71,** 136
  Tibia-1 pathway studies in, **71,** 172–191
  toxicologic effect of lead on, **71,** 290
Axon regeneration, Purkinje neuronal culture studies of, **71,** 91
Axon size, **71,** 8–9
Axonal and dendritic length, neural-enriched cultures for study of, **71,** 143
Axonal development
  chick forebrain neuron studies of, **71,** 53–62
  examining with fluorescent tracing, **71,** 132–133
  in hippocampal and cortical neurons, **71,** 111
  in long-term cortical and hippocampal cultures, **71,** 124
  observing in *Heliosoma* neurons, **71,** 157
  slow in cortical neuronal cultures, **71,** 119–121
  speeded by glia-conditioned medium, **71,** 120
  studies in *Xenopus* spinal neurons, **71,** 130
  variable according to substrate, **71,** 146–147
Axonal growth cones, properties of, *vs.* dendritic growth cones, **71,** 2
Axonal halo, and peripheral neuron explant cultures, **71,** 29
Axonal morphogenesis, chick forebrain neurons, **71,** 52
Axonal retraction experiments, peripheral neuron cultures and, **71,** 25
Axonal transport, **71,** 18
  live-cell imaging in cultured neurons, **71,** 305–321
  observing, **71,** 314–321
Axons, outgrowth
  DiI microinjections, **69,** 355
  embryo slices, **69,** 355–356
  overview, **69,** 353
  slice preparation, **69,** 353
Axotomy, peripheral neurons and, **71,** 18
Azure C, nucleolus staining, **53,** 305

B

2B4.11 cell line, **66,** 418, 420–423, 425, 426–427
  transfection, **66,** 426–427
B lymphocytes, cultures, **63,** 390–392
BA:BB, *see* Benzyl alcohol:benylbenzoate solution
BAC, *see* Bacterial artificial chromosome
Background approximation, in CRM, **63,** 589
Background binding, nonspecific antibodies, **63,** 283–284
Backreflection, instrumental noise, **55,** 60, 152–153
Backscattered light reflection, CLSM enhancement, **70,** 280–281
Bacmid DNA, Sf9 cell transfection, **67,** 15
Bacteria
  antibiotic effects, **64,** 558
  cell cycle, **64,** 540
  cell viability, **64,** 521–524
  detection, **64,** 514–516
    by artificial neural networks, **64,** 520–521
  DNA content, **64,** 516–517
  DNA staining, **64,** 541–542
  dye efflux, **64,** 543
  enzymatic activity, **64,** 523–524
  fixed, staining, **64,** 543–544
  flow cytometer setup for, **64,** 512–513
  flow cytometric measurement, **64,** 547–548
  flow cytometry, fluorescent dyes for, **64,** 554–555
  Gram stain, **64,** 525–526
  immunofluorescence, **64,** 517–520
  light scatter measurements, **64,** 516
  membrane potential, **64,** 522–523
  rapidly growing, drug effects, **64,** 558–562
  sample preparation, **64,** 513–514
  size and shape, **64,** 541
  spores and sheath fluid, **64,** 546
  viable, FISH identification, **64,** 524–525
  vital cells, fluorescent dye uptake, **64,** 555–558
  vital staining, **64,** 544–546

Bacterial artificial chromosome
comparison with other artificial chromosome systems, **60,** 236–237
complementation by injection of genomic clones, **60,** 275–276
overview of cloning system, **60,** 238, 272–273
positional cloning
candidate gene isolation
direct cDNA selection, **60,** 248
direct hybridization of clones to cDNA libraries, **60,** 249, 279, 281–282
direct sequencing, **60,** 249–250, 282–284, 278–280
exon trapping, **60,** 248
gene density in genome, **60,** 247
validation of genes, **60,** 250, 280–281
chromosomal walking
end-isolation of inserts, **60,** 244–246
large-insert clone characterization, **60,** 246–247
overview, **60,** 244, 273
genetic mapping, **60,** 243–244
overview of steps, **60,** 243
zebrafish genomic libraries
availability, **60,** 239–241
hybridization-based screening, **60,** 241, 273, 276–277
polymerase chain reaction-based screening, **60,** 241
Bacterial cells, making efficient recombinant adenovirus vectors in, **71,** 390
Bacterial fusion proteins, antibodies against, **67,** 131–132
Bacterial overexpression
recombinant caspases, **66,** 7–9
Bacterial pore, single protomer unzipping
AFM studies, **68,** 287
entire pore unzipping, **68,** 287–289
Bacterial spheroblasts
annexin V purification, **66,** 344
Bacterial sphingomyelinase assay
metabolic labeling, **66,** 142
Bacteriophage T5, STM
general characteristics, **68,** 6
indium tin oxide, **68,** 5
instrumentation, **68,** 5
tail structures, **68,** 6
Bacteriorhodopsin
AFM studies, membrane protein assembly, **68,** 285
EF loop, polypeptide, AFM imaging
end removal, **68,** 267–270
force-induced conformational changes, **68,** 272–274
loop removal, **68,** 267
loop replacement, **68,** 265–267
surface, AFM topograph–atomic model comparison, **68,** 276–278
Baculovirus
insect cells
tumor necrosis factor-related ligands, **66,** 503–504
Baculovirus expression systems
Apaf-1 purification
caspase assay, **66,** 21–25
Baculovirus, for γTuSC expression, **67,** 14–16
Baculovirus, pCOCO, **69,** 75–76
Bad
protein interactive cloning, **66,** 35
yeast, **66,** 464
Baffles, use in limiting blast damage in biolistics, **71,** 359
Bak (Bcl-2 homologous antagonist killer)
domains, **66,** 32–33
overexpression, **66,** 464
PCR, **66,** 34
Banker method, **71,** 112–121
Barley aleurone protoplasts
GA, **66,** 442
Barrel design, improvements in biolistics, **71,** 366
Barrier-free path length, plasma membranes, **55,** 186–187
Basal bodies, **62,** 32, 34, 53–54, 68, 70, 82, 88, 335
Basal lamina
adhesive role of, **71,** 177
establishment, in Tibia-1 pathway, **71,** 181
in grasshopper limb bud studies, **71,** 165
Base hydrolysis
ceramide, **66,** 148
Base-unpairing region (BUR)
binding proteins, *see* Base-unpairing region binding protein
functions in cell, **53,** 324–325
identification in matrix attachment regions
chemical cleavage of modified DNA, **53,** 326
chloracetaldehyde modification of plasmids, **53,** 325
end labeling and purification, **53,** 325–326
gel electrophoresis, **53,** 326–327
matrix attachment region component, **53,** 324
Base-unpairing region binding protein
ammonium sulfate precipitation, **53,** 334

Base-unpairing region binding protein (*cont.*)
buffers in purification, **53,** 333
DNA affinity chromatography
column preparation
oligonucleotide annealing and phosphorylation, **53,** 330
oligonucleotide coupling, **53,** 332
oligonucleotide ligation, **53,** 330–331
oligonucleotide purification, **53,** 329–330
oligonucleotide sequences, **53,** 327, 329
Sepharose activation, **53,** 331–332
running conditions, **53,** 334–335
DNA ligand isolation and cloning
amplification and cloning of cross-linked DNA
isolation of cloned inserts, **53,** 347–348
linker ligation, **53,** 345–346
linker removal, **53,** 346–347
polymerase chain reaction, **53,** 346, 350
subcloning, **53,** 347, 350
binding assay, **53,** 348
chromatin cross-linking
cross-linking, *in vivo*, **53,** 342–343
genomic DNA labeling with tritiated thymidine, **53,** 342
restriction digestion, **53,** 343
urea gradient purification, **53,** 343
isolation of cross-linked DNA
digestion of DNA, **53,** 345
immunoprecipitation, **53,** 343, 345
reversal of cross-links, **53,** 345
sequencing, **53,** 350–351
expression library screening
hybridization, **53,** 341
plating of library, **53,** 341
probe preparation, **53,** 340
second round of screening, **53,** 341
solutions, **53,** 339–340
extraction
cultured cells, **53,** 333
tissue, **53,** 333
gel mobility shift assay, **53,** 338–339, 348
phosphocellulose chromatography, **53,** 333–334
probe preparation for assays, **53,** 335–336
Southwestern blotting assay, **53,** 336–338
types, **53,** 351
Basement membrane proteins
native
conditioned media, **69,** 115–116
EHS components, **69,** 115
fragments, **69,** 116–117
placental components, **69,** 115–116
recombinant
fragments, **69,** 118–120
heterotrimeric laminins, **69,** 120–124
single-subunit glycoproteins, **69,** 118–120
Basement membranes
3D matrix for cell culture
materials, **69,** 408
matrigel, **69,** 408
overview, **69,** 407
ECM, **63,** 563
formation, **69,** 112
formation *in vivo*, **69,** 114
intact, preparation from tissues, **63,** 565
laminin role, **69,** 112–114
macromolecular components, **69,** 112
models, **69,** 114
self-assembly
affinity chromatography, **69,** 124–125
affinity coelectrophoresis, **69,** 125–127
affinity retardation chromatography, **69,** 125
cell–surface contributions, **69,** 135–136
electron microscopy, **69,** 133–135
equilibrium gel filtration, **69,** 128
genetic and molecular analyses, **69,** 137–138
polymerization assays, **69,** 129–130
solid-phase binding, **69,** 128–129
solution binding analysis, **69,** 125
Baseplate, Aufderheide rotocompressor device, **70,** 349
Bax (Bcl-2 associated X protein), **66,** 30, 455
coimmunoprecipitation, **66,** 34
domains, **66,** 32–33
overexpression, **66,** 464
Bax-induced cell death inhibition
yeast, **66,** 465–466
BBMV, *see* Brush border membrane vesicles
BCECF, pHi measurement, confocal microscopy
dual-excitation LSCM, **70,** 420
overview, **70,** 418–419
Bcl-XL/Bcl-XS
hybridization, **66,** 33–34
structure, **66,** 33
Bcl-2 associated X protein, **66,** 30, 455
coimmunoprecipitation, **66,** 34
domains, **66,** 32–33
overexpression, **66,** 464
B50 cell lines, compared with B35 cell lines, **71,** 287–288
Bcl-2 cleavage assay, **66,** 44–45, 44

Bcl-2 family, **66,** 30, 455–456
amino acid sequences, **66,** 31
anti-Fas (CD95/Apo-1) induced cytotoxicity assay, **66,** 40
carboxy-terminal transmembrane (TM) domain, **66,** 31
cell death assays, **66,** 38–40
coimmunoprecipitation, **66,** 41–42
IL-3 withdrawal assay, **66,** 38–39
interactions, **66,** 40–42
posttranslational modifications, **66,** 42–45
proapoptotic, **66,** 30
schematic diagrams, **66,** 32
subgroups, **66,** 32–33
protein detection
antibodies, **66,** 37
regulation, **66,** 36–44
schematic diagrams, **66,** 32
subfamilies, **66,** 30
TNF-alpha cytotoxicity assay, **66,** 39–40
transfections, **66,** 36–38
Bcl-2 family cloning, **66,** 33–36
coimmunoprecipitation, **66,** 34
differentiation-induced genes, **66,** 36
DNA analysis, **66,** 33–34
EST screen, **66,** 34
PCR, **66,** 34
protein interactive cloning, **66,** 35
protein-protein interactions, **66,** 34–36
yeast two-hybrid, **66,** 35–36
Bcl-2 homologous antagonist killer
domains, **66,** 32–33
overexpression, **66,** 464
PCR, **66,** 34
Bcl-2 homology (BH) domain, **66,** 31–33
Bcl-2 interacting killer
domains, **66,** 33
yeast two-hybrid, **66,** 35
Bcl-2 phosphorylation assay, **66,** 42–44, 43
Bcl-2 protein, apoptotic effects, **63,** 471
Bcl-2 related ovarian killer
domains, **66,** 32–33
yeast two-hybrid, **66,** 35
Bcl-w
PCR, **66,** 34
Bead coating, cell membrane mechanics, **55,** 160–161
Beads
brightness, **64,** 298, 300
fluorescence, IFI value, **64,** 88–90
for instrument performance verification, **64,** 294–295
mixtures, in flow cytometry, **63,** 128
quantum simply cellular, **63,** 333
Rainbow
cross-calibration, **63,** 324–325
emission spectra, **63,** 319, 323–324
Beads, DNA, preparation, **67,** 245–246
Beam
dimensions, **63,** 24–26
focusing lenses, **63,** 23–24
illuminating, particle centering in, **63,** 26–33
Beam-pointing, instrumental noise fluctuations, **55,** 59, 62, 153
Beam scanning confocal microscope
slow-scan CLSM
Bio-Rad Laboratories, **70,** 26–27
Carl Zeiss, **70,** 27–29
Leica microsystems, **70,** 31–33
Nikon, **70,** 33–36
Olympus, **70,** 37–39
overview, **70,** 25–26
video-rate CLSM
Bio-Rad laboratories, **70,** 43
NORAN instruments, **70,** 39–42
overview, **70,** 39
Beam-splitting prisms
high-resolution multimode digital microscopy, **56,** 193–194
video-enhanced differential interference contrast light microscopy, **56,** 174
Beckman airfuge, **71,** 132
Beckman Instruments, **71,** 137–138
Beef heart mitochondria, **65,** 5–13
broken submitochondrial particles, **65,** 9
CoQ, **65,** 12–13
coupled submitochondrial particles, **65,** 7–9
cytochrome c, **65,** 9–12
oxidative phosphorylation system, **65,** 6
small-scale preparation, **65,** 5–6
Behavior, **62,** 363–375
Benzamidine hydrochloride, proteoglycans, **69,** 56–57
Benzyl alcohol:benylbenzoate solution, *Xenopus* oocytes and eggs
clearing procedures, **70,** 394–395
mounting procedures, **70,** 395–397
photobleaching, **70,** 397
Beta-actin neoepitope
detection, **66,** 408
Beta-galactosidase cell death assays
Drosophila S2 cells, **66,** 334–335
BFP, *see* Blue fluorescent protein

BH3-only subfamily, **66,** 30
domains, **66,** 32–33
Bias retardation, video-enhanced differential interference contrast light microscopy, **56,** 160–161
Bicinchoninic acid assay, signal transduction, **69,** 370
Bid
protein interactive cloning, **66,** 35
yeast, **66,** 464
Bifunctional cross-linking reagents, **65,** 226
Bik (Bcl-2 interacting killer)
domains, **66,** 33
yeast two-hybrid, **66,** 35
Bik-like-killer
domains, **66,** 33
EST screen, **66,** 34
Bim_L
domains, **66,** 33
protein interactive cloning, **66,** 35
BIM1–green fluorescent protein, fusion protein studies in yeast, **58,** 99
Binding assay
for *Spisula* MT nucleation, **67,** 158–161
for SPB proteins, **67,** 91–92
Binding force, intact cell measurements, **68,** 178–179
Binding kinetics, antibody, **63,** 220–224
Binding map analysis, AFM analysis, **68,** 179–180
Binding pocket, ligand–receptor recognition, **68,** 125
Binding probes
heparan sulfate
fixed cell monolayers, **69,** 99
heparin binding of HS, **69,** 98–99
specificity, **69,** 100–101
tissue sections, **69,** 101–102
regulatory HS identification
chimeric proteins, **69,** 97–98
FGF production, **69,** 95–97
Binding strength, vascular endothelial cadherin, **68,** 128–129
Binning
high-resolution multimode digital microscopy, **56,** 203
low-light microscopy, **56,** 54, 61
Biochemical procedures, **71,** 11
Bio-Rad Laboratories
confocal imaging resolution parameters, **70,** 321, 323–324
slow-scan CLSM, **70,** 26–27
video-rate CLSM, **70,** 43
Bioartificial membranes, preparation and characterization, **69,** 177–179
Biofilms, confocal objective lens selection, **70,** 275–278
Bioinformatics
web site, **66,** 249
Biolistic bombardment, *see also* Transformation, **62,** 501–511
Biolistics, **71,** 353–367
advantages over chemical transfection or viral infection methods, **71,** 354
compared with nuclear injection, **71,** 310–314
disadvantages of, **71,** 354
future improvements to, **71,** 365–367
parameters of, **71,** 355
untested parameters of, **71,** 365
Biolistics device, **71,** 358
Biological activity, retained by PTP-delivered cargo, **71,** 329
Biological ballistics, *see* Biolistics
Biological membrane solubilization
BN-PAGE, **65,** 235–236
Biological properties, of neurons, **71,** 8–10
Biological responsiveness, as criterion for choice of neuronal cultural culture, **71,** 2
Biological safety cabinet, **71,** 398, 401
Biomarkers
clinical, cytometric quantitation, **64,** 449
expression, correlation with proliferation data, **64,** 452
Biomechanics, AFM studies, **68,** 184
Biosafety, blood sample handling, **64,** 569
Biosafety level 2+ safety procedures, for adenovirus work, **71,** 405–406
Biospherix, **71,** 12
Biotin, **65,** 319
Biotin/avidin, protein labeling, **63,** 191–193
Biotin, labeled DNA: visualization, **64,** 65–66
5-biotinamido-pentylamine (X-biotin), **66,** 120
2′,7′-Bis(carboxyethyl)-5-carboxyfluorescein, intracellular ion ratio imaging, properties, **56,** 239–244
incorporation into intact living cells, **56,** 242–243
pitfalls, **56,** 243–244
signal-to-noise ratio, **56,** 243
spectral characteristics, **56,** 239–242
Bivariate analysis, cyclins, **64,** 433
Bivariate analysis, PCNA expression, **63,** 424–425
Bivariate distributions, mitotic cells, **63,** 348
Bivariate profiles, hemoglobin

effect of TGF-β, **64,** 145–147
flow data, **64,** 143–144
isolation of fetal nucleated red cells, **64,** 144–145
problems and limitations, **64,** 149–151
study of sickle-cell erythropoiesis, **64,** 148–149
BJ5183 *Escherichia coli,* **71,** 394
preparing electrocompetent vector containing pAdEasy-1, **71,** 394
Black-and-white video signals, *see also* Video monitors
basic adjustments, **56,** 2–4, 10–12
Blank standard, defining, **63,** 317
Blast cells
and monocytes: intermediate cell population between, **64,** 353–354
populations, surface antigens, **64,** 351–352
studied within bone marrow, **64,** 344
Blast damage, in biolistics, **71,** 359
Blastomere injections
compared with lipofection, **71,** 136
with *Xenopus* cultures, **71,** 131, 133–135
BLAST search
web sites, **66,** 494
Blasticidin, **66,** 485–486
Blastomeres
abnormal, **64,** 111
preparation for
chromosome translocation study, **64,** 102–103
PRINS, **64,** 58–59
Bleaching
CSU-10 fluorescence, **70,** 101–104
Xenopus oocytes and eggs, **70,** 391
Blk (Bik-like-killer)
domains, **66,** 33
EST screen, **66,** 34
Block, cell cycle, *see* Synchronization, cell cycle
Blocking
Fc receptors with IgG prior to staining, **63,** 229–232
with IgG, **63,** 239–240
Blood
collection, transport, and storage, **63,** 208–209
peripheral, erythrocyte lysis protocols, **63,** 214
Blood cells
early counters, **63,** 5–6
fixation and preservation, **63,** 210
preparation, **63,** 207–214
staining, **63,** 212
viability, **63,** 211–212
Blood cultures, maternal, isolation of fetal nucleated red cells, **64,** 144–145
Blood gene expression, **59,** 313–333
Bloomington stock center (http://flystocks.bio.indiana.edu/), **71,** 202
Blot overlay assays, **65,** 218–219
Blue fluorescent protein
flow cytometry of expressing cells, **58,** 330
green fluorescent protein fluorescence resonance energy transfer
applications, **58,** 103–104, 294, 300–301, 307–309
controls for microscopy, **58,** 303
design of constructs, **58,** 301
fluorescence microscopy imaging system
acquisition and processing of images, **58,** 306–307
deconvolution of images, **58,** 307
detectors, **58,** 305–306
filters, **58,** 305
microscope, **58,** 303–305
functional assays, **58,** 301–302
Pit-1 dimerization, **58,** 307–309
transient expression techniques, **58,** 302–303
multiple organelle labeling, **58,** 81
Blue fluorophores, **70,** 177–178
Blue-native electrophoresis, **65,** 232–237
buffers, **65,** 232–233
gel types, **65,** 233–234
protein assembly, **65,** 208–209
sample preparation, **65,** 234–237
Blue-native gels, **65,** 231–242
protein complex isolation, **65,** 231–242
Blue native-polyacrylamide gel electrophoresis (BN-PAGE)
biological membrane solubilization, **65,** 235–236
buffers, **65,** 233
chromatographically prepurified complexes, **65,** 241–242
future, **65,** 242
gel types, **65,** 233
membrane protein solubilization, **65,** 234
native protein electroelution, **65,** 236–237
native protein semidry electroblotting, **65,** 237
OXPHOS, **65,** 238
plant mitochondria, **65,** 67–68
protein load, **65,** 234–235
running conditions, **65,** 236
Blurring, reduction in confocal microscopy, **63,** 94
BMP signalling pathway, zebrafish, **59,** 163

BN-PAGE
see Blue native-polyacrylamide gel electrophoresis
B35 neuroblastoma cell lines, **71,** 3, 8
antisense RNA use in, **71,** 295
as models for axon growth and cell migration, **71,** 291–292
cell death studies using, **71,** 289–290
compared with B50 cell lines, **71,** 287–288
detecting migration and expression via immunocytochemistry, **71,** 301
developmental properties of, **71,** 5
differentiation studies, **71,** 289–290
intracellular signaling mechanisms in, **71,** 292
protocols for working with, **71,** 295–302
research applications for, **71,** 289–290
toxicology studies using, **71,** 289–290
BO-PRO-3, *Xenopus* oocytes and egg chromosome staining, **70,** 393–394
BODIPY, *see* Boron dipyrromethene difluoride
Bok (Bcl-2 related ovarian killer)
domains, **66,** 32–33
yeast two-hybrid, **66,** 35
Bond lifetime, ligand–receptor recognition, **68,** 124
Bone marrow
for MRD detection in leukemias, **64,** 386
light microscopic examination, **64,** 347–348
sample processing, **64,** 345
Bone marrow stem cells, **66,** 224
Borohydride reduction, *Xenopus* oocytes and eggs, **70,** 391
Boron dipyrromethene difluoride, **70,** 171–173
Bouton development, in *Drosophila* studies, **71,** 228
Bovine corneal endothelial cell basement membrane, glioma culture
biological effects, **57,** 161–162
cell attachment to culture dishes, **57,** 151, 161
Bovine heart mitochondria
giant multiprotein complexes, **65,** 240
Bovine serum albumin, gradient preparation, **57,** 172–173
Bovine serum albumin, supported bilayer substrate, **68,** 233–234
Boyden chamber assay, cell migration, **63,** 554–557
Brace–Kohler compensator retardance, **61,** 182
Bradford assay, signal transduction, **69,** 369
Brain mitochondria, **65,** 20–22
biochemical parameters, **65,** 22
respiratory chain activities, **65,** 22
Brain tissue, confocal objective lens selection, **70,** 273–274
Brain, slice preparation, **69,** 349
BrdU protocol
growth assays, **66,** 221–222, 222
BrdUrd, *see* 5′-Bromodeoxyuridine
BrdUrd, *see* Bromodeoxyuridine
Breast cancer
BrdUrd labeling, **64,** 463
cell lines, under serum-free conditions, **63,** 618–619
endocrine therapy, **64,** 376–377
JC-1 probe role, **63,** 478–479
proliferation markers and prognosis, **64,** 368–370
Brefeldin A, effect on protein expression, **63,** 227
Brightness, video microscopy
basic adjustments, **56,** 13
digital image analysis, **56,** 86–90
low-light microscopy, **56,** 47
Broken submitochondrial particles (SMP)
respiratory chain activities, **65,** 11
5′-Bromodeoxyuridine
as tumor radiosensitizer, **64,** 450
in early clinical studies, **64,** 453
incorporated
immunostaining of, **64,** 276
measurement by flow cytometry, **64,** 136–137
labeling of
cells, **64,** 131
RNA, **64,** 134–135
subcellular distribution, **64,** 132–133, 135
substituted RNA, **64,** 130–131
Bromodeoxyuridine
and general laboratory reagents, **63,** 359–360
anti-BrdUrd fluorescence signal, **63,** 364
anti-BrdUrd primary antibodies, **63,** 358
as radiation enhancer, **63,** 11
as tumor radiosensitizer, **64,** 450
cells labeled by, **63,** 356–357, 368–369
content, *vs.* DNA content: data displayed as, **63,** 365–366
in combination with CFSE labeling, **63,** 383–384
in early clinical studies, **64,** 453
incorporated
immunostaining of, **64,** 276
measurement by flow cytometry, **64,** 136–137
labeling of
cells, **64,** 131
RNA, **64,** 134–135
subcellular distribution, **64,** 132–133, 135
substituted RNA, **64,** 130–131

Bromodeoxyuridine, *see* DNA replication site, mapping *in situ*
Brownian motion
hydrodynamic drag, **55,** 139–140
overview, **55,** 129–131, 136, 154
power spectrum, **55,** 136–138
trap calibration, **55,** 138–139
Brush border membrane vesicles, dynamic force microscopy, **68,** 131
BSA adhesion assay
deposited ECM, **69,** 35–36
Mabs for instructive ECM identification, **69,** 45
BSA fraction V, **71,** 84
BSA, *see* Bovine serum albumin
BSL, *see* Backscattered light reflection
B27 supplement, **71,** 12
Buccal ganglion, of freshwater snail *Heliosoma,* **71,** 3
Budding cells
mitochondrial movements, **65,** 348
Buffers
BN-PAGE, **65,** 233
caspases, **66,** 9t–10
for baculovirus-expressed γTuSC, **67,** 17
for cell division analysis with CFSE, **63,** 378
for centrosome overproduction in mammalian cells, **67,** 257
for centrosome protein analysis, **67,** 262
for *Dictyostelium* centrosome, **67,** 339, 342, 348–350
for embryo extract preparation, **67,** 4, 7
for hydroxyurea-arrested CHO cell extract preparation, **67,** 260
for γTuSC and γTuRC, **67,** 18–19
for MT nucleation reconstitution, **67,** 142–143
in cell adhesion assay, **63,** 610
lysis, cell–matrix adhesion assays, **69,** 367–368
microtiter-based integrin binding assays, **69,** 252
optimal, for protein labeling, **63,** 187–189
Buffer composition, as electroporation parameter, **71,** 349
Buffers, SFM of DNA, **68,** 370–371
Bufo japonicus, cell-free chromatin remodeling
decondensation and remodeling, incubation conditions, **53,** 507
frog and extract preparation, **53,** 504
overview, **53,** 498
Bulk culture, lymphocyte subsets, **63,** 448–449
Bulked segregant analysis, positional cloning project initiation, **60,** 263–264
BUR, *see* Base-unpairing region
Bureaucracy, in neuronal culture systems, **71,** 14

C

C-banding, **61,** 58
C-Jun N-terminal kinases (JNKs), **66,** 187–188
apoptosis, **66,** 187–193
detection, **66,** 190–193
evaluation, **66,** 193
in vitro kinase assay, **66,** 190–192
endogenous JNK, **66,** 190–192
transfected JNK, **66,** 191–192
isoforms, **66,** 188–189, 188
signaling pathway, **66,** 189
Western Blotting
antiphospho-JNK antibody, **66,** 192
C-mitosis, **61,** 278
C. elegans
TGase glutamyl donor substrate proteins, **66,** 123–124, 124
C_0t curve, analysis of DNA classes, **60,** 100
C_2C_{12} cell line, **66,** 419, 420 t–423 t, 425 t, 430–431
transfection, **66,** 431
C32 strain, characteristics, **60,** 358
Cadherin, VE cells, dynamic force microscopy
binding strengh, **68,** 128–129
Ca^{2+} dependence, **68,** 127–128
conformation, **68,** 126
overview, **68,** 125–126
thermodynamics and structural parameters, **68,** 129–130
trans association, **68,** 127–128
Caged Ca^{2+}, **71,** 152–154
in PC12 cell line studies, **71,** 278–282
Caged-FITC dextrans, **71,** 132
Calcium
and exocytosis, **62,** 45–47
gradients, **62,** 364
intracellular
prostate adenocarcinoma, **66,** 241–242
measurements
aequorin, **65,** 358
Calcium-binding
calcium dyes, **66,** 236
Calcium binding proteins, **62,** 67, 455–464
Calcium dyes
calcium-binding, **66,** 236
fluorescence properties, **66,** 236
Calcium dynamics
imaging and manipulating in *Drosophila* neurons, **71,** 224–226
in *Drosophila* studies, **71,** 206–207
neural-enriched cultures for study of, **71,** 143

Calcium flux
apoptosis, **66,** 229–242
early studies, **66,** 229–230
calcium imaging, **66,** 237–242
calcium-specific probes, **66,** 234–237
compartmentalization, **66,** 230–233
intracellular pools, **66,** 234
nuclear, **66,** 233–234
Calcium Green-1, pCa_i, **70,** 456
Calcium imaging
aequorin, **66,** 239–240
calcium flux, **66,** 237–242
FACS, **66,** 239
fluorescence microscopy, **66,** 240–242
luminescence measurements, **66,** 239–240
spectrofluorimetric measurements, **66,** 237–239
Calcium indicator dextrans, **71,** 131
Calcium imaging, by confocal microscopy, **63,** 102
Calcium ion imaging
intracellular, *see* Intracellular calcium
live cell and fixed tissues, **70,** 271–272
multiphoton, Ca^{2+} dynamics, **70,** 141–142
Calcium ions, Ca^{2+} waves, study with JC-1 probe, **63,** 478
Calcium ions, intracellular ratio imaging, **56,** 237–251
experimental chamber, **56,** 249–250
fluorescence signal calibration, **56,** 245–247
imaging workstation components, **56,** 247–249
indicator properties, **56,** 239–244
incorporation into intact living cells, **56,** 242–243
pitfalls, **56,** 243–244
signal-to-noise ratio, **56,** 243
spectral characteristics, **56,** 239–242
structural characteristics, **56,** 239–242
overview, **56,** 237–238, 251
perfusion system, **56,** 249–250
rationale, **56,** 238
Calcium, measurements in protoplasts, **63,** 520, 522
Calcium phosphate, transfection with, **71,** 310
Calcium signaling, examining with dextran tracing, **71,** 132
Calcium-specific probes
aequorin, **66,** 236–237
calcium flux, **66,** 234–237
cameleons, **66,** 237
synthetic fluorescent dyes, **66,** 234–236
Calcium uptake, **65,** 357–358
Calcium, vascular endothelial cadherin, **68,** 127–128
Calcofluor, **66,** 483
stalk cells, **66,** 484
Calibrated measurements, applications of, **63,** 330–332
Calibration
common cytometry measurements, **63,** 312–330
fluorescence axes, problems with, **64,** 149–150
light scatter, **63,** 327–328
material for, **63,** 311
results of telomere length measurement, **64,** 85–86
Calmidazolium, **66,** 264
Calmodulin, **62,** 38, 54
Calmodulin, **66,** 264
Calpain, **66,** 251–255
cell-permeable inhibitors, **66,** 256–258
Calpeptin, **66,** 252, 263
Calyx terminals, applications of chick ciliary neuron, **71,** 38–42
Cameleons
calcium-specific probes, **66,** 237
Camera
fluorescence, **65,** 344
Camera, immunofluorescence microscopy, **57,** 314, 323–324
Cameras
photon counting, **65,** 367
Cameras, *see specific types*
Cameras, CCD
in flow cytometric detection, **63,** 122–123
in LSCM, **63,** 52, 57
Camptothecin
apoptosis induction, **64,** 429–430
cytotoxicity, **64,** 177
Cancerous tissue, multiphoton imaging *in vivo*, **70,** 141
Candidate gene approach
isolation from large-insert clones
direct cDNA selection, **60,** 248
direct hybridization of clones to cDNA libraries, **60,** 249, 279, 281–282
direct sequencing, **60,** 249–250, 282–284
exon trapping, **60,** 248
gene density in genome, **60,** 247
validation of genes, **60,** 250, 280–281
limitations, **60,** 260–261
overview, **60,** 260
Cantilevers, AFM probe microscopy, **69,** 167–168
CAP2–green fluorescent protein, fusion protein studies in yeast, **58,** 102–103
Capping assay, for γTuSC and γTuRC, **67,** 20–21

Carbocyanine dyes, symmetrical and asymmetric, **64,** 119–120
Carboxy-terminal transmembrane (TM) domain
  Bcl-2 family, **66,** 31
Carboxyfluorescein diacetate succinimidyl ester, *see* CFDA-SE
Carboxyfluorescein succinimidyl ester, *see* CFSE
Carboxyseminaphthorhodafluor-1 (SNARF1), **65,** 372
Carboxyseminapthofluorescein (SNAFL), **65,** 372
CARD, **66,** 4, 454
Cardiolipin dye, for mitochondria, **64,** 120, 124–125
Carl Zeiss, slow-scan CLSM, **70,** 27–29
Carolina Biological (www.carolina.com), **71,** 70
Carotid artery, bifurcation of rat, **71,** 26
Caryonide, *see* Karyonide
Caspase-2, **66,** 366
  mitochondria released, **66,** 377
Caspase-3, **66,** 182
  apoptosis, **66,** 456
    flow cytometry, **66,** 63
  detection, **66,** 408
Caspase-7
  detection, **66,** 408
Caspase-8, **66,** 369
  enzymatic activity, **66,** 16
  yeast growth, **66,** 463
Caspase-9, **66,** 366
  Apaf-1 apoptosome assay, **66,** 18–20
  depletion
    S-100 extracts, **66,** 25
  detection, **66,** 408
  mitochondria released, **66,** 377
Caspase-10
  yeast growth, **66,** 463
  yeast-induced cell death, **66,** 463
Caspase-associated recruitment domain (CARD), **66,** 4, 454
Caspase-induced cell death inhibition
  yeast, **66,** 465–466
Caspases
  activation, **66,** 4–5
    cell-free systems, **66,** 181–183, 182
  activation, identification with fluorogenic substrates, **63,** 495, 497
  assays, **66,** 5–25, 175–181
    affinity labeling assay, **66,** 176
    affinity labeling caspases, **66,** 176–177, 178
    Apaf-1 purification
      baculovirus expression systems, **66,** 21–25
    caspase labeling, **66,** 177–178
    caspase processing, **66,** 179–181
    caspase purification
      inclusion bodies, **66,** 12–14
    cloning, **66,** 6–7
    fluorometric assays, **66,** 15
    materials, **66,** 5–6
    methods, **66,** 6–25
    PARP cleavage assay, **66,** 175–176
    protease assay, **66,** 178–179
    protein substrate assay, **66,** 16–17
    radiolabeled substrate cleavage, **66,** 17–18
    recombinant caspase expression, **66,** 7–9
    S-100 extract preparation, **66,** 14
  buffers, **66,** 9–10
  cell death responses to, **63,** 488
  cell-permeable irreversible inhibitors, **66,** 253–254
  cleavage sequences, **66,** 9
  flow cytometry
    apoptosis, **66,** 50
  inhibition, followed by CD95 induction of Jurkat cells, **63,** 490
  labeled inhibitor analysis, **66,** 93–97
  labeling, **66,** 177–178
  mitochondria released, **66,** 377
  oxidants, **66,** 313–315
  pathways, **66,** 454
  processing, **66,** 179–181
  prodomains, **66,** 4
  purification, **66,** 12–14
    recombinant caspase expression, **66,** 9–12
    soluble proteins, **66,** 10–12
  redox state, **66,** 315–318
  sequence alignment, **66,** 3
  structure and organization, **66,** 2–4, 4
  transformed yeast, **66,** 459
  zymogens, **66,** 3
Cassette purification
  yeast, **65,** 261
Catharanthus roseus
  TUNEL staining, **66,** 443
Cathepsin G, **66,** 5
Cationic dyes, Nernstian redistribution
  methods, **70,** 437–441
  plasma membrane potential, **70,** 434–436
  single mitochondria membrane potential, **70,** 436–437
Caudal cilium, **62,** 88
CB, *see* Coiled body
Cbs, *see* Chromosome breakage

CCD camera, **61,** 162, 184, 375, 378
CCD camera, *see* Charge-coupled device camera
CCD camera, *see* Cooled charge-coupled device camera
CCD, *see* Charge-coupled devices
CD10, expression patterns, **64,** 390
CD11b, coexpression with CD15, **64,** 156
CD11b, conjugated to phycoerythrin, **63,** 80–81
CD15, coexpression with CD11b, **64,** 156
CD15, conjugated to FITC, **63,** 80–81
CD3, and CD28, simultaneous binding, **63,** 435
CD34
  expression on immature HP, **64,** 154
  lost by developing HP, **64,** 158
CD45
  as gating reagent, **64,** 573
  B and T cells positive for, **64,** 333
  expression, AMLs showing original features of, **64,** 354
  intensity, combined with linear side scatter, **64,** 344
CD45/SSC gating, **64,** 345, 348–356
CD61, and cell differentiation, **64,** 159
CD69, identification of activated lymphocytes, **63,** 449, 451
CD95, induction of Jurkat cells, **63,** 500–501
  caspase inhibition followed by, **63,** 490
cDNA clone tagging
  applications
    candidate gene cloning, **60,** 232
    expression profiles, **60,** 223, 225
    genome mapping, **60,** 187, 190, 230–232
    homolog identification, **60,** 225–226
    isolation of full-length cDNA clones, **60,** 226–226
    marker identification, **60,** 230
    promoter isolation, **60,** 228, 230
  cDNA library
    construction, **60,** 215–216
    insert length determination, **60,** 216
  management of tagged clones
    documentation, **60,** 220–221
    plasmid storage, **60,** 220
    redundant clone minimization, **60,** 221–223
    tissue libraries, **60,** 223
  overview of expressed sequence tag generation, **60,** 214–215
  prescreening hybridization, **60,** 219–220, 222
  sequencing
    homology analysis by BLAST *vs.* FASTA, **60,** 217–219
    manual sequencing, **60,** 217
    plasmid preparation, **60,** 216–217
CED-3, **66,** 290
ced-3 gene, required for programmed cell death, **63,** 487–488
CED-9, **66,** 30
Cell adhesion assay
  adhesive substrata preparation, **69,** 283–284
  basement membrane self-assembly, **69,** 135–136
  blockage of nonspecific adhesion, **69,** 284–285
  cell preparation, **69,** 285–286
  glycosaminoglycans, **69,** 298–299
  materials, **69,** 299
  overview, **69,** 299
  procedure, **69,** 300–302
  reporter cells to instructive ECM, **69,** 47
  stimulatory signals, **69,** 261–262
Cell adhesion, determination techniques, **68,** 91–92
Cell adhesion force spectroscopy
  cell–cell adhesion force
    research results, **68,** 105–109
    single cell immobilization, **68,** 104–105
  Cell-cell attractions/repulsions, **71,** 3, 10
  cell layer measurements
    monolayer immobilization, **68,** 99–100
    research results, **68,** 100–104
  cell–surface adhesion force
    functionalized sensors, **68,** 98
    passivated force sensors, **68,** 95
    research results, **68,** 95–99
    sphere immobilization, **68,** 94
  instrumentation, **68,** 92–93
Cell adhesion molecules, endocytosis of, in B35 cells, **71,** 294–295
Cell adhesion receptors, AFM studies, **68,** 307–308
Cell adhesion studies
  aggregation assays, **63,** 600–603
  by confocal microscopy, **63,** 102
  monolayer cell–cell adhesion assays, **63,** 603–605
  reversed centrifugation for, **63,** 605–606
Cell adhesion, specificity of, **63,** 608–609
  optimization of experimental conditions for, **63,** 609–611
Cell-assembled extracellular matrix
  advantages, **69,** 152
  cells and proteins, **69,** 152–153
  disadvantages, **69,** 152
  DOC-soluble and -insoluble material, **69,** 154
  immunofluorescence staining, **69,** 153

serum starvation, **69,** 154
time course, **69,** 154
troubleshooting, **69,** 155
Cell attachment assay
cell–matrix interactions, **69,** 287–288
counting quantitation, **69,** 286–287
fluorescent dyes, **69,** 287
incubation time, **69,** 287
radioactivity, **69,** 287
washing step, **69,** 287
Cell behavior, and choice of neuronal culture, **71,** 2–5
Cell biology
confocal microscopic applications, **63,** 100
importance of extracellular matrix to, **63,** 561–562
Cell biological mechanisms
correlating with growth cone motility, **71,** 157
in *Drosophila* studies, **71,** 216
monoamine biogenesis, **71,** 267
neurotrophin action, **71,** 267
protein trafficking, **71,** 267
secretory vessel dynamics, **71,** 267
Cell biopsy, and fixation, **64,** 111
Cell body size, **71,** 8–9
Cell–cell adhesion force
research results, **68,** 105–109
single cell immobilization, **68,** 104–105
Cell cloning, for centrosome duplication assay, **67,** 290–291
Cell counter
Coulter counter, **57,** 8
maintenance, **57,** 8–9
Cell counts
CD4 absolute, **64,** 573–574
for erythropoiesis study, **64,** 142–143
Cell cracking methods, **71,** 272
Cell crawling and navigation, **71,** 2–3
Cell culture, **62,** 34–35, 187–206
cell division asynchrony in, **63,** 384
3-D ECM substrates for, **63,** 563–565
Dictyostelium, **66,** 473–476
embryonal zebrafish, **59,** 29–35
establishment of 3-D systems, **63,** 574–575
for centrosome duplication assay, **67,** 290–291
for MRD detection, **64,** 397
for yeast SPB permeabilization, **67,** 167–168
hanging drop culture, **62,** 285–286
integrity: quality control, **64,** 178–179
long-term storage, **62,** 200–206, 211–215, 524
mammalian cells, **67,** 181, 187
NADH, **66,** 309
nutritional requirements, **62,** 34–35
primary fibroblast, **59,** 39–42
caudal fin amputation, **59,** 40–41
cell lines, **59,** 41
cryopreservation, **59,** 41
introduction, **59,** 39–40
results and discussions, **59,** 41–43
tissue culture techniques, **59,** 40
zebrafish strains, **59,** 40
single cell isolation, **62,** 198, 231–237
SM labeling, **66,** 138
Cell culture, *see* Animal cell culture, Invertebrate cell culture
Cell cultures
3D matrix, basement membrane, **69,** 407–408
3D matrix, collagen I, **69,** 405–407
Dictyostelium, **68,** 110
HEC/RL cells on coverslips, **68,** 109
JAR cells on cantilever, **68,** 109–110
mammary epithelial, adenovirus infection, **69,** 420–422
mouse mammary epithelial cells, **69,** 409
Cell cycle, **62,** 68–79
and RNA synthesis, **64,** 134–135
bacteria, **64,** 540
difficulty in defining phases, **57,** 247–248
doubling time, **62,** 192–194
effect of TGF-β, **64,** 146
histone H_3 phosphorylation during, **63,** 343–344
kinases, **62,** 78–79
kinetics, estimation by BrdUrd incorporation, **63,** 357–372
phase duration estimation
flow cytometry, **57,** 237
importance in synchronization studies, **57,** 231–232
labeling index determination, **57,** 234–235
mean generation time, **57,** 232
mitotic index determination, **57,** 233, 235–237
proliferative fraction determination, **57,** 232–233
pulse-chase labeling, **57,** 235–237
regulation
cyclins and kinases, **57,** 230–231
growth-dependent *vs.* regulatory events, **57,** 247–248
related expression of DNA topoisomerases, **64,** 180
simultaneous analysis with apoptosis using flow cytometry
annexin V/cell cycle analysis

Cell cycle (*cont.*)
filters, **57,** 273
performance of assay, **57,** 273–274
preparation of cells, **57,** 272
principle, **57,** 267–268
staining of cells, **57,** 272–273
TUNEL/cell cycle analysis
filters, **57,** 269
fixation of cells, **57,** 268
labeled nucleotides, **57,** 275–276
performance of assay, **57,** 269–272
principle, **57,** 259, 267
staining, **57,** 268–269, 271–272
stages and transitions, **57,** 230
synchronization, *see* Synchronization, Cell cycle
synchronization, plant chromosomes, **64,** 19, 21
Cell cycles-dependent pericentrin recruitment, **67,** 58–60
Cell cycles, embryonic zebrafish, **59,** 11–25
Cell damage, affecting biolistics performance, **71,** 365–367
Cell death
in B35 and B50 cell line studies, **71,** 289–291
regulated by CyP-A, **71,** 291
target-driven *vs.* programmed, **71,** 38
Cell death
associated changes in scattering properties, **63,** 512, 514
flow cytometric analysis, **63,** 528
HIV-induced, **64,** 583
incidence, correlation with apoptotic index, **63,** 529–531
nuclear protein-induced, **64,** 270
programmed, *ced-3* required for, **63,** 487–488
responses to specific caspases, **63,** 488
Cell death, *see also* Apoptosis
assays
Drosophila S2 cells, **66,** 334–336
Bcl-2 family, **66,** 38–40
incidence
apoptotic index, **66,** 104–105
morphological features, **66,** 395–397
yeast-induced, **66,** 463
Cell death-resistant mutants, **66,** 491–494
course, **66,** 493
selection protocol, **66,** 491–494
Cell debris, in cortical neuronal cultures, **71,** 119
Cell density
as biolistics parameter, **71,** 354
as electroporation parameter, **71,** 345–346
as parameter of biolistics, **71,** 355
Cell-derived extracellular matrix, **69,** 157–158
Cell diameter, affecting survival of electroporation, **71,** 340
Cell differentiations, and choice of neuronal culture, **71,** 2–5
Cell disruption, **62,** 357
Cell dissociation solution, **71,** 77
in spinal and motor neuron cultures, **71,** 82
Cell division, **62,** 68–78
asynchrony, in cultures, **63,** 384
CFSE distribution during, **63,** 376–377
Cell enucleation, for centrosome duplication assay, **67,** 291–292
Cell–extracellular matrix interactions, **69,** 281
Cell extraction medium, in *Heliosoma* cultures, **71,** 160
Cell fate, in chick embryos, **71,** 68
Cell-free chromatin remodeling, *see also* Cell-free nuclear assembly
Bufo japonicus system
decondensation and remodeling, incubation conditions, **53,** 507
frog and extract preparation, **53,** 504
overview, **53,** 498
Drosophila system
decondensation and remodeling, incubation conditions, **53,** 507
extract preparation
eggs, **53,** 504
embryos, **53,** 504–505
preblastoderm embryos, **53,** 510–511
NURF role, **53,** 500
overview, **53,** 499
p22 role, **53,** 499–500
transcriptional remodeling, incubation conditions, **53,** 511
micrococcal nuclease digestion and chromatin structure analysis, **53,** 508–509
polyacrylamide gel electrophoresis of chromatin-associated proteins, **53,** 507–508
system selection, **53,** 510
Xenopus system
decondensation and remodeling, incubation conditions, **53,** 507
extract preparation, **53,** 500–504
frog selection and preparation, **53,** 501, 503
nucleoplasmin role, **53,** 498–499
overview, **53,** 498
sperm chromatin preparation, **53,** 505–507
Cell-free nuclear assembly
Drosophila system
disassembly

monitoring, **53,** 410–412
nuclei preparation, **53,** 409–410
stage 14 oocyte exctract preparation, **53,** 408–409
embryo extract preparation, **53,** 398–399
incubation conditions for assembly, **53,** 399
lamin assembly
disassembly assays, **53,** 412–413
interphase *vs.* mitotic/meiotic isoform assembly, **53,** 405–407
monitoring of aggregation and polymerization, **53,** 405
purification of native and recombinant proteins, **53,** 402–404
regions in head-to-tail binding, **53,** 406
regulation by lamin isoforms, **53,** 407–408
monitoring of assembly
immunofluorescence microscopy, **53,** 401
light microscopy, **53,** 400
transmission electron microscopy, **53,** 402
Western blot analysis, **53,** 402
sperm chromatin preparation, **53,** 399
egg maturation stage and male pronuclei assembly, **53,** 418
mammalian system, Chinese hamster ovary cells
collection of cells, **53,** 359–360
homogenate preparation, **53,** 360
immunodepletion studies of protein function, **53,** 363–364
immunofluorescence microscopy, **53,** 362–363
inhibition by ATP, **53,** 358, 363
nuclear envelope and lamina assembly, **53,** 361–362
nuclear reassembly, **53,** 360–361
solution preparation, **53,** 364–365
synchronization of cell cycle, **53,** 359
Western blot analysis, **53,** 363
overview of systems, **53,** 358
sea urchin system for male pronuclei assembly
chromatin decondensation
incubation conditions, **53,** 429–431
microscopic observation, **53,** 431–432, 446–449
stages, **53,** 430
egg extract preparation
egg collection and fertilization, **53,** 423–424
lysis, **53,** 424–425
membrane-free extracts, **53,** 426
membrane vesicle isolation and characterization, **53,** 426–429
lamina assembly and pronuclear swelling
antibodies for detection, **53,** 438–439
electron microscopy, **53,** 437–438
immunodepletion studies, **53,** 441–443
immunofluorescence microscopy, **53,** 439–440
induction, **53,** 436–438
Western blotting of lamins, **53,** 440–441
membrane protein analysis
N-ethylmaleimide treatment of vesicles, **53,** 444–445
extraction of proteins, **53,** 445–446
nuclear envelope preparation, **53,** 443–444
trypsinization of vesicles, **53,** 444–445
nuclear envelope formation
fluorescent dyes and membrane labeling, **53,** 432–433
membrane vesicle binding to decondensed chromatin, **53,** 433–434
membrane vesicle fusion and assays, **53,** 434–436
sperm
permeabilization, **53,** 419–422
collection and storage, **53,** 420
lipophilic structure isolation, **53,** 422
surf clam system for male pronuclei assembly
chromatin decondensation
incubation conditions, **53,** 431
microscopic observation, **53,** 431–432
egg extract preparation
egg collection and fertilization, **53,** 424
lysis, **53,** 425–426
membrane-free extracts, **53,** 426
lamina assembly and pronuclear swelling
immunofluorescence microscopy, **53,** 440
induction, **53,** 438
sperm
collection and storage, **53,** 420
lipophilic structure isolation, **53,** 422
permeabilization, **53,** 419–422
Xenopus system
advantages, **53,** 358–359, 370
egg extract preparation
buffer preparation, **53,** 376
centrifugation, **53,** 376–377
dejellying, **53,** 375–376
laying of eggs, **53,** 374–375
stability, **53,** 378
immunofluorescence microscopy
antibodies, **53,** 382
nuclei recovery, **53,** 382–383

Cell-free nuclear assembly (*cont.*)
troubleshooting, **53,** 383–384
nuclear envelope assembly
electron microscopy, **53,** 381–382
incubation conditions, **53,** 378, 390
membrane staining, **53,** 381
model, **53,** 391–393
morphology and DNA staining, **53,** 378, 380
protein import assay, **53,** 380–381
nuclear envelope disassembly
cyclin response, **53,** 384–385
metaphase extract response, **53,** 385
nuclear envelope precursor fraction A, isolation, **53,** 386–389
nuclear envelope precursor fraction B
isolation, **53,** 386–387, 389–390
sucrose flotation of vesicles, **53,** 390
S_{200} fraction
isolation, **53,** 386–387
properties, **53,** 387–388
sample preparation for electron microscopy, **53,** 134–137
sperm permeabilization
buffers, **53,** 371
lysolecithin treatment, **53,** 373
recovery from testes, **53,** 371–373
storage, **53,** 373–374
Cell-free nuclear import, *see* Protein import, nucleus, RNA transport, nucleus
Cell-free proteins
protein substrate assay
in vitro, **66,** 17
Cell-free systems, **66,** 167–183
caspase activation, **66,** 181–183, 182
extract preparation, **66,** 167–171
adherent cells, **66,** 170–171
protein concentrations, **66,** 171
requirements, **66,** 168–169
suspension cells, **66,** 169–170
in vitro apoptosis assays, **66,** 171–181
caspase assays, **66,** 175–181
nuclei assays, **66,** 171–175
mitochondrial cytochrome *c* release, **66,** 375–377
Cell fusion, laser micromanipulation, **55,** 85–87
Cell heterogeneity, detection by comet assay, **64,** 237
Cell hybridization, *see* Hybridization, cells, Hybridoma
Cell invasion assay
glycosaminoglycans, **69,** 298–299
invading cell percent, **69,** 305–307
invasion depth, **69,** 304
materials, **69,** 302–303
overview, **69,** 302
procedure, **69,** 303–304
Cell junctions, instructive ECM studies with Mab, **69,** 29–32
Cell labeling, integrin-associated proteins, **69,** 239
Cell layers, proteoglycan anion-exchange chromatography, **69,** 58–59
Cell–ligand adhesion
flow measurements, **63,** 607–608
interference reflection microscopy, **63,** 607
monolayer assays, **63,** 606–607
Cell line models
in vitro apoptosis, **66,** 417–432
Cell line, *see also* Animal cell culture, Glioma, Testicular cell
authentication, *see* Authentication, Cell lines
commercial utility, **57,** 37–38
development by American Type Culture Collection, **57,** 38–40
embryonic stem cell line establishment, **57,** 284–285
government regulation, **57,** 35
immortalization, *see* Immortalization, Cell lines
nomenclature, **57,** 36–37
repositories, **57,** 32–36, 44–45
source species selection, **57,** 32–33
Cell lineages, bone marrow, identification, **64,** 348
Cell lines
CHO, *see* Chinese hamster ovary cells
hybridoma, and *Dictyostelium* centrosomes, **67,** 351–352
protein complex isolation, **65,** 240–241
Sf9, *see* Sf9 cells
Cell lines, for studies of drug efflux and MDR, **64,** 196–197
Cell lines, transfection, recombinant laminin generation, **69,** 121–123
Cell loss factor, and potential doubling time, **64,** 472
Cell lysis, in SPB precipitation, **67,** 85–86
Cell lysis, optimization of conditions, **64,** 261
Cell–matrix adhesion assay
drug development, **69,** 18–19
fields of study, **69,** 19
integrin $\alpha V\beta 3$, **69,** 17–18
integrin VLA-4, **69,** 18
signal transduction studies
adhesion assay, **69,** 364–365
applications, **69,** 361

cell lysates, **69,** 366–370
cell preparation, **69,** 363–364
coimmunoprecipitation, **69,** 375–376
controls, **69,** 365–366
drug inhibitors, **69,** 376–378
GTPases, **69,** 378
immunoprecipitation, **69,** 370–372
kinase assays, **69,** 375
matrix-coated plates, **69,** 362–363
molecular genetics, **69,** 379–382
phospho-specific antibodies, **69,** 374–375
Western immunoblotting, **69,** 372–374
Cell lines, **71,** 3
with stable expression of transgenes, **71,** 8
Cell–matrix interactions
best studied, **69,** 8
cell attachment assays, **69,** 287–288
dynamic reciprocity, **69,** 8
ECM role, **69,** 9–10
external protein meshwork, **69,** 7
thrombospondins, **69,** 8–9
Cell membrane mechanics
bead coating, **55,** 160–161
head position tracking, **55,** 165
laser tweezers
bound *vs.* unbound receptor molecules, **55,** 189–192
cell fusion, **55,** 85–87
fence effects, **55,** 188–192
laser calibration, **55,** 179–181
maximal force calibration, **55,** 161–162
membrane skeleton protein transport, **55,** 188–189
optoporation, **55,** 85–87
overview, **55,** 174–176
receptor molecule dragging obstacles, **55,** 181–185
representative experiments, **55,** 181–183
trajectory analysis, **55,** 183–185
single-particle tracking
elastic intercompartmental barriers, **55,** 185–188
future research directions, **55,** 192
gold labeling particle preparation, **55,** 176–178
protein motion, **55,** 176
video microscopy, **55,** 178–179
tether effects, **55,** 188–192
trap calibration, **55,** 162–165
membrane mechanical properties, **55,** 167–168
membrane–skeleton fence model, **55,** 187–188
membrane tension, **55,** 168–170
overview, **55,** 157–160
tether force measurement, **55,** 166–167
Cell migration
biological processes, **69,** 341–342
Boyden chamber assays, **63,** 554–557
multistep process, **63,** 549
somites
appropriate migration, **69,** 347–348
embryo media preparation, **69,** 343
embryo slicing and embedding, **69,** 345–346
embryo staging, **69,** 343–344
imaging, **69,** 346
microinjections, **69,** 345
microscopy dish preparation, **69,** 346
migratory cycle, **69,** 348
overview, **69,** 342–343
Rac, **69,** 348
slice cultures, **69,** 347
slice environment, **69,** 348
stimulatory signals, **69,** 261–262
subventricular zone to olfactory bulb
brain slice preparation, **69,** 349
DCC, **69,** 351–353
neuronal precursor cells, **69,** 351
overview, **69,** 349
synthetic provisional matrix, **69,** 156
time-lapse microscopic measurement, **63,** 550–554
Cell migration assay
agarose drop assay, **69,** 291–292
overview, **69,** 290–291
wound healing assay, **69,** 292–293
Cell monolayers, fixed, HS binding probe, **69,** 99
Cell morphology, for identification of apoptotic cells, **63,** 539, 541–542
Cell motility
B35 cells as models for, **71,** 291–292
examining with fluorescent tracing, **71,** 132
study via ultraviolet photolysis, **71,** 152–154
Cell organelles, *see* Organelles
Cell organization, **62,** 32–34
Cell-permeable inhibitors
calpain, **66,** 256–258
Cell plate, **61,** 414, 418, 427
Cell plating, for centrosome duplication assay, **67,** 291
Cell plating, micropatterned culture surfaces, **69,** 393, 398
Cell polarity, **62,** 79–88
development of, **71,** 4
in hippocampal and cortical neurons, **71,** 112

Cell populations
fluorescence histograms and plots, **64,** 291–293
intermediate
between blast cells and monocytes, **64,** 353–354
between lymphocytes and blast cell, **64,** 353
Cell position, LSCM list mode feature, **63,** 71–76
Cell preparation
analysis of RNA synthesis, **64,** 131–132
and immunochemical staining, **63,** 422–423
and staining
for flow cytometry, **63,** 137–138
procedures, **63,** 237–244
B and T lymphocytes, **63,** 378
blood cells, **63,** 207–214
universal precautions, **63,** 207–208
by solid phase LSCM, **63,** 80–82
cell suspensions, **63,** 394–395
for identification of leukocytes, **63,** 217–250
mitochondria fluorescent detection, **64,** 122
study of erythropoiesis, **64,** 141–142
Cell production rates
and therapy, **64,** 469–470
clinical studies with thymidine analogs, **64,** 452–474
in normal tissues, **64,** 473
studies in nonmalignant epithelium, **64,** 464–468
Cell proliferation
analysis with PCNA, **63,** 427–428
effect on migration assays, **63,** 552
Cell regulation, ECM
micropatterned culture surfaces, **69,** 386–387
molecular basis, **69,** 385–386
shape control, **69,** 386
structural form, **69,** 386
Cell responses
extracellular matrix, **69,** 3
reporter cells, characterization, **69,** 48–49
reporter cells, instructive ECM
adhesion assays, **69,** 37–40
coculturing, **69,** 37
deposited ECM, **69,** 35–36
ECM testing, **69,** 34–35
overview, **69,** 33–34
soluble ECM, **69,** 36–37
Cell shape, and gene expression, as mechanism for contact effect, **64,** 221–222
Cell size
apoptosis
flow cytometry, **66,** 52–57
propidium iodide, **66,** 54–57
as criterion for choice of neuronal culture, **71,** 2
as parameter of biolistics, **71,** 355
Cell size, forward scatter reflecting, **64,** 289–290
Cell sorting
from multicell culture systems, **64,** 214–215
methodology, **64,** 215
Cell sorting, *see also* Presorting
added to early flow cytometers, **63,** 8
flow cytometric, **63,** 44–48
fluorescent activated, **63,** 619–626
on CFSE peaks, **63,** 384, 386
Cell spreading assay, protocol, **69,** 288–289
Cell–surface adhesion force, living cells
functionalized sensors, **68,** 98
passivated force sensors, **68,** 95
research results, **68,** 95–99
sphere immobilization, **68,** 94
Cell surface exposed-phosphatidylserine (PS)
annexin V injection, **66,** 349–351
detection
paraffin sections, **66,** 352
ultrastructural level, **66,** 353
whole mounts, **66,** 351–352
in vivo detection, **66,** 349–354
Cell survival, in complex systems, predication by DNA damage, **64,** 244–246
Cell suspension temperature, as electroporation parameter, **71,** 347
Cell suspensions
acetone fixation, **63,** 405
adherent cells, **63,** 610–611
paraformaldehyde/methanol fixation, **63,** 406
preparation, **63,** 394–395
single-cell, for aggregation assay, **63,** 601
Cell synchronization, for centrosome duplication assay, **67,** 291
Cell viability, bacteria
enzymatic activity, **64,** 523–524
membrane integrity, **64,** 521–522
membrane potential, **64,** 522–523
Cell wall, bacteria, **64,** 542
Cell wall, digestion, in yeast SPB permeabilization, **67,** 168–169
Cells
activation, time application, **63,** 180–181
AFM applications
actin filaments, **68,** 78–80
cell immobilization, **68,** 53–54
cell viability, **68,** 56
cytoskeleton elasticity, **68,** 76–77
elasticity, **68,** 74–76

homogeneity, **68,** 83–84
imaging consequences, **68,** 80–82
limitations, **68,** 82
sample thickness, **68,** 82–83
tip shape, **68,** 82
tip, probe surfaces, **68,** 121
apoptosis-associated changes
measurement, **66,** 102
autofluorescence measurements, **65,** 92–93
birth rate and death rate, **63,** 530–531
blood, *see* Blood cells
constituents, LSC values, **63,** 59–61
control, *see* Control cells
cryopreservation, **63,** 406–407
cultured
labeling and fixation *in vitro* and *in vivo*, **63,** 360–361
staining procedure, **63,** 361–362
dead
considered in data interpretation, **63,** 224
tracking fate of, **63,** 386
fixed, structure, **63,** 279–281
flow rate past laser beam, **63,** 30–32
interaction with light, in flow cytometry, **63,** 110–118
isolation
MEFs, **66,** 200–201
layers, adhesion force
monolayer immobilization, **68,** 99–100
research results, **68,** 100–104
live, *see* Live cells
living
confocal microscopic applications, **63,** 101–102
engulfing apoptotic bodies, **63,** 538
reactive oxygen species, **65,** 91–92
mitotic, *see* Mitotic cells
necrobiology
flow cytometry, **66,** 70–72
number per division, estimation, **63,** 388–390
permeabilization
mitochondrial ATP synthesis, **65,** 135
permeabilized
antibody-antigen kinetics, **63,** 282–287
structure, **63,** 274–281
positive: subtracting, **63,** 231
processing for antibody titration, **63,** 245
removal, **66,** 340
structure, **63,** 274
transfection
Drosophila S2 cells, **66,** 333–334
Cellular dynamics, AFM studies, **68,** 84–86
Cellular fractionation
cytochrome *c*, **66,** 369–372
Cellular mechanics, actin filaments, AFM studies, **68,** 78–80
Cellular potentials
TMRE, **65,** 90
Cellular proteins
apoptosis
protein substrate assay, **66,** 16–17
Cellular proteins, dynamic force microscopy
Na^+/D-glucose cotransporter
epitope mapping, **68,** 131–133
membrane-sidedness and functionality, **68,** 131
overview, **68,** 130–131
vascular endothelial cadherin
trans association, **68,** 127–128
binding strength, **68,** 128–129
Ca^{2+} dependence, **68,** 127–128
conformation, **68,** 126
overview, **68,** 125–126
thermodynamics and structural parameters, **68,** 129–130
Cellular response, HS sulfation, **69,** 91
Cellular retinaldehyde-binding protein
fluorescence confocal microscopy, **70,** 327
thick tissue sections fixation, **70,** 309, 317–319
Cellular uptake
promoted by noncovalently associated peptides, **71,** 327–328
promoted by peptide sequences, **71,** 326–327
CEN3 probe, **61,** 146–148, 151–152
CENP-B–green fluorescent protein, fusion protein studies of centromere dynamics
centromere binding, **58,** 184
confocal microscopy
cell culture in perfusion chamber, **58,** 186–187
data storage and transport, **58,** 189–190
focal planes, **58,** 185–186
image collection, **58,** 188–189
morphometric analysis, **58,** 193–194
motion analysis, **58,** 192–193
optics, **58,** 187–188
quantitative analysis, **58,** 191–192
time-lapse imaging, making and playing movies, **58,** 190–191
fixed cell visualization, **58,** 185
subcellular localization, **58,** 184–185
Centering, particles in illuminating beam, **63,** 26–33

Central motor neurons
B35 neuroblastoma cells as models of, **71,** 287–288
recording in *Drosophila,* 230
Central nervous system neurons, **71,** 3, 90
limitations of, in transfection, **71,** 288
quantity of yield from, **71,** 11
transfection via electroporation, **71,** 288
Centrifugal elutriation, cell cycle synchronization, **57,** 244–246
Centrifugation
antibodies, for aggregate removal, **63,** 221–223
differentiating
yeast mitochondria isolation, **65,** 40–43
plant mitochondria, **65,** 55–56
repeated, leading to cell loss, **63,** 537
reversed, quantitation of cell adhesion, **63,** 605–606
Centrifuge, animal cell culture, **57,** 9
Centrin, **62,** 441
Centriole assembly pathway, centrosomes, **67,** 220
Centromere
CENP-B–green fluorescent protein
cell culture in perfusion chamber, **58,** 186–187
centromere binding, **58,** 184
confocal microscopy (focal planes), **58,** 185–186
data storage and transport, **58,** 189–190
fixed cell visualization, **58,** 185
image collection, **58,** 188–189
morphometric analysis, **58,** 193–194
motion analysis, **58,** 192–193
optics, **58,** 187–188
quantitative analysis, **58,** 191–192
subcellular localization, **58,** 184–185
time-lapse imaging, making and playing movies, **58,** 190–191
dynamics in interphase, **58,** 194–195
Centromere fragmentation
concept, **61,** 72
mitotic cells with unreplicated genomes (MUGs)
techniques
caffeine, **61,** 73
Chinese hamster ovary cells, **61,** 73
hydroxyurea incubation, **61,** 73
mitotic index, **61,** 73
types, **61,** 72
uses, **61,** 73
Centromere–kinetochore complex
C-banding, **61,** 58
composition, **61,** 58
electron microscopy structural analysis, **61,** 58
plant sources
CREST immunofluorescence protocol, **61,** 77–78
EK anti-sera staining, **61,** 77
silver enhancement
bright-field microscopy, **61,** 78
protocol, **61,** 78
Centromere stretching
centromere–kinetochore complex features
alpha-satellite DNA, **61,** 76–77
CREST-positive segments, **61,** 76
interphase and metaphase chromatin mapping, **61,** 77
degree of stretch, **61,** 76
limitations and precautions, **61,** 76
procedures
centrifugation, **61,** 75
centromere distention gradient, **61,** 75
hypotonic stretching, **61,** 75
lysis, **61,** 75
metabolic block, **61,** 76
mitotic shake-off, **61,** 75
nonintercalating DNA-binding agent facilitation, **61,** 76
purpose, **61,** 75
Centromeric markers, zebrafish, **60,** 361–363
Centrosomal aster formation, **67,** 243
Centrosome
biochemical composition
evolutionary conservation
gamma-tubulin homologs, **61,** 25–26
main function, **61,** 24–25
sequence tags, **61,** 26
genuine chromosomal protein
amino acid sequence, **61,** 27
centrosome-associated proteins, **61,** 27
definition, **61,** 27
examples, **61,** 27
polyacrylamide gel electrophoresis
keratin contamination, **61,** 24
protein profile, **61,** 24
biogenesis
assay features *in vitro*, **61,** 29–30
computational tomography, **61,** 30
genome sequencing, **61,** 30
Xenopus oocyte parthenogenetic assay, **61,** 28–29
centrosome function study *in vivo*

calmodulin and spindle-pole bodies (SPB), **61,** 30
centrin, **61,** 30
GFP-tagged centrosomal proteins, **61,** 30
composition, **61,** 14
immunological probes
autoimmune sera, **61,** 20
immunogen preparation, **61,** 19
microtubule affinity chromatography, **61,** 19
monoclonal antibody categories, **61,** 20
isolation
Drosophila embryo
advantages, **61,** 2
buffers and solutions, **61,** 3
equipment, **61,** 3
preparation of embryo extract, **61,** 3–4
sucrose gradient isolation, **61,** 4–5, 18
yield, **61,** 2
Spisula solidissima oocytes, **61,** 18
human lymphoblastic cultured cells
drug pre-treatment, **61,** 18
immunofluorescence analysis, **61,** 16–17
materials, **61,** 15
procedure, **61,** 15–16
yield, **61,** 18
thymus, **61,** 17
microtubule nucleation assays
aster spin-down assay
calculation of yield, **61,** 9
electron microscopy preparation, **61,** 9
steps, **61,** 27–28
tubulin stocks, **61,** 8
buffers and solutions, **61,** 7, 24
equipment, **61,** 7
on-glass microtubule nucleation assay, **61,** 9–10
organization of microtubules, **58,** 224
pericentrin–green fluorescent protein fusion protein
construction of vector, **58,** 225
controls following imaging
cell viability assay, **58,** 237
functional assay of centrosomes, **58,** 236–237
microtubule integrity assays, **58,** 237
movements of cells or cytoplasm, **58,** 238
structural comparisons of centrosomes in fixed cells, **58,** 236
subcellular localization by immunofluorescence, **58,** 235–236
expression in COS-7 cells
confirmation, **58,** 226
electroporation, **58,** 226
Lipofectamine transfection, **58,** 225–226
high-speed fluorescence microscopy
charge-coupled device camera, **58,** 230–231
deconvolution of images, **58,** 231–234
dynamic properties of centrosomes, **58,** 234–235
heated cell chamber, **58,** 231
image capture, **58,** 234
lasers, **58,** 229
overview of instrumentation, **58,** 228–229
piezoelectric focus drive, **58,** 229
subcellular localization, **58,** 227–228
spindle pole body, **61,** 11
ultrastructural analysis
centriole pair and centrosomal matrix organization, **61,** 21
ring complexes, **61,** 21
role of centrioles, **61,** 21
immunolocalization of centrosomal antigens
centrin, **61,** 23–24
techniques, **61,** 22–23
structural preservation
role of divalent cations, **61,** 20
role of high sucrose concentrations, **61,** 21
Centrosome-associated proteins
cDNA cloning
bacterial fusion protein antibodies, **67,** 131–132
clone characterization, **67,** 129–130
expression library immunoscreening, **67,** 128–129
mammalian cell expression, **67,** 132–133
Cep135, **67,** 133–137
Centrosome protein kinases
functional analysis, **67,** 313–314
genomics-based isolation, **67,** 311–312
mammalian cell duplication assay, **67,** 314–315
proteomics-based isolation, **67,** 312–313
recruitment assay, **67,** 317–318
sperm aster formation assay, **67,** 316
substrate identification, **67,** 318–319
Xenopus egg extract assay, **67,** 315–316
Centrosome proteins
colocalization with pericentrin, **67,** 57–58
functional analysis, **67,** 261–264
recruitment assay, **67,** 317–318
Centrosomes
active *vs.* inactive centrosome, **67,** 218–219
antibodies against, **67,** 126–127
anticentrosome antibodies, **67,** 126–127
centriole assembly pathways, **67,** 220

Centrosomes (*cont.*)
defect detection, **67,** 67–68
Dictyostelium discoideum
electron microscopy, **67,** 346–347
indirect immunofluorescence microscopy, **67,** 342–343
isolation, **67,** 338–342
monoclonal antibody generation, **67,** 347–351
MT nucleation assay, **67,** 344–345
screening, **67,** 351–352
SDS-gel electrophoresis, **67,** 345–346
yield and purity estimation, **67,** 343–344
Drosophila, **67,** 114–122, 141–142
duplication, **67,** 297–298
from quiescent cells, **67,** 215–217
from synchonized cells, **67,** 215–217
immobilized recovery assay, **67,** 153–155
in mammalian cells, **67,** 257–260
independent spindle assembly, **67,** 245–247
injection and egg activation, **67,** 215
isolation, **67,** 124–125, 150–151
parthenogenesis, **67,** 217–220
pericentrin enrichment, **67,** 65
pericentrin localization, **67,** 54–58
pericentrin recruitment, **67,** 58–63
preparation, **67,** 149–150
protein kinases, localization, **67,** 310–311
reproduction
assay monitoring, **67,** 300–301
aster duplication *in vitro* studies, **67,** 281–283
duplication assay logic, **67,** 289–290
duplication assay preparation, **67,** 290–292
duplication assay steps, **67,** 292–294
manipulation *in vitro*, **67,** 283
polarization microscopy studies, **67,** 271–275
spindle assay, **67,** 294–295
video microscopy system, **67,** 275–276
Xenopus egg S-phase extract preparation, **67,** 270
salt-inactivated, MT nucleation reconstitution, **67,** 142–145
salt-stripped, **67,** 261–264
starting, **67,** 295–297
structures containing, **67,** 124–125
Centrosomes, *Xenopus* oocytes, **70,** 412
Ceolomocytes, JC-1 staining technique applied to, **63,** 480
Cep135, **67,** 133–137
Ceramidase
signaling regulation, **66,** 159–160
Ceramide, **66,** 136
acid hydrolysis, **66,** 147–148
analysis, **66,** 137
base hydrolysis, **66,** 148
enzyme assays, **66,** 149–161
fluorescence spectroscopy, **66,** 147–149
HPLC, **66,** 147–149
metabolic labeling, **66,** 136–142
OPA derivatives, **66,** 148–149
Ceramide-diacylglycerol (DAG) kinase assay
enzymatic quantitation, **66,** 144–147
kinetics, **66,** 146–147
lipid extraction, **66,** 144–145
materials, **66,** 144
method, **66,** 145–146
reagents, **66,** 145
Ceramide synthase, **66,** 155–157
materials, **66,** 155–156
molecular probes, **66,** 157
principle, **66,** 156
procedure, **66,** 156–157
signaling regulation, **66,** 155
Cerebellum, **71,** 90
CFDA-SE
bright and dim cells, **64,** 160, 167
diminished fluorescence, **64,** 157
halving, **64,** 159–160
in HP maturation studies, **64,** 155–156
CFDA-SE, tracking of human lymphocytes, **63,** 453, 455
CFSE
application to *in vitro* lymphocyte cultures, **63,** 390–392
detection of intracellular components, **63,** 383
distribution during cell division, **63,** 376–377
labeling of
cell surface markers, **63,** 381–382
cells, **63,** 378–379
monitoring of
lymphocyte responses *in vivo*, **63,** 392–395
position of undivided group, **63,** 386–387
peaks, cell sorting on, **63,** 384, 386
protein labeling, **63,** 191
staining intensity, appropriate levels, **63,** 379–380
Chaetopterus pergamentaceous, see Spindle birefringence
Channel microarray scans
criteria, **65,** 441
Channels
ion, voltage-sensitive, AFM–path–clamp combination, **68,** 12–13
transduction, OHC, gating, **68,** 161–165

Channels per decade, and data accuracy, **63,** 305–309
Chaotic flow, at point of analysis, **63,** 173–174
Chaperonins, AFM imaging, **68,** 219
CHAPS, *see* 3-[(3-Cholamidopropyl) dimethylammonio]-1-propanesulfonate
Charge-coupled device camera, fluorescence microscopy detection, **58,** 36–39, 52–53, 70–71, 230–231, 256, 266, 287, 289, 305–306
Charge-coupled device cameras
basic adjustments
black level, **56,** 10–12
coordinate monitor adjustments, **56,** 14–17
gain, **56,** 5–9
gamma, **56,** 6
linearity, **56,** 6
specimen illumination intensity, **56,** 9–10
features, **56,** 35–43
color cameras, **56,** 39–40
fast-scan cameras, **56,** 37
intensified cameras, **56,** 37–39
monochrome cameras, **56,** 35–36
slow-scan cameras, **56,** 36–37
fluorescence lifetime-resolved imaging, **56,** 287–289
high-resolution multimode digital microscopy
background reference image subtraction, **56,** 205–206
camera features, **56,** 199–200
cooling advantages, **56,** 203–205
dark reference image subtraction, **56,** 205–206
linearity, **56,** 203–205
noise, **56,** 203–205
overview, **56,** 186–187
quantum efficiency, **56,** 200–202
readout, **56,** 202–203
shading, **56,** 206
signal-to-noise ratio, **56,** 206–207
low-light microscopy
advanced technology, **56,** 62
camera noise, **56,** 26–28
dynamic range, **56,** 32–33, 58–59
fill factor, **56,** 24–26
frame rate, **56,** 33–34, 62
gain, **56,** 33
geometric distortion, **56,** 60
imaging rate, **56,** 34–35
linearity, **56,** 31–32
overview, **56,** 46–47
quantum efficiency, **56,** 23–26, 54, 62
sensitivity
definition, **56,** 47–48
high sensitivity enhancement, **56,** 49–54
intensified cameras compared, **56,** 56–57
parameters, **56,** 34–35
quantum efficiency, **56,** 23–26
signal-to-noise ratio, **56,** 26–28, 47, 54–56
shading, **56,** 60
spatial resolution, **56,** 28–31, 34–35, 60
spectral response, **56,** 23
temporal resolution, **56,** 60
time response, **56,** 32
uniformity, **56,** 31–32
usability, **56,** 61–62
Charge-coupled device, in halo–comet assay, **64,** 265
Charge-coupled devices
in flow cytometric detection, **63,** 122–123
in LSCM, **63,** 52, 57
Chariot PTP, **71,** 328
delivering proteins and peptides into neurons with, **71,** 335–336
Charles River Laboratories, **71,** 113
Cheek cells, video-enhanced differential interference contrast light microscopy performance test, **56,** 180–181
CHEF, *see* Contour-clamped homogeneous electric field electrophoresis
Chemical compounds, screening in *Xenopus* egg, **67,** 250–251
Chemical mutagens
advantages of, **71,** 201–202
point mutations induced by, **71,** 202
Chemical transfection, disadvantages compared with biolistics, **71,** 354
Chemokine receptors, **64,** 583
Chemokinesis assays, **63,** 555
Chemoreceptors, **62,** 365–366
Chemotactic assays, for cell migration, **63,** 552
Boyden chamber, **63,** 555
Chemotherapeutic agents
activity patterns in spheroids, **64,** 222–223
contact effect, **64,** 218–219
Chemotherapy
correlation with TLI, **64,** 373, 375–376
neoadjuvant, **64,** 375–376
Chick ciliary neurons
advantages of, **71,** 38
applications of, **71,** 38
culturing methods, **71,** 46–47
developmental properties of, **71,** 5
disadvantages of, **71,** 41
dissection and culturing of, **71,** 37–49

Chick ciliary neurons (*cont.*)
endocytosis and recycling in, **71,** 5–8
materials for dissection of, **71,** 42
materials for dissociation and culturing of, **71,** 42–43
methods of dissecting and dissociating, **71,** 43–47
neuroanatomy and development of, **71,** 38
neuronal types of, **71,** 38
synapses formed in, **71,** 10
Chick dorsal root ganglia (DRG)
ease of dissection of, **71,** 18
endocytosis and recycling in, **71,** 5–8
quantity of yield from, **71,** 11
relative infectivity of, **71,** 408
short-term cultures of, **71,** 32–33
Chick embryo extract, **71,** 8
Chick embryos
as sources of spinal motor neurons, **71,** 67–68
infecting by retroviral injection, **71,** 379–382
obtaining and maintaining, **71,** 57
preparation of, **71,** 71–73
Chick forebrain neurons, **71,** 4
axonal development studies in, **71,** 53
axonal morphogenesis, **71,** 52
contrasting with rat hippocampal neurons, **71,** 54–56
culture conditions for, **71,** 57–63
culturing, **71,** 51–64
developmental properties of, **71,** 5
dissection of, **71,** 13, 57–58
dissociation of, **71,** 59
elongation behaviors in, **71,** 54
future improvements in culturing, **71,** 63–64
growth and development characteristics of, **71,** 53–56
in toxicological assays, **71,** 53
isolation of single, **71,** 56–59
plating efficiency and culture density of, **71,** 63
pyramidal shape of, **71,** 59
relative purity of, **71,** 11
small size of, **71,** 8–9
Chick motor neurons, as basis of injury models, **71,** 69–70
Chick peripheral neurons, expression of transgenes via retroviral infection, **71,** 369–385
Chick primary neurons, electroporation protocol for, **71,** 344–345
Chick Purkinje neuron cultures, **71,** 90
Chick sensory ganglia, dissection of, **71,** 25
Chick sensory neurons, **71,** 17
developmental properties of, **71,** 5
organelle transport in, **71,** 8
Chick serum, **71,** 377
Chick spinal motor neurons, **71,** 4
quantity of yield from, **71,** 11
relative purity of, **71,** 11
Chick sympathetic neurons, **71,** 4
axonal-somatic synapses in, **71,** 10
developmental properties of, **71,** 5
endocytosis and recycling in, **71,** 5–8
Children, perinatal transmission of HIV, **64,** 574
Chimeric proteins, regulatory HS binding probes, **69,** 97–98
Chinese hamster ovary cells, **69,** 157–158
hydroxyurea-arrested, **67,** 260–261
recruitment of pericentrin to centrosomes, **67,** 58–60
Chloramphenicol, **65,** 448–449
vs. untreated yeast
log–transformed expression ratios, **65,** 450
Chlorate treatment
HS–growth factor interaction, **69,** 89–90
HS sulfation, **69,** 90–91
Chlorodeoxyuridine, *see* DNA replication site, mapping *in situ*
Chlorophyll fluorescence
plant senescence, **66,** 449–450
Chloroplasts
protein complex isolation, **65,** 238–244
CHO cells, *see* Chinese hamster ovary cells
Cholera toxin β-subunit, AFM studies, **68,** 237–238
3-[(3-Cholamidopropyl) dimethylammonio]-1-propanesulfonate, **69,** 55
Cholinergic motor neurons, **71,** 140
Chondroitin sulfate, GAG–growth factor interaction, **69,** 93
Choroid neurons, of chick ciliary ganglia, **71,** 38
Chromatic aberration
multicolor LSCM, **70,** 235–237
objective lens, **70,** 262–263
Chromatic aberrations, ratio imaging, **56,** 227–230
Chromatin, **62,** 137, 144–148, 379–381
condensation
in apoptosis, **63,** 507–509
measurement in apoptotic plant cells, **63,** 514–516
condensed, mitotic cells, **63,** 351
Chromatin condensation
assays, **66,** 172–174
LSC, **66,** 77–89

Chromatin packaging, as mechanism for contact effect, **64,** 220–221
Chromatin remodeling protein 1 (CRP1)
functions in cell-free systems, **53,** 500
mass measurement with scanning transmission electron microscopy, **53,** 145–146, 156
Chromatin, *lac* repressor in visualization of large-scale organization
applicability to different species, **58,** 220
binding to operator sequence, **58,** 204
chromonema fiber visualization, **58,** 219
dynamics of amplified region, **58,** 218
gene amplification and cell cloning, **58,** 212–214, 220
green fluorescent protein–repressor–nuclear localization sequence fusion protein
construction, **58,** 211–212
perfusion chamber for microscopy, **58,** 215–216
phototoxic effects, **58,** 216–218
operator introduction into Chinese hamster ovary cells
operator repeat
construction, **58,** 206–208
manipulation for stability, **58,** 208–210
overview, **58,** 205
repressor–nuclear localization sequence fusion protein
construction, **58,** 210–211
immunostaining in operator repeat detection, **58,** 214–215
Chromatin, *see also* Miller chromatin spreading
alignment with nuclear envelope, **53,** 80
atomic force microscopy, **53,** 182
compaction ratio, **53,** 99–100
remodeling, *see* Cell-free chromatin remodeling, Chromatin remodeling protein 1
fluorescence image microscopy, **53,** 62–63, 66–68
micrococcal nuclease digestion structure analysis, **53,** 508–509
polyacrylamide gel electrophoresis of associated proteins, **53,** 507–508
preparation from *Xenopus* sperm, **53,** 505–507
remodeling, *see* Cell-free chromatin remodeling, Chromatin remodeling protein 1
transmission electron microscopy
in situ
buffer system selection, **53,** 169
cryoultramicrotomy, **53,** 175
data collection, **53,** 175–176
low-temperature embedding and ultraviolet polymerization, **53,** 172
Lowicryl resin handling and sectioning, **53,** 170–173
overview, **53,** 168
progressive lowering of temperature embedding, **53,** 169–171
rapid freezing and freeze substitution, **53,** 169–172
staining of sections, **53,** 173–175
in vitro
adhesion to substrate, **53,** 178–179
cryoelectron microscopy, **53,** 180–183
fixation, **53,** 177–178
heavy metal shadowing, **53,** 178
negative staining, **53,** 180
nuclei isolation, **53,** 176
overview, **53,** 168
positive staining, **53,** 178–180
soluble chromatin preparation, **53,** 176–177
Chromatographically prepurified complexes
BN-PAGE, **65,** 241–242
Chromatography
affinity, **65,** 219
basement membrane self-assembly, **69,** 124–125
DE52
actin-binding assay, **65,** 165
DNase I affinity
actin-binding assay, **65,** 163
integrin-associated proteins, **69,** 237–239
integrins, **69,** 232–234
mono Q
plant mitochondria, **65,** 67
proteoglycans, **69,** 57–63
solubilized integrins, **69,** 228–233
Chromophore
caspase, **66,** 15
Chromophore, green fluorescent protein
Förster cycle, **58,** 10–12
mutation, mechanisms of spectra alterations, **58,** 12–14, 22–26, 299, 317–319
oxidation, **58,** 6–7, 22
pH effects on spectra, **58,** 9, 33
photobleaching, **58,** 32–33, 36, 166
photoisomerization, **58,** 156
protein folding and autocatalytic formation, **58,** 3–6, 21–22, 33
screening of enhanced mutants, **58,** 27–29
sequencing, **58,** 2–3

Chromophore, green fluorescent protein (*cont.*)
stability in protein, **58,** 140–141, 284
tautomers and absorption maxima, **58,** 9–12, 63, 143
water interactions, **58,** 14–15
Chromosomal aberrations, inducing, **71,** 202
Chromosomal walking, *see* Positional cloning
Chromosome, **62,** 263
breakage, **62,** 138
macronuclear, **62,** 138
micronuclear, **62,** 134
staining, **62,** 293–296
Chromosome analysis
fluorescence *in situ* hybridization, **57,** 209, 211
Giemsa staining, **57,** 209
quinacrine mustard staining, **57,** 209, 211
trypsin–Giemsa banding, **57,** 209–210
Chromosome translocation
detection in interphase nuclei, **64,** 98–99
PRINS–painting of, **64,** 63–64
probe preparation
applications, **64,** 112
buffers and other solutions, **64,** 101–102
critical aspects, **64,** 111–112
discussion of techniques, **64,** 109–111
instruments, **64,** 102
materials, **64,** 99–101
protocols, **64,** 102–108
Chromosome, *see also* Centromere, Chromatin
identity maintenance during interphase, **58,** 183–184
in situ hybridization, **58,** 204–205
segregation visualization, **58,** 203–204
Chromosome, *see* Chromatin; Fluorescence *in situ* hybridization; Nucleosome
Chromosomes
border, overlaid with telomere border, **64,** 87–88
human, application of PRINS, **64,** 56
painting, combined with PRINS, **64,** 61–64
plant
flow cytogenetics problems, **64,** 4–5
mitotic, **64,** 5
preparation of suspensions, **64,** 14, 19, 21–22
sorted fraction purity, **64,** 17–18, 20–21, 25–26
sorting, **64,** 17, 20, 24–26
staining, **64,** 14–15
prematurely condensed, **64,** 398
protective cap provided by telomeres, **64,** 70
segmentation, **64,** 82, 84–85, 92–93
Chromosomes, absence in mitotic spindle-like structure assembly, **67,** 298
Chromosomes, laser surgery, **55,** 81–83
Chromosomes, *Xenopus* oocytes and eggs, staining, **70,** 393–394
Chromosomes, zebrafish
C-banding, **60,** 324, 328, 334
chromomycin A3 banding, **60,** 324, 328
karyotyping by fluorescence *in situ* hybridization
analysis of results, **60,** 332–334
antibody detection, **60,** 332
counterstaining, **60,** 332
denaturation, **60,** 331
hybridization, **60,** 331
pretreatment of slides, **60,** 331
probe synthesis, **60,** 329–330
washes, **60,** 331–332
metaphase spread preparation from hamster/zebrafish radiation hybrids, **60,** 296–297
mouse/zebrafish somatic cell hybrid characterization, **60,** 310–313
Nucleolar Organizer Regions, **60,** 335
orthologous chromosomes, **60,** 149, 152
preparation
embryo cell culture, **60,** 326
embryo extract, **60,** 326–327
fin culture, **60,** 324–325
harvesting for spreads, **60,** 327
subculture, **60,** 325–326
replication banding, **60,** 328–329, 334
tagging with *neo* resistance gene, **60,** 306–307
trypsin G-banding, **60,** 328
Cilia, **62,** 34, 52–55, 79, 82, 84
basal bodies, **62,** 32, 34, 52–55, 68, 70, 82, 88
caudal cilium, **62,** 88
ciliary reversal, **62,** 364, 370–371
ciliary rows and patterning, **62,** 79–83
dynein, **62,** 284–287, 433–439
isolation, **62,** 433–435, 442–443
mutants, **62,** 279–287
Ciliary neurons, **71,** 3
Circuits
hybrid, pulse characterization with, **63,** 150–154
peak and hold, **63,** 37
Circular Airy diffraction pattern, **63,** 91
Citrate synthase 1-green fluorescent protein (CS1-GFP), **65,** 279
fluorescence patterns, **65,** 280
Clam sperm, CSU-10 studies, **70,** 97, 100
CLASSIF1 algorithm
asthma, **64,** 494, 499
data classification, **64,** 488–492
HIV infection, **64,** 503

Classification coincidence factor, **64,** 492
asthma, **64,** 502
melanoma, **64,** 494
Classifier masks, **64,** 492–494
Cleavage and blastula period, zebrafish, **59,** 212–215
Cleavage and blastula period, zebrafish, **59,** 212–215
Cleavage characterization
PARP-1, **66,** 299–300
cleavage site analysis, **66,** 300
protein sequencing, **66,** 299–300
Cleavage period, embryonic zebrafish, **59,** 14–16
Cleavage sequences
caspases, **66,** 9
Clinical oncology, future of cytometry in, **64,** 474–478
Clogging, flow channel, **63,** 5–6, 28
Clone-specific markers, in PCR, **64,** 404–406
Clone-specific PCR, **64,** 408
Clone-specific primers, in limiting dilution, **64,** 410–411
Clones, selection from library, **64,** 111–112
Clonigenic survival assays, **66,** 465
Cloning
caspase assay, **66,** 6–7
Dictyostelium centrosomes, **67,** 351–352
expression, *see* Expression cloning
for centrosome duplication assay, **67,** 290–291
genes into two-hybrid vectors, **67,** 75–76
Mabs for instructive ECM identification, **69,** 47
molecular, *see* Molecular cloning
plant γ-tubulin cDNA and genes, **67,** 196–200
Cloning applications, **71,** 11
Drosophila mutant analyses, **71,** 222–223
Cloning sites, retroviral vectors, **71,** 371–378
Cloning strategy, zebrafish genes, **59,** 229–234
Cloning, animal cell culture, **57,** 16, 98–99
Cloning, centromere DNA
method strategy
cell homogenization, **61,** 73
centrifugation fractionation, **61,** 73
isopycnic gradient banding, **61,** 75
MUG (mitotic cells with unreplicated genomes) procedure, **61,** 73
CLSMs, *see* Confocal laser scanning microscopes
Clumping, of neuronal cell bodies, **71,** 24
Cluster analysis, **65,** 445
Cluster analysis, statistical analysis of particle distribution, **53,** 95–96
CM5 chips, SPR of integrin ligand binding, **69,** 255–256
CMXRos
atractyloside, **65,** 155
CMXRosamine, probe for mitochondria, **64,** 119, 125–126
CO_2-dependent or -independent media, culturing chick forebrain neurons with, **71,** 52
Co-stimulation, **62,** 91, 93
Coated pits, **62,** 42–43, 48
Coaxial mixing device, **63,** 174–175
Cochlea
hair cells, AFM combination techniques, **68,** 19
organ of Corti, **68,** 143–144
organ of Corti preparation, **68,** 146–147
stereocilia, **68,** 144–146
Cockroach flight muscle mitochondria, **65,** 24
COCO system, proteins
aggregate formation, **69,** 80
antigenicity, **69,** 79–80
expression failure, **69,** 80
oligomerization, **69,** 78
posttranslational modifications, **69,** 79
protein variety, **69,** 78
yields, **69,** 78
Coculture systems
in Purkinje neuronal cultures, **71,** 93
Xenopus spinal neurons, **71,** 145
Codon usage, **62,** 150–152, 565–566
Coefficient of variation
DNA content, **63,** 278–279, 293
fluorescence intensity measurement, **63,** 313–314
for repeated samples, **63,** 287
LSCM, **63,** 63
measurement of resolution, **63,** 309–310
Coenzyme Q
beef heart mitochondria, **65,** 12–13
Cognitive impairment, **71,** 291
Coiled body (CB), components, **53,** 312
Coimmunoprecipitation
assay, **66,** 41–42
ATPase oligomers, **65,** 223–225
Bax, **66,** 34
Bcl-2 family, **66,** 34, 41–42
digitonin recrystalization, **65,** 223
protein–protein interactions, **65,** 222–225
Coimmunoprecipitation, cell–matrix adhesion assays, **69,** 375–376
Colcemid, cell cycle synchronization, **57,** 239–240
Colchicine, *see* Drug role in analyzing mitosis
colgate, mutation effects in development, **60,** 81

Collagen, **71,** 21
as culture substrate, **71,** 11–12
as substrate for explant cultures, **71,** 29–30
rat tail, as substrate, **71,** 17
Collagen
characterization, **69,** 283
component of ECM, **63,** 562–563
degradation, **69,** 171
fibrillar collagen, **69,** 187–188
fibrillogenesis, **63,** 590
invasion assays, **69,** 203–204
moldable gel, **63,** 564
preparation of purified collagen gel, **63,** 568–569
Collagen I
AFM imaging, **69,** 171–174
ECM self-assembly paradigm, **69,** 111–112
real-time AFM imaging, **69,** 172–174
3D matrix for cell culture, **69,** 405–407
Collagen III, **69,** 174
Collagenase treatment, in neural-enriched *Xenopus* cultures, **71,** 144
Collection
blood, prior to cell preparation, **63,** 208–209
data, for Boyden chamber assays, **63,** 556–557
fluorescent light, **63,** 32
light, optics for, **63,** 114–115
light signals from particles, **63,** 33–37
Colloidal gold, *see* Scanning transmission electron microscopy, Transmission electron microscopy
Colocalization studies, by confocal microscopy, **63,** 102–104
Color code, objective lens, **70,** 258
Color compensation, fluorochromes, **63,** 318–319
Color scanning
plant senescence, **66,** 449
Color video cameras
low-light microscopy, **56,** 39–40
two-camera imaging, **56,** 125
Colorectal adenocarcinomas, BrdUrd labeling, **64,** 462–463
Colorectal cancer, proliferation markers and prognosis, **64,** 370–372
Colorimetry
plant senescence, **66,** 447
Colorimetry, application to Boyden chamber assay, **63,** 557
Comet assay
alkaline method (by Olive), **64,** 256–257
alkaline method (by Singh), **64,** 253–254
application to clinical samples, **64,** 247
development, **64,** 236–237
neutral method (by Olive), **64,** 257–258
neutral method (by Singh), **64,** 254–256
types of DNA damage detected by, **64,** 242–246
Comets
antibody-stained, **64,** 237
image analysis, **64,** 240–242
preparation steps, **64,** 238–240
Commercial detection kits, for apoptosis, **63,** 538–539
Common acute lymphoblastic leukemia antigen, *see* CD10
Communication, intracellular, as mechanism for contact effect, **64,** 220
Comparator, element in trigger circuit, **63,** 151–152
Compartmentalization
calcium flux, **66,** 230–233
Compensation
color, fluorochromes, **63,** 318–319
fluorescence, and logarithmic amplification, **63,** 124–127
in immunophenotyping of activated lymphocytes, **63,** 443–444
phycoerythrin emission requiring, **63,** 235–236
Compensation matrix, manual application, **64,** 307
Compensation standard, files for, **64,** 304–305
Competitive PCR assay, **64,** 409
Competitive polymerase chain reaction
mtDNA, **65,** 418–419
Complementary DNA
centrosome-associated protein
bacterial fusion protein antibodies, **67,** 131–132
clone characterization, **67,** 129–130
expression library immunoscreening, **67,** 128–129
mammalian cell expression, **67,** 132–133
expression cloning, **69,** 213
plant γ-tubulin, cloning, **67,** 196–200
recovery by Hirt supernatant, **69,** 216–217
Complementary metal-oxide semiconductor imagers, features, **56,** 41–42
Complementation, **62,** 237, 252–258
Complex I, **65,** 101–104
assays, **65,** 76–78
NADH-cytochrome c reductase, **65,** 78
NADH-ferricyanide reductase assay, **65,** 77–78
NADH-Q reductase, **65,** 76
skeletal muscle mitochondria
membrane disruption, **65,** 102

Complex II, **65,** 104–105
assays, **65,** 78–79
Complex III, **65,** 105–110
assays, **65,** 79–80
ubiquinol-cytochrome c reductase, **65,** 79–80
deficiency
diagnosis, **65,** 114
dodecyl maltoside, **65,** 108
inhibitors, **65,** 105–106
protocol, **65,** 107
reagents, **65,** 106–107
Complex IV, **65,** 110
assays, **65,** 80
dedecyl maltoside, **65,** 111
Computer hardware, confocal microscopy, **70,** 22–23
Computer software, confocal microscopy, **70,** 23, 370
Computers
curve fitting programs, **63,** 388–389
digital, data storage and display with, **63,** 161–167
LSC, drawing of contours around cell event, **63,** 58–61
Computers, multiple trap synthesis, laser position control, **55,** 212–213
Con A binding protein, **62,** 94
Condensers
light microscopy, **56,** 138
video-enhanced differential interference contrast light microscopy, **56,** 173–174
Conditioned media
native basement membrane proteins, **69,** 115–116
proteoglycan anion-exchange chromatography, **69,** 58–59
Confocal analysis, morphogenetic movements, **59,** 196
Confocal digital imaging
considerations, **70,** 248
electronic scan zoom, **70,** 252
infinity *vs.* finite lens systems, **70,** 254–256
objective lens, *see* Objective lenses
on-axis *vs.* off-axis, **70,** 247–248
optical resolution, **70,** 251
point-spread function, **70,** 253
resolution, **70,** 252
signal-to-noise ratio, **70,** 253–254
wide-field, optical resolution, **70,** 249–250
Confocal fluorescence lifetime imaging microscopy
characteristics, **69,** 272–273
signaling molecules, **69,** 262
Confocal fluorescence microscopy
Aufderheide rotocompressor device
assembly, adjustment, repair, **70,** 351–352
background information, **70,** 347–349
lubrication and maintenance, **70,** 353
parts description, **70,** 349–351
use, **70,** 352–353
biological studies, **70,** 166
Ca^{2+} in perfused afferent arteriole, **70,** 423–425
CSU-10, low bleaching, **70,** 101–104
cultured cell cytoskeletal organization, **70,** 302–303
development, **70,** 130
drawbacks, **70,** 327–328
intact tissue problems, **70,** 303–304
intact tissue solutions, **70,** 304–308
maximum resolution parameters, **70,** 321–326
overview, **70,** 12–13
pH and Ca^{2+} in living cells, **70,** 417–418, 426
pHi in cortical collecting tubule, **70,** 422–423
pHi with BCECF, **70,** 418–420
retinal cell biology applications, **70,** 326–327
Ringlien differential screw thread compressor
construction details, **70,** 354–356
operating instructions, **70,** 356–357
overview, **70,** 353–354
Tetrahymena thermophila
overview, **70,** 338
reagents, **70,** 340–342
sample preparation, **70,** 342–344
vital fluorescent dyes, **70,** 357–358
Tetrahymena thermophila rDNA FISH
nick translation reaction, **70,** 346–347
probe synthesis, **70,** 346
sample preparation, **70,** 344–346
thick tissue preparation
agarose embedding, **70,** 330
antibody penetration, **70,** 315–319
fixation, **70,** 308–313, 329
fluorochromes, **70,** 319–321
future prospects, **70,** 328–329
labeling procedures, **70,** 330–331
mounting sections, **70,** 331–332
Vibratome sectioning, **70,** 314–315, 330
Xenopus oocytes and embryos
cytoskeletal and nuclear probes, **70,** 410–413
keratin filaments, **70,** 386–388
microfilaments, **70,** 388–390
microtubule fixation, **70,** 383–386
overview, **70,** 380–382
recipes and reagents, **70,** 409–410

Confocal fluorescence microscopy (*cont.*)
Yokogawa CSU-10 design, **70,** 91–92
Confocal fluorescence microscopy, *see* Microscopy, Light
Confocal imaging, embryos, **59,** 181
Confocal laser-scanning microscope, **71,** 96, 152
in *Xenopus* live-cell imaging studies, **71,** 151–152
Confocal laser scanning microscopes, *see also* Multicolor laser scanning confocal microscopy
backscattered light reflection, **70,** 280–281
biofilms, **70,** 275–278
computer software, **70,** 23
development, **70,** 129–131
Drosophila embryos
autofluorescence, **70,** 368–369
bleed-through, **70,** 369–370
collection and permeabilization, **70,** 363
fluorescent probes, **70,** 364
hard copy images, **70,** 372
image archiving, **70,** 372
image collection, **70,** 366–368
image processing, **70,** 370–372
image transfer, **70,** 370
microscope details, **70,** 365–366
mounting, **70,** 364–365
multiple labels, **70,** 372–375
overview, **70,** 361–363
interference reflection contrast, **70,** 281
light source characteristics, **70,** 17–19
membrane potential, nonlinear optical imaging
methods, **70,** 446–448
second harmonic generation, **70,** 441–444
second harmonic generation enhancement, **70,** 444–446
two-photon excited fluorescence, **70,** 441–444
multiple-label confocal fluorescence imaging, **70,** 278–279
Nernstian redistribution of permeant cationic dyes
methods, **70,** 437–441
plasma membrane potential, **70,** 434–436
single mitochondria membrane potential, **70,** 436–437
pHi measurement with BCECF, **70,** 420
scanning transmission differential interference contrast, **70,** 282
scanning transmission phase contrast, **70,** 281–282
ultrawide low-magnification confocal imaging, **70,** 282, 287
video-rate
Bio-Rad laboratories, **70,** 43
NORAN instruments, **70,** 39–42
overview, **70,** 39
Xenopus oocyte
brightness and contrast extremes, **70,** 402
digital image processing, **70,** 403–404
image brightness optimization, **70,** 400–402
image scale optimization, **70,** 399
objective choice, **70,** 398–399
optical sectioning, **70,** 399–400
presentation, **70,** 404–408
Xenopus oocytes and eggs, **70,** 392–393
X-Z and 3D imaging, **70,** 279–280
Confocal microscopy, **61,** 82, 378, 427
AFM integration, **68,** 176–177
AFM settings, **68,** 177
calcium imaging, **63,** 102
cell adhesion studies, **63,** 102
cell biology applications, **63,** 100
colocalization studies, **63,** 102–104
3-D imaging, **63,** 585–586
development, **63,** 90–92
fluorescence recovery after photobleaching, **63,** 104–105
fluorescent probes for, **63,** 96–99
image formation, **63,** 92–96
living cells, **63,** 101–102
Confocal microscopy core facility
ancillary services, **70,** 485
core issues, **70,** 483–484
data storage, analysis, handling, **70,** 482–483
instrument care, **70,** 481–482
operator and user training, **70,** 478–481
overview, **70,** 475–476
planning, **70,** 476–477
resources, **70,** 485
scheduling, **70,** 477
Confocal microscopy overview
advantages, **70,** 3–4
airy disks, **70,** 9–10
applications, **70,** 52–55
basic concepts, **70,** 2–3, 88, 90, 149–150
computer hardware, **70,** 22–23
computer software, **70,** 23
3D reconstruction, **70,** 60–66
4D reconstruction, **70,** 66–68
deconvolution alternative, **70,** 68–69
depth discrimination, **70,** 156–158
design overview, **70,** 23–24
detector systems, **70,** 159–161
disadvantages, **70,** 4–7
fluorescence imaging, **70,** 12–13

general suppliers, **70,** 76–77
lateral and axial resolution, **70,** 10–11
lens selection, **70,** 155–156
light microscopy comparison, **70,** 7–9
light source, **70,** 16–19
microscope, **70,** 14–15, 49–51, 74–76, 90–91
multiphoton microscopy alternative, **70,** 69–73
multiple-label imaging, **70,** 56
objective lens, **70,** 15–16
personal and multiuser systems, **70,** 51–52
photodetector, **70,** 20–22
real lens behavior, **70,** 158
refractive index effects, **70,** 150–155
scan head, **70,** 20
single-label imaging, **70,** 55–56
spatial filters, **70,** 9
spatial sampling effects, **70,** 161–163
specimen preparation, **70,** 52
stage scanning, **70,** 24–25
stereo imaging, **70,** 58–59
web sites, **70,** 77–80
z-series, **70,** 56–57
Confocal microscopy, *see* Laser scanning confocal microscopy
Confocal microscopy, DNA–drug interactions, **64,** 184–185
Confocal reflection microscopy, ECM and ECM–cell interactions, **63,** 587–591
Confocal scanning laser microscopy
antifading agents, **53,** 55
DNA replication site mapping, *see* DNA replication site, mapping *in situ*
fluorophore selection
DNA-specific dyes, **53,** 54
long-wavelength fluorophores, **53,** 53–54
rhodamines *vs.* fluorescein, **53,** 53
nuclear pore complex
antibody labeling, **53,** 87–88, 96
cell culture, **53,** 86
imaging parameters, **53,** 88
model system
correction for photobleaching, **53,** 84–85
electron microscopy verification, **53,** 85
optimization of imaging conditions, **53,** 83–84
point spread function approximation, **53,** 84
single particle localization, **53,** 85–86
smoothing, **53,** 85
specimen preparation, **53,** 82–83
permeabilization
digitonin, **53,** 87
mechanical method, **53,** 87
statistical analysis of particle distributions
cluster analysis, **53,** 95–96
nearest-neighbor distribution function, **53,** 89, 91, 93–95
numerical computation of estimators, **53,** 92
pair correlation function, **53,** 91–95
out-of-focus blur removal, **53,** 51–53
principle, **53,** 50–52
spherical aberration, **53,** 71–72
Conformation, AFM studies
individual proteins, **68,** 291
membrane proteins, force-induced changes, **68,** 272–274
native membrane proteins
HPI layer changes, **68,** 280–282
OmpF porin, **68,** 282–285
surface structure changes, **68,** 278–280
vascular endothelial cadherin, **68,** 126
Conformation, antigen, **63,** 262
Conjugation, *see also* Mating, **62,** 91–98, 128–133
exconjugant, **62,** 133–134
exconjugant isolation, **62,** 221–222, 235–237
Conjugation chemistry, protein labeling, **63,** 186–187
Constitutively active (CA) mutant proteins, in *Xenopus* cultures, **71,** 131
Contact effect
chemotherapeutic agents, **64,** 218–219
ionizing radiation, **64,** 217–218
proposed mechanisms for, **64,** 219–222
Contact mode atomic force microscopy
cell imaging, **68,** 57–58
operation, **68,** 35
overview, **68,** 260–261
Contact mode photonic force microscopy, **68,** 208–209
Contamination, cell culture
chemical contamination, **57,** 53–55
cross-contamination, **57,** 41, 43
mycoplasma, **57,** 41–43, 55–63
physical contamination, **57,** 52–53
prevention, **57,** 50–52
types, **57,** 49–50
Contour-clamped homogeneous electric field electrophoresis (CHEF), *see* Pulsed-field gel electrophoresis
Contour plots, for cell populations, **64,** 292–293
Contouring, critical step in LSCM, **64,** 318–319

Contours
around cell event, LSC
computer-drawn, **63,** 58–61
two-parameter histograms, **63,** 164–166
Contractile vacuole, **62,** 51–52, 81–82, 86–87
visualization, **62,** 314
Contrast, *see also* Video-enhanced differential interference contrast light microscopy
definition, **56,** 77
digital image enhancement
differential interference contrast images, **56,** 209–210
gray value, *see* Gray value
methods, **56,** 77–79, 101–104
processors, **56,** 177–178
fluorescence microscopy, phase contrast imaging, **56,** 190, 193
Köehler illumination, **56,** 143–146
monitor controls, **56,** 13–14
Control cells, **63,** 262
mock-transfected, for adhesion studies, **63,** 602
positive/negative, **63,** 289–290
Controls, to confirm cell adhesive specificity, **63,** 608–609
Convolution kernel, CSU-10
neighborhood-based processing, **70,** 107
Convolution, *see also* Deconvolution
digital image processing, **56,** 108–109
digital mask filters, **56,** 109–114
edge detection, **56,** 112–113
Laplacian filters, **56,** 113–114
line detection, **56,** 112–113
point detection, **56,** 112
Cooled charge-coupled device camera
charge coupling, **53,** 57
comparison to other cameras
home video camera, **53,** 57–59
SIT, **53,** 57
cost, **53,** 57–58
quantum efficiency, **53,** 57, 224
read-out rate, **53,** 57–58, 224
Copurification
protein–protein interactions, **65,** 219
Core human mtRNA, **65,** 420
Core yeast mitochondrial RNA polymerase, **65,** 420
Corneal damage, from adenovirus, **71,** 415–416
Cornified envelope proteins
ε-(γ-glutamyl)-lysine cross link density, **66,** 117
Correction collar, objective lens, **70,** 259, 265–266
Correlated list mode data
acquisition and analysis, **64,** 290–293
multiparameter analysis produced by, **64,** 311
Cortex, **62,** 314, 317
cortical alveoli, **62,** 32, 48
cortical cytoskeleton, **62,** 39, 65–67, 79–88, 443–446
during conjugation, **62,** 98
patterning, **62,** 85
Cortical collecting tubule, pHi, **70,** 422–423
Cortical neurons, **71,** 3
acquisition of, **71,** 113–117
culture characteristics, **71,** 119–121
culturing, **71,** 119–121
long-term culture of, **71,** 121–126
short-term culture methods, **71,** 117–121
Cosedimentation assay, *Spisula* MT nucleation, **67,** 158–161
Co-Stimulations, **62,** 91, 93
Cosmids, **61,** 366
Cost, of obtaining/maintaining neurons, **71,** 2
Cotransformants, **65,** 385
Coulter counter, **63,** 5
Coulter counter, *see* Cell counter
Counting, cell attachment assays, **69,** 286–287
Coupling
fluorescent probes to proteins, **63,** 186–187
principle, and use of heterobifunctional reagents, **63,** 195–201
Cover glass, objective lens, **70,** 259
Coverage Plus disinfectant, **71,** 405
Coverslip method, **65,** 272–274
Coverslips
Aufderheide rotocompressor device, **70,** 350–351
cleaning of, **71,** 118
growing hippocampal neurons on, **71,** 112
plating of hippocampal and cortical neurons on, **71,** 117–119
Ringlien differential screw thread compressor, **70,** 357
Coverslips, FITC–gelatin coated, **63,** 623–625
Coverslips, HEC/RL cell culture, **68,** 109
COX, *see* Cytochrome *c* oxidase
[^{14}C]putrescine method
Tgase assay, **66,** 114–115
CRALBP, *see* Cellular retinaldehyde-binding protein
CREST immunofluorescence, **61,** 59, 60, 77–78
Creuzfeldt-Jakob disease, association with Purkinje neurons, **71,** 191
Crithidia lucilae, **65,** 29
CRM, *see* Confocal reflection microscopy
Cross-correlation function, alignment of images

fluorescence microscopy, **53,** 70
transmission electron microscopy, **53,** 113
Cross-linkers
cell fixation accompanied by, **63,** 254, 256
fixatives: properties, **63,** 277–279
heterobifunctional, **63,** 197, 200–201
photoreactive, **65,** 226
Cross-linking
antibodies
protein–protein interactions, **65,** 222
radiolabeled precursor protein
import machinery component, **65,** 227–228
reagents, **65,** 226
properties, **65,** 226
recombinant involucrin protein, **66,** 128
TGase substrate proteins
location, **66,** 126–129
Cross-linking, DNA, **64,** 182
Cross-linking, integrin ligand binding pocket, **69,** 248–250
Cross-reactivity, as limitation of *Heliosoma* antigens, **71,** 169
Crosslinker binding, AFM tips, **68,** 119
CRP1, *see* Chromatin remodeling protein 1
Cryo-atomic force microscopy
deep etching, **68,** 252–253
future research, **68,** 253–254
general design, **68,** 244–245
individual molecules, **68,** 248–250
initial characterizations, **68,** 246–248
instrumentation, **68,** 245–246
large assembly surface details, **68,** 250–252
Cryopreservation
blood cells, **63,** 210
single cell suspensions, **63,** 406–407
Cryopreservation, *see* Freezing
Cryopreservation, primary fibroblast, **59,** 41
Cryostat sections, tissue, **69,** 37–40
Cryosubstitution
ITEM, **65,** 250–252
Cryotechniques
mitochondrial imaging, **65,** 251
Cryptostome, **62,** 90
Crystal violet
viability assays, **66,** 215–216
CS1-GFP, **65,** 279
fluorescence patterns, **65,** 280
CsCI gradient purification of adenovirus, **71,** 401–402
CSF, *see* Cytostatic factor
CSU-10, *see* Yokogawa CSU-10
CTB, *see* Cholera toxin β-subunit
Culture
MEFs, **66,** 201–202
Culture density, chick forebrain neurons, **71,** 63
Culture dishes
for spinal and motor neuron cultures, **71,** 82–85
for superior cervical ganglia, **71,** 309–310
plating of hippocampal and cortical neurons on, **71,** 117–119
Culture media, **62,** 192–197
bacterized, **62,** 194–196
Cerophyl, **62,** 196
chick forebrain neurons, **71,** 49–60
defined, **62,** 35, 194–195
Dryl's, **62,** 196
for superior cervical ganglia, **71,** 306–308
Neffs, **62,** 193, 212
phagocytosis-deficient cells, **62,** 197
PP210, **62,** 241
PPYS, **62,** 400
proteose peptone, **62,** 192–193
PYG, **62,** 459–460
simplified for *Heliosoma* cultures, **71,** 158
SPP, **62,** 490
starvation media, **62,** 196, 212
Culture methods
chick ciliary neurons, **71,** 46–47
chick forebrain neurons, **71,** 60–64
cocultures, **71,** 145
dissociated mixed cultures of *Xenopus* spinal neurons, **71,** 141–143
neural-enriched cultures, **71,** 143–145
substrata, for *Xenopus* cultures, **71,** 145–148
Culture properties, **71,** 11–12
Culture substrates, **71,** 11–12
in *Xenopus* neuronal cultures, **71,** 139–148
Culture system
3-D: cellular and environmental heterogeneity, **64,** 212–213
multicell spheroids, **64,** 213–215
chemotherapy in, **64,** 222–225
Cultured cell mitochondria
isolation, **66,** 376
Cultured cells
cytochrome *c* release, **66,** 369–372
mitochondria, **65,** 30–31
mitoplasts, **65,** 30–31
transmitochondrial cybrid generation, **65,** 402–405
Cultured cells, cytoskeletal organization, **70,** 302–303

Cultured cells, mitotic extract preparation, **67,** 226–228
Cultured neurons, experimental properties of, **71,** 6–7
Culturing, *see* Cell culture
Cupric sulfate charring
SM mass levels, **66,** 143–144
Current
converting optical signals to, **63,** 118–123
converting to voltage, **63,** 123–124
Curve-fitting analysis, estimating number of cells per division, **63,** 388–389
Cy3-labeled tubulin, **67,** 45–46
Cyanine fluorophores, immunofluorescence histochemistry, **70,** 176
Cyanobiphenyl monomolecular layers, STM studies, **68,** 6
Cyclin A
mitotic cells negative for, **63,** 349
plus H_3 phosphorylation, immunocytochemical detection, **63,** 345–346
Cyclin B1, maximally expressed during mitosis, **63,** 349
Cyclin D1, MAbs reacting with, **63,** 288–289
Cyclin-dependent kinase inhibitors, **64,** 378
Cyclin E, immunoreactivity pattern, **63,** 286–287
Cyclins, **62,** 78
Cyclins, immunocytochemical detection, **63,** 347
Cyclins, staining for, **64,** 432–433
Cycloheximide, **62,** 470, 510, 520
Cyclophosphamide, plus adriamycin and 5-fluorouracil, **64,** 374
cyclops, gamma-ray mutagenesis, **60,** 67
CyP-A, regulating death and survival in neuronal cells, **71,** 291
Cysteine aspartic acid-specific proteases, *see* Caspases
Cysteine proteinase inhibitors, **66,** 251
Cystolic calcium
spectrofluorimetric measurement
prostate adenocarcinoma cells, **66,** 238–239
Cystolic NADPH, **65,** 92–93
Cytocentrifugation
LSC, **66,** 76
vs. touch preparation, **66,** 75
Cytochalasin B-induced enucleation, **65,** 408
cytochrome *c*, **66,** 366
Cytochrome *c* oxidase (COX), **65,** 110, 314–317
cytochemical staining, **65,** 115
genetics, **65,** 316
histochemistry, **65,** 314–317, 326
immunolocalization
KSS, **65,** 319
KCN titration, **65,** 122–123
integrated, **65,** 123–124
isolated, **65,** 125–126
visualization, **65,** 316
Cytochrome *c* oxidase (COX) 1 gene
mtDNA
histochemistry, **65,** 325
PCR, **65,** 325
Cytochrome *c*, release without simultaneous drop in $\Delta\Psi$, **63,** 471
Cytochrome c
beef heart mitochondria, **65,** 9–12
staurosporin, **65,** 153
Cytochrome c–depleted and reconstituted mitochondria
oxidative and phosphorylating properties, **65,** 12
Cytochrome c oxidase (COX) II immunoprecipitation
in organelle translation, **65,** 220–222
antibody coupling, **65,** 220
lysis, **65,** 220
radiolabeled translation products, **65,** 220
Cytogamy, **62,** 156–158, 257
Cytogenetics
FISH, **64,** 398, 400–401
for MRD detection, **64,** 397–401
Cytograms, bivariate halogenated thymidine analog-DNA, **63,** 366
Cytokeratin, and surface antigen MC5: simultaneous staining, **63,** 264–266
Cytokeratin-18 neoepitope
detection, **66,** 408
Cytokines
differentiation driven by, HP subsets during, **64,** 157–167
intracellular, labeling, **64,** 582–583
Cytokinesis, **62,** 73–76
Cytomegalovirus immediate early promoter, **66,** 204
Cytometers, bench-top, **63,** 22
Cytometry
fluorescence lifetime image, **63,** 133
image, analysis of PCNA, **63,** 425–426
quantitative fluorescence, **63,** 311–312
Cytoplasm, *see also* Organelles
cytoplasmic streaming, laser tweezer manipulation, *in vivo*, **55,** 199–202
Cytoplasmic fluorescence
LSC, **66,** 74

Cytoplasts, MTs
Cy3-labeled tubulin preparation, **67,** 45–46
enucleation, **67,** 47–48
image acquisition, **67,** 48–49
image analysis, **67,** 49–51
microinjection, **67,** 47–48
Cytoproct, **62,** 33, 56
Cytosine arabinoside, reducing nonneuronal contamination with, **71,** 30
Cytosine beta-arabinofuranoside (AraC), **71,** 10, 92, 121
Cytoskeletal dynamics, zebrafish, **59,** 209–212
Cytoskeletal organization, **71,** 18
Cytoskeletal polymers, transport of, **71,** 305
Cytoskeletal probes, *Xenopus* oocytes, **70,** 410–413
Cytoskeletal proteins, fluorescently tagged, **71,** 39
Cytoskeletal transport and dynamics, **71,** 5, 9
Cytoskeleton, **62,** 52–67, 87, 95, 441–446
14-nm filaments, **62,** 59, 96, 98
actin, **62,** 63–64
antibodies, **62,** 291–310
apical filament ring, **62,** 57
basal bodies, **62,** 32, 34, 53–54, 68, 70, 82, 88, 335
calcium-binding protein, **62,** 67, 455–464
centrin, **62,** 441
conjugation, **62,** 95–98
cortical organization, **62,** 39, 41
cultured cells, **70,** 302–303
dynein, **62,** 64–65
fenestrin, **62,** 70, 93, 96, 98
fimbrin, **62,** 57
hsp, **62,** 61, 441, 450
intermediate filament, **62,** 443
kinesin, **62,** 64–65
kinetids, **62,** 52, 82
live cell and fixed tissue confocal applications, **70,** 272
microtubules, **62,** 51–57, 95
microtubules, longitudinal bands, **62,** 81
myosin, **62,** 63
oral filaments, **62,** 57–60
profilin, **62,** 57
striated rootlets, **62,** 57
TCBP, **62,** 455–464
tetrins, **62,** 54, 58, 442–446
tubulin, **62,** 54, 60–63
visualized by immunofluorescence, **62,** 291–310, 459–461
Xenopus oocyte, CLSM
brightness and contrast extremes, **70,** 402
digital image processing, **70,** 403–404
image brightness optimization, **70,** 400–402
image scale optimization, **70,** 399
objective choice, **70,** 398–399
optical sectioning, **70,** 399–400
presentation, **70,** 404–408
Xenopus oocytes and embryos, CLSM
keratin filaments, **70,** 386–388
microfilaments, **70,** 388–390
microtubule fixation, **70,** 383–386
Cytoskeleton, elasticity, AFM studies, **68,** 76–77
Cytoskeleton, Inc., **71,** 132
Cytosol, green fluorescent protein targeting
chimera construction and expression, **58,** 76, 78
expression in primary cultures, **58,** 80–83
fluorescence microscopy, **58,** 79–80, 83–84
multiple organelle labeling with blue-shifted protein, **58,** 81
transfection, **58,** 83
Cytosol, pCa$_i$ measurement, **70,** 461
Cytosolic proteins
immunocytochemical detection in PC12 cell lines, **71,** 284
transport of, **71,** 314
Cytosolic S-100 extracts
Apaf-1 apoptosome, **66,** 23
Cytospin chamber, in LSCM of human tumors, **64,** 424, 428–429
Cytospin, specimen preparations, **64,** 317
Cytostatic factor, **61,** 386–387, 393, 395
Cytotoxicity
anthraquinones, **64,** 177
Hoechst, **64,** 33342, 215, 217
Cytotoxicity assay, **66,** 311

D

D011.10 cell line, **66,** 418, 420 t–423 t, 425 t, 427–428
transfection, **66,** 427–428
D1
gene cloning, **60,** 92, 94–95
retrovirus-mediated insertional mutagenesis, **60,** 92
DALP-MC
TGase glutamyl donor substrate proteins, **66,** 120–121
DANA
DANA/mermaid polymerase chain reaction, **60,** 317–318
zebrafish elements, **60,** 107–108

Dandelion pollen grain, CSU-10 studies, **70,** 93
Danio rerio, see Zebrafish
Dansylcadaverine, **66,** 120
DAPI, **65,** 338–339
  apoptosis assays, **66,** 213
  flow cytometry
    DNA fragmentation assay, **66,** 88
  mtDNA, **65,** 415–417
  staining of plant chromosomes, **64,** 14–15
  stock solution, **64,** 10
DAPI, **62,** 291–310
DAPI dye, *Drosophila* embryo confocal imaging, **70,** 373
Dark conditions, PECY5 reagents kept in, **63,** 236
Dark noise
  high-resolution multimode digital microscopy, reference image subtraction, **56,** 205–206
  low-light microscopy, sensitivity enhancement, **56,** 52, 54
DASPMI, **65,** 340
Data collection, for Boyden chamber assays, **63,** 556–557
Data display
  analog, **63,** 154–155
  as histograms or by list mode, **63,** 176–178
  logarithmic, and dynamic range, **63,** 302–304
Data file
  continuum to information, **63,** 41–43
  from light signals to, **63,** 37–41
Data management, HIV patients
  database, **64,** 574–575
  problems with immunophenotyping, **64,** 577–578
  proficiency testing programs, **64,** 576–577
  quality control, **64,** 575–576
  reference ranges, **64,** 575
Data pattern classification
  asthma, **64,** 497–502
  CLASSIF1 algorithm for, **64,** 491–493
  for diagnosis, **64,** 506
  melanoma, **64,** 493–494
Data storage
  flow cytometry standard format, **63,** 161
  formats for, **63,** 175–176
Date–time generator, video-enhanced differential interference contrast light microscopy, **56,** 178
Daunomycin, cellular retention, **64,** 203, 205
DCC, *see* Deleted in colorectal carcinoma
DCFDA assay, **65,** 81–83
DD, **66,** 454, 501
D, see Diffusion constant
DE52 chromatography
  actin-binding assay, **65,** 165
dead eye
  gene cloning, **60,** 92, 94–95
  retrovirus-mediated insertional mutagenesis, **60,** 92
Death domain (DD), **66,** 454, 501
Death effector domain (DED), **66,** 454
Death receptors, **66,** 368–369
Decay, heterogeneous fluorescence, **63,** 144–145
Deconvolution
  images, **58,** 39–40, 231–234, 307
  microscopy, **58,** 354–355
Deconvolution microscopy
  confocal microscopy alternative, **70,** 68–69
  development, **70,** 130
  web sites, **70,** 78
Deconvolution microscopy, **61,** 82
Deconvolution, *see also* Convolution
  digital fluorescence images, **56,** 305–315
    constrained iterative deconvolution, rationale, **56,** 306–307
    evaluation, **56,** 309–314
    future research directions, **56,** 314–315
    nearest-neighbor deconvolution
      implementation, **56,** 308–309
      rationale, **56,** 307–308
    overview, **56,** 305–306
DED, **66,** 454
Deep etching, cryo-AFM, **68,** 252–253
Defined medium, for spinal and motor neuron cultures, **71,** 84
Deflectors, multiple trap synthesis
  acousto-optic deflectors, **55,** 208–209
  mechanical deflectors, **55,** 208
Degradation
  collagen, probe microscopy, **69,** 171
  multicomponent ECM, **69,** 197–198
  proteoglycans, **69,** 55–57
Degradation, membrane, invadopodia activity as measurement, **63,** 615–619
Dehydration, solutions, DNA on mica, **68,** 360
Deleted in colorectal carcinoma, **69,** 351–353
Deleted-mtDNA
  single-fiber PCR, **65,** 325
Deletions, chromosomal, **71,** 202
$\Delta\Psi$, *see* Mitochondrial membrane potential
Denaturation, DNA, **63,** 357
Dendrites
  in long-term cortical and hippocampal cultures, **71,** 124

organelle traffic in, **71,** 8
production of, **71,** 10
Dendritic growth cones, properties of, *vs.* axonal growth cones, **71,** 2
Dendritic spines, in long-term cortical and hippocampal cultures, **71,** 120
Dense-core secretory granules, **62,** 32, 45–47, 349–350
Dense fibrillar component, *see* Nucleolus
Density gradient centrifugation, **71,** 79–80
Density gradient centrifugation, *Dictyostelium discoideum*, **67,** 340–342
Density gradient purification
plant mitochondria, **65,** 57–61
Percoll gradients, **65,** 57–59
sucrose gradients, **65,** 57
transgenic line screening, **65,** 57–59
Density plot, two-parameter histograms, **63,** 164–166
Density, cell, effect on migration assays, **63,** 550
Density, of neuron yield in culture, **71,** 11
Dental cell biology, confocal objective lens selection, **70,** 273
Deoxycholate, cell-assembled ECM, **69,** 154
Deposited extracellular matrix
antigen identity, **69,** 41–42
reporter cell response, **69,** 35–36
Depressions, AFM studies
fusion pores, **68,** 40, 43
plasma membranes, **68,** 39–40
transient fusion, **68,** 43–47
Depth correction, confocal microscopy, **70,** 151–152
Depth cues, CSU-10 digital signal processing, **70,** 115
Depth discrimination, confocal microscopy, **70,** 156–158
Detection threshold, **63,** 314–317
Detectors
double-labeled specimen cross-talk, **70,** 213–217
multicolor laser scanning confocal microscopy, **70,** 190–191
Detector systems, confocal microscopy, **70,** 159–161
Detergents
cell membrane dissolved with, **63,** 272
for permeabilization, **63,** 256, 258–259
nonionic, **63,** 276–277
Deterioration, chronic, in optical system, **64,** 295
Deubiquitinase, **66,** 263
Development, **62,** 88–98
Development, embryonic zebrafish, **59,** 11–25
Developmental mutant screening, zebrafish, *see also* Early pressure screens, zebrafish, Haploid screen, zebrafish, Maternal-effect genes, gynogenesis-based screening in zebrafish
advantages of zebrafish system, **60,** 22, 214
allele screens
effective carrier frequencies, **60,** 31–33
fused somites, **60,** 31–33
rationale, **60,** 31–32
sonic you, **60,** 31–33
unplugged, **60,** 31–33
you-too, **60,** 31–33, 38
breeding scheme, **60,** 26
dominant screens, **60,** 34
ethylnitrosourea mutagenesis of premeiotic male germ cells
dose dependence, **60,** 24–25
safety, **60,** 24
F_2 screening, calculation of genome number, **60,** 27–28
line selection for mutagenesis, **60,** 22–23
management of outcomes
complementation analysis, **60,** 36
criteria for keeping mutants, **60,** 35–36
database management, **60,** 36
mapping of mutations, **60,** 36
saturation level calculation, **60,** 36–37
mutagen selection, **60,** 23–24
mutation frequency dependence, **60,** 25
progeny tracking, **60,** 25–26, 28
touch response screen, **60,** 29
transgenic fish, **60,** 38
visual screens, **60,** 28–29, 35
whole mount staining in screening, **60,** 29, 38
DFF45, **66,** 300–303
cleavage roles, **66,** 302–303
endogenous cleavage, **66,** 301
in vitro cleavage, **66,** 301–302
DHFR, **65,** 225
Developmental properties, **71,** 10–11
of cultured neurons, **71,** 5
Developmental stages, advantageous in study of Tibia-1 pathway, **71,** 173
Dextrans and proteins, preparing injection samples of, **71,** 132–133
Diagnosis
acute leukemias, **64,** 355–356
with data pattern classification, **64,** 506
Diamidino-2-phenylindole (DAPI), **65,** 338–339

Diamidino-2-phenylindole (DAPI) (*cont.*)
mtDNA, **65,** 415–417
4,6-Diamidino-2-phenylindole, **63,** 10
Diatoms, video-enhanced differential interference contrast light microscopy performance test, **56,** 179–180
DiBAC$_4$(3), membrane potential sensitive, **64,** 562
Dibucaine, **62,** 350–352
DIC microscopy, optics, **68,** 202
DIC, *see* Differential interference contrast
DIC, *see* Differential interference contrast microscopy
5,6-Dichloro-β-D-ribofuranosylbenzimidazole (DRB)
inhibition of nucleolar function, **53,** 316
segregation of nucleolus, **53,** 315–316
Dichlorotriazinylaminofluorescein, protein labeling, **63,** 191
Dichroic mirror
in wavelength selection, **63,** 117–118
long-pass and short-pass, **63,** 36–37
Dichroic mirror, **71,** 153
Dichroic mirrors, multiply tuned
advantages, **53,** 55–56
applications with filter wheel, **53,** 56–57
spectral characteristics, **53,** 55
Dictyostelium discoideum, see Spindle birefringence
Dictyostelium discoideum
axenic strains, **66,** 473–476
cell culture, **66,** 473–476
cloning, **66,** 476
media, **66,** 474
routine, **66,** 474–475
storage, **66,** 475–476
centrosomes
electron microscopy, **67,** 346–347
indirect immunofluorescence microscopy, **67,** 342–343
isolation, **67,** 338–342
monoclonal antibody generation, **67,** 347–351
MT nucleation assay, **67,** 344–345
screening, **67,** 351–352
SDS-gel electrophoresis, **67,** 345–346
yield and purity estimation, **67,** 343–344
development, **66,** 471–472
differentiation, **66,** 471
freezing, **66,** 475
genetic manipulation, **66,** 485–491
cell death-resistant mutants, **66,** 491–494
DNA introduction, **66,** 485–486
gene rescue, **66,** 487–490
homologous recombination, **66,** 490–491
insertional mutagenesis, **66,** 487–490, 489
selectable markers, **66,** 485–486
PCD, **66,** 469–494
characteristics, **66,** 480–485
induction, **66,** 476–480
molecular mechanisms, **66,** 472–473
morphotype, **66,** 472
rationale, **66,** 470–471
regrowth assay, **66,** 480–481
staining methods, **66,** 481–485
stalk cell differentiation, **66,** 477–478
thawing, **66,** 475
ultrastructure, **66,** 473
wild-type strains, **66,** 478–480
Dictyostelium discoideum, CSU-10 studies, **70,** 96
Dictyostelium
cell–cell adhesion force measurements, **68,** 105–109
cell culture, **68,** 110
web sites, **66,** 494
Dideoxynucleotides, blocking with, **64,** 61, 66
Dielectric sphere, single-beam gradient laser trap force calculations, **55,** 1–25
gradient trap force, **55,** 9–16
arbitrary trap location, **55,** 11–16, 23–25
Y axis trap focus, **55,** 10–11, 22–23
Z axis trap focus, **55,** 9–10
index of refraction effects, **55,** 19–20
mode profile effects, **55,** 16–20
overview, **55,** 1–4, 20–21
ray force, **55,** 21–22
ray optics regime, light forces, **55,** 4–8
transmission electron microscope mode profile effects
$_{00}$-mode profile, **55,** 16–17
$_{01}$-do-nut-mode profile, **55,** 17–19
Diethylaminoethyl cellulose (DE52) chromatography
actin-binding assay, **65,** 165
Difference filter, chromosomes, **64,** 84–85
Differential interference contrast
objective lens application markings, **70,** 260
scanning transmission DIC, **70,** 282
Differential interference contrast microscopy (DIC), **61,** 114, 161, 301
Differential interference contrast microscopy, *see* Video-enhanced differential interference contrast light microscopy

Differential nonlinearity, problem in ADCs, **63,** 157
Differentiation of neurons, epigenetic factors in, **71,** 102–104
Differentiation-induced genes
  Bcl-2 family, **66,** 36
Differentiation map analysis, data processing for, **63,** 390
Differentiating centrifugation
  yeast mitochondria isolation, **65,** 40–43
Difficulty of preparation, **71,** 6–7
Diffusers, light microscopy, **56,** 138
Diffusion constant, calculation in fluorescence recovery after photobleaching of membrane proteins, **58,** 271–272, 276–278
Digestion, cell wall, in yeast SPB permeabilization, **67,** 168–169
Digital fluorescence microscopy, cell cytoplasts
  Cy3-labeled tubulin preparation, **67,** 45–46
  enucleation, **67,** 47–48
  image acquisition, **67,** 48–49
  image analysis, **67,** 49–51
  microinjection, **67,** 47–48
Digital frame memory, confocal microscopy, **70,** 54
Digital image analysis, for QDFM, **64,** 46
Digital image processing
  contrast enhancement
    differential interference contrast images, **56,** 209–210
    gray value, *see* Gray value
    methods, **56,** 77–79, 101–104
    processors, **56,** 177–178
    video-enhanced differential interference contrast light microscopy, **56,** 165–167
  convolution, *see* Convolution
  digital masks
    convolution filter, **56,** 109–114
    noise reduction, **56,** 99
  digitization process, **56,** 92–94
  high-resolution multimode digital microscopy
    computer control, **56,** 207–208
    image analysis, **56,** 208–209
    image processing, **56,** 208–209
    software, **56,** 209
  intensity quantification, **56,** 94–95
  noise reduction, **56,** 95–101
    arithmetic removal, **56,** 97–99
    digital mask, **56,** 99
    filters, **56,** 100–101
    spatial methods, **56,** 96–101
    temporal averaging, **56,** 95–96
  overview, **56,** 91–92, 115
  ratio imaging, **56,** 233–234
  transforms, **56,** 105–108
  video microscopy optimization, **56,** 63–90
    analog processing compared, **56,** 70–73
    analysis
      brightness analysis, **56,** 86–90
      image accuracy, **56,** 83–86
      image processor, **56,** 89–90
      intensity quantification, **56,** 94–95
      optical system characterization, **56,** 87–88
      shading correction, **56,** 88–89
    fundamentals, **56,** 63–70
      conventional microscopy compared, **56,** 65–66
      image sampling, **56,** 68–69
      imaging process, **56,** 66–67
      overview, **56,** 63–64
      pixels, **56,** 67–68
      small object visualization, **56,** 69–70
      voxels, **56,** 67–68
    mechanics, **56,** 73–83
      contrast enhancement, **56,** 77–79
      detail enhancement, **56,** 82–83
      image averaging, **56,** 81
      image integration, **56,** 81–82
      image subtraction, **56,** 79–80
      processor components, **56,** 74–76
Digital mask
  convolution filter, **56,** 109–114
    edge detection, **56,** 112–113
    Laplacian filters, **56,** 113–114
    line detection, **56,** 112–113
    point detection, **56,** 112
  spatial noise reduction, **56,** 99
Digital microscopy, *see* Digital image processing, Multimode digital microscopy
Digital resolution, confocal microscopy, **70,** 252
Digital signal processing
  3D and 4D image viewing, **70,** 115–116
  3D projection and stereo image generation, **70,** 113
  arithmetic precision, **70,** 111–112
  image stacks to projections, **70,** 115
  image storage requirements, **70,** 116–117
  neighborhood-based processing, **70,** 107–108
  overview, **70,** 105–107
  sharpening, **70,** 108–110
  unsharp masking, **70,** 108–110
  unsharp masking significance, **70,** 112–113
  Xenopus oocyte cytoskeleton, **70,** 403–404

Digital signal processing, pulse characterization by, **63,** 159–161
Digital still cameras, features, **56,** 40
Digital-to-analog converter, **63,** 156–157
Digitalis purpurea, **66,** 371
Digitization, of electrophysiological data, **71,** 234
Digitonin, **65,** 137
membrane permeabilization, **53,** 87, 289–290, 521–522
mitochondria, **66,** 371
recrystalization
coimmunoprecipitation, **65,** 223
solubilization, **65,** 203
stock solution preparation, **53,** 520–521
titration, **65,** 127,135
Digitonin, **63,** 490–491
Digoxigenin, labeled DNA: visualization, **64,** 65
Dihexyloxacarbocyanine iodide
mitochondrial membrane potential, **65,** 291–292
Dihydroethidium, **66,** 382
Dihydrofolate reductase (DHFR), **65,** 225
Dikinetids, **62,** 82
Dil microinjections, axonal outgrowth, **69,** 355
Dimensions, of neurons, **71,** 8–9
Dimerization
aequorin and green fluorescent protein, **58,** 19–20
green fluorescent protein mutants, **58,** 21
DioC_6, **65,** 88,340
Diphenylene iodinium (DPI), **66,** 308
Direct binding assay
basement membrane self-assembly, **69,** 124–129
method, **69,** 289–290
Direct labeling, for γTuSC and γTuRC, **67,** 21–23
Direct-view high-speed confocal scanner
3D and 4D image viewing, **70,** 115–116
3D projection and stereo image generation, **70,** 113
arithmetic precision, **70,** 111–112
clam sperm, **70,** 97, 100
dandelion pollen grain, **70,** 93
design, **70,** 91–92
image stacks to projections, **70,** 115
image storage requirements, **70,** 116–117
low bleaching of fluorescence, **70,** 101–104
mechanical and optical performance, **70,** 117–118
neighborhood-based processing, **70,** 107–108
overview, **70,** 105–107
potentials, **70,** 118–122
Schizosaccharomycetes pombe, **70,** 97
sharpening, **70,** 108–110
slime mold, **70,** 96
Tetrahymena, **70,** 96–97
unsharp masking, **70,** 108–110
unsharp masking significance, **70,** 112–113
very-high-speed full-frame confocal imaging, **70,** 104–105
Directional cues, **71,** 172
Disaggregation, tissue for primary cell culture, **57,** 14–15, 148–150, 175–177, 180
Disease screening, expression cloning, **69,** 219–220
Dissecting microscope, **71,** 42
Dissection
ease of, **71,** 13
in Tibia-1 pathway studies, **71,** 172
enzymatic, of intact spinal cords or ventral halves, **71,** 77–79
intact chick spinal cords, **71,** 73–77
isolation of chick ciliary ganglion, **71,** 43–44
mechanical, **71,** 77–79
of chick ciliary ganglion neurons, **71,** 37–49
of chick forebrain neurons, **71,** 57–58
of *Drosophila* embryos and larvae for immunolabeling, **71,** 218–220
of grasshopper embryos, **71,** 186–187
of hippocampal and cortical neurons, **71,** 113–117
of peripheral neurons, **71,** 25–27
of retrovirally infected chick embryos, **71,** 382–385
Disinfectants, exposure to, **64,** 527, 529
Disk recorders, video-enhanced differential interference contrast light microscopy, **56,** 178–179
Dissociated cultures
biolistic transfection of, **71,** 354
chick ciliary ganglia, **71,** 46
chick forebrain tissue, **71,** 57–59
Drosophila retina, **71,** 236
Heliosoma ganglia, **71,** 166
hippocampal and cortical neurons, **71,** 117
in Purkinje neuron studies, **71,** 90
of *Xenopus* spinal neurons, **71,** 140–143
preparation of, **71,** 27–28
transfection of, **71,** 333
Distance, particles separated by, **63,** 30–31
1,2-Distearoyl-*sn*-glycero-3-phosphatidylethanolamine, **69,** 177–178
Distortion, low-light microscopy, **56,** 60
Division-furrow filaments, **62,** 57
diwanka, mutation effects in development, **60,** 79
DNA
adsorption

charged molecules in solution, **68,** 364
surface treatments, **68,** 365
altered stainability, **63,** 351
and dsRNA, simultaneous measurement, **63,** 143
Bacmid, **67,** 15
bacteria
staining, **64,** 541–542
structure, **64,** 542
base composition, **62,** 18, 150–151
bead preparation, **67,** 245–246
cancer cells, propidium iodide-stained, **63,** 72
complementary, *see* Complementary DNA
complementary, mapping, **64,** 51
covalent, anchoring for SFM, **68,** 365–366
curvature and flexibility
overview, **68,** 379–380
SFM studies, **68,** 380–385
DNA–enzyme processes, AFM imaging, **68,** 219
DNA–protein interactions, AFM imaging, **68,** 219
DNase digestion, AFM studies, **68,** 227
electrostatic adsorption
λ-phage, B–S transtition, **68,** 332
ethidium bromide-bound, **63,** 135–136
fluorescence staining, **63,** 9–10
fragmentation
in estimation of apoptosis frequency, **63,** 531–532
studied by LSC, **63,** 543
fragmentation
and cell lysis assay, **64,** 178
calculation, **64,** 179
high molecular weight
isolation, **66,** 440
high-molecular-weight, purification, **64,** 38–40
histograms, *vs.* BrdUrd histograms, **63,** 371
immunolocalization
respiratory chain, **65,** 317–318
integrity, drug-induced disruption, **64,** 174–175
intercalating agents, **65,** 399
isolation, **62,** 239–250, 271
labeled, visualization, **64,** 65–66
labeling, primed *in situ*, **64,** 17–18
ladder
detection with gel electrophoresis, **63,** 494–495, 499
indicator of nDNA fragmentation, **63,** 516–517
libraries, chromosome-specific, **64,** 27–28
loops, relaxed, **64,** 265–266
maximum pixel values in LSCM, **63,** 64
metabolism, inhibitors, **64,** 175
methylation, **62,** 144, 497–498
mica
multivalent cation activation, **68,** 361–362
silane treatment, **68,** 360–361
solution dehydration, **68,** 360
thermodynamic equilibrium, **68,** 363–364
nuclear, fragmentation, **63,** 506–509
measuring techniques, **63,** 516–520
overview, **68,** 361–362
soluble cation effects, **68,** 362–363
percentage in comet tail, **64,** 241
prepared from YAC clones, **64,** 42–43, 104–107
preparing injection samples of, **71,** 132–134
probes for centrosome-associated protein DNA, **67,** 130
purification from carbohydrates, **62,** 477
rearrangement, **62,** 134–140
repeat sequences, detection with oligonucleotide primers, **64,** 59–60
replication, **62,** 141–144
SFM
AC modes, **68,** 367
electron microscope legacy, **68,** 359
fragment ablation, **68,** 379
mobility modulation and solution dynamics, **68,** 371–374
molecular motion, **68,** 374–376
organic solvents, humidity issue, **68,** 366–367
protein and ligand interactions, **68,** 368–369
protein interactions, **68,** 376–377
secondary structure, **68,** 369–370
sharper probes, **68,** 367–368
single-molecule dissection, **68,** 377–379
single-molecule stretching, **68,** 388
solid-state sizing, **68,** 370
ss-DNA structure, **68,** 388
supercoiled DNA, salt and surface effects, **68,** 385–388
water and buffers, **68,** 370–371
single-fiber PCR, **65,** 328–329
staining, **63,** 266, 383–384
strand breaks
apoptotic cells, **63,** 352
flow cytometric analysis, **63,** 493–494
supercoiling, **64,** 252–254
synthesis, **62,** 71–73, 76
as *Xenopus* culture injections, **71,** 132
DNA affinity chromatography, *see* Base-unpairing region binding protein

DNA amplification, *in vitro*, generation of probes by, **64,** 43–44
DNA analysis
Bcl-2 family cloning, **66,** 33–34
DNA
as *xenopus* culture injections, **71,** 132
preparing injections samples of, **71,** 132–134
DNA-binding dyes, **65,** 338–339
DNA-binding protein, *see* Base-unpairing region binding protein
DNA circularization, **61,** 365–366
DNA coating density
and future biolistics improvements, **71,** 365
as parameter of biolistics, **71,** 355–356
DNA constructs, species and, **71,** 9
DNA content
analytic methods for, **64,** 131–134
bacteria, **64,** 516–517, 558–559, 561
fractional, **63,** 533–534
in human tumors, LSC applications, **64,** 423–425
in multiparametric analyses, **63,** 292–293
nuclear quantification in S phase, **64,** 363–364
vs. BrdUrd content: data displayed as, **63,** 365–366
DNA damage
electrophoretic detection, **64,** 236–237
measurement on cell-by-cell basis, **64,** 253–266
prediction of cell survival in complex systems, **64,** 244–246
radiation-induced, **64,** 251–252
types detected by comet assay, **64,** 242–244
DNA damage, affecting biolistics performance, **71,** 365
DNA–drug interactions
confocal microscopy, **64,** 184–185
flow cytometric analysis of
drug uptake, **64,** 184
Hoechst 33342–DNA binding, **64,** 185–188
ligand characteristics, **64,** 183
spectrofluorimetry, **64,** 183–184
DNA fibers
genomic, gene mapping by hybridization onto, **64,** 48, 50
preparation for QDFM, **64,** 40–41
DNA fingerprinting
databases, **57,** 214
embryonic stem cells, **57,** 287
polymerase chain reaction amplification of polymorphisms, **57,** 212–213
principle in cell line authentication, **57,** 211
restriction fragment length polymorphism analysis, **57,** 212
DNA fragmentation
gel electrophoresis, **66,** 438–443
DNA isolation, **66,** 438–439
nuclease activity, **66,** 441–442
protoplasts, **66,** 439–441
reagents, **66,** 442–443
tissues, **66,** 441
TUNEL assay, **66,** 90–93
vs. poly(ADP-ribose) polymerase, **66,** 99
DNA fragmentation assays
flow cytometry, **66,** 87–88
fractional DNA content, **66,** 87–90
LSC, **66,** 88
materials, **66,** 87
nuclei assays, **66,** 174–175
DNA fragmentation factor II (DFF45), **66,** 300–303
endogenous cleavage, **66,** 301
DNA fragments
and ethanol fixation, **64,** 426
cloned, probes generated from, **64,** 41–43
mapped by FISH, **64,** 34
DNA histograms
and gating, **64,** 142
bacteria, **64,** 540
standards and controls, **64,** 548–550
in bivariate profiles of hemoglobin, **64,** 144
DNA index, **64,** 425
in monitoring MRD, **64,** 402
DNA laddering, apoptosis analysis, **57,** 259–261
DNA plasmids, transfection of chick ciliary neurons with, **71,** 39
DNA repair, analysis with PCNA, **63,** 426
DNA repair, chromatin structure effect, **64,** 220–221
DNA replication
alteration after heat shock, **64,** 276–278
patterns, **64,** 281–282
DNA replication site, mapping *in situ*
clustering of replicons, **53,** 456–457
digoxigenin deoxynucleotide method
cell culture on cover slips, **53,** 463
controls, **53,** 464
fixation, **53,** 460, 464
fluorescent labeling with antibody, **53,** 464
materials, **53,** 462–463
mounting of coverslips, **53,** 464
permeabilization of cells, **53,** 463–464
principle, **53,** 456–457, 462
run-on labeling, **53,** 464
troubleshooting, **53,** 465
dual labeling at different time points
cell culture on cover slips, **53,** 466

controls, **53,** 468
fixation, **53,** 467
fluorescent labeling with antibody, **53,** 467–468
materials, **53,** 466
mounting of coverslips, **53,** 468
principle, **53,** 457, 465–466
pulse labeling with halodeoxyuridines, **53,** 467
troubleshooting, **53,** 468
halogenated deoxynucleotide method
cell culture on cover slips, **53,** 460
controls, **53,** 461
DNA denaturation for antibody access, **53,** 456, 460
fixation, **53,** 460
fluorescent labeling with antibody, **53,** 460–461
materials, **53,** 458–460
mounting of coverslips, **53,** 461
principle, **53,** 455–458
pulse labeling, **53,** 458, 460
troubleshooting, **53,** 461–462
DNA strand breaks, labeling, **64,** 427–428
DNA topoisomerase I, inhibitors, **64,** 177, 429–430
DNA topoisomerase II
drug-induced trapping, **64,** 181–182
inhibitors, **64,** 177
DNA topoisomerase II, nucleic acid interactions during interphase and mitosis/meiosis, **53,** 413
DNA topoisomerases, cell cycle related expression, **64,** 180
DNA, *Tetrahymena thermophila*, FISH for confocal microscopy
nick translation reaction, **70,** 346–347
probe synthesis, **70,** 346
sample preparation, **70,** 344–346
DNA, complementary
expression cloning, **69,** 213
recovery by Hirt supernatant, **69,** 216–217
DNase I, **71,** 83, 92
DNase I affinity chromatography
actin-binding assay, **65,** 163
DNase I column
actin-binding assay, **65,** 160–161
DNase I, digestion of nuclei, **53,** 15–16, 27, 42
DNase, DNA digestion, AFM studies, **68,** 227
DOC, *see* Deoxycholate
Dodecyl maltoside
complex III, **65,** 108
complex IV, **65,** 111
Dominant-negative (DN) mutant proteins, in *Xenopus* cultures, **71,** 131
$DoOC_6(3)$, mitochondrial dye, **64,** 119
Dorsal appearance, zebrafish, **59,** 157–158
Dorsalized mutants, zebrafish, **59,** 162–163
Dorsal root ganglia, advantages of, **71,** 18
Dorsal-ventral patterning, zebrafish, **59,** 159–165
Dorsal-ventral polarity, zebrafish, **59,** 158–159
Dosage-dependent suppressor screen, SPBs, **67,** 97–100
Dose-response studies, **71,** 326
Dot blotting
TNF ligands, **66,** 508–509
Dot plot
flow cytometric: simple display, **63,** 161–164
two-dimensional, **63,** 43
Dot plots
for cell populations, **64,** 293
two-parameter, **64,** 198, 200
DOTAP liposomal transfection reagent, **71,** 137
Double crossovers
homologous
mtDNA sequences, **65,** 388–389
Double fluorescein diacetate/propidium iodide staining
stalk cells, **66,** 483
Double-labeled specimens
detector cross-talk in LSCM, **70,** 213–217
two-color LSCM with krypton–argon ion laser, **70,** 225–228
Double-sensor-stabilized atomic force microscopy, protein unfolding, **68,** 328–331
Dow Corning, **71,** 119
Down syndrome cell adhesion molecule, **71,** 214
Doxorubicin
cellular retention, **64,** 203, 205
fluorescence, **64,** 219
penetration problem, **64,** 224
plus 5-fluorouracil and cyclophosphamide, **64,** 374
plus vincristine, **64,** 373–374
DPI, **66,** 308
DPI-induced apoptosis
exogenous NADH, **66,** 313
Drag, micromechanical measurement, **55,** 139–140
DRB, *see* 5,6-Dichloro-β-D-ribofuranosylbenzimidazole
Drilled dishes, in spinal and motor neuron cultures, **71,** 83
Drops, formation and delay time, **63,** 45–47
Drosophila, **53,** **66,** 321–327
AO staining
embryos, **66,** 325

Drosophila (*cont.*)
imaginal discs, **66,** 325–326
apoptosis detection, **66,** 323–330
cell-free chromatin remodeling
decondensation and remodeling, incubation conditions, **53,** 507
extract preparation
eggs, **53,** 504
embryos, **53,** 504–505
preblastoderm embryos, **53,** 510–511
NURF role, **53,** 500
overview, **53,** 499
p22 role, **53,** 499–500
transcriptional remodeling, incubation conditions, **53,** 511
cell-free nuclear assembly
disassembly
monitoring, **53,** 410–412
nuclei preparation, **53,** 409–410
stage 14 oocyte exctract preparation, **53,** 408–409
embryo extract preparation, **53,** 398–399
incubation conditions for assembly, **53,** 399
lamin assembly
disassembly assays, **53,** 412–413
interphase *vs.* mitotic/meiotic isoform assembly, **53,** 405–407
monitoring of aggregation and polymerization, **53,** 405
purification of native and recombinant proteins, **53,** 402–404
regions in head-to-tail binding, **53,** 406
regulation by lamin isoforms, **53,** 407–408
monitoring of assembly
immunofluorescence microscopy, **53,** 401
light microscopy, **53,** 400
transmission electron microscopy, **53,** 402
Western blot analysis, **53,** 402
sperm chromatin preparation, **53,** 399
centrosomes, analysis during embryogenesis, **67,** 114–122
embryo extracts
γ-tubulin-containing complexes, **67,** 6–11
γTuRC, **67,** 3–6
γTuSC, **67,** 4–6
embryo fixation
transmission EM, **66,** 330
embryos, centrosomes, **67,** 141–142
eye discs
double labeling, **66,** 328
embryo embedding, **66,** 328–330
lamin purification
histidine-tagged recombinant proteins, **53,** 404
immunoaffinity chromatography of native proteins, **53,** 402–404
phagocytosis, **66,** 330–332
polyacrylamide gel electrophoresis of chromatin-associated proteins, **53,** 507–508
sample preparation for whole-mount fluorescence *in situ* hybridization
buffer preparation, **53,** 199–200
egg chamber
manual dissection, **53,** 207–208
mass preparation, **53,** 203–207
formaldehyde fixation
cells cultured on coverslips, **53,** 210
duration, **53,** 199
egg chambers, **53,** 203–208
embryo fixation and methanol devitellinization, **53,** 201–203
formaldehyde preparation from paraformaldehyde, **53,** 200–201
tissues on slides, **53,** 208–209
methanol fixation/devitellinization of embryos, **53,** 203
squashes, **53,** 210–211
TUNEL staining
embryos, **66,** 326–327
eye discs, **66,** 327–328
whole-mount hybridization reactions
equipment, **53,** 212, 215
reagents and solutions, **53,** 211–212
slide preparations, **53,** 215–217
squash preparations, **53,** 217
tube preparations, **53,** 212–215
Drosophila embryo centrosome isolation, *see* Centrosome
Drosophila embryos, CLSM
autofluorescence, **70,** 368–369
bleed-through, **70,** 369–370
collection and permeabilization, **70,** 363
fluorescent probes, **70,** 364
hard copy images, **70,** 372
image archiving, **70,** 372
image collection, **70,** 366–368
image processing, **70,** 370–372
image transfer, **70,** 370
microscope details, **70,** 365–366
mounting, **70,** 364–365
multiple labels, **70,** 372–375
overview, **70,** 361–363
Drosophila melanogaster, **71,** 3, 196

antibodies found in, **71,** 208–212
as genetic model system for molecular neurobiology, **71,** 199–203
background references for, **71,** 196–199
cell biology techniques in, **71,** 216–222
central motor neuron recording, **71,** 232–234
clonal techniques for mutant analysis, **71,** 222–223
coimmunoprecipitation to identify interacting proteins, **71,** 213–214
compared with *Xenopus* cultures, **71,** 130
dissection of embryos and larvae for immunolabeling, **71,** 206–213
electrophysiology in, **71,** 228–238
embryo, small size of, **71,** 8
excitatory junctional current (EJC) recordings, **71,** 231
functional analysis techniques, **71,** 224–228
giant fiber system of, **71,** 234–235
head extracts, **71,** 203
identifying the phenotype of, **71,** 203
imaging and manipulating calcium levels, **71,** 224–226
immunolabeling nervous system cells, **71,** 217–218
isolation and extraction of heads, **71,** 205–206
life cycle, **71,** 216
lipid analysis in, **71,** 214–215
monitoring membrane dynamics with stryl dyes, **71,** 226–227
monitoring protein dynamics in, **71,** 227–228
nervous system biochemistry, **71,** 203–215
neuromuscular junction recording in, **71,** 230–232
online database, **71,** 198
optic lobe of fly brain, **71,** 203
phospholipid metabolism in the eye of, **71,** 203
recording
from adult brain, **71,** 237
from cultured neurons, **71,** 228–229
from sensory systems, **71,** 236–238
shibire mutation in, **71,** 204
species advantages in gene expression, **71,** 10
synaptic transmission studies, **71,** 203
temperature dependency in, **71,** 216
transgenic protein expression studies in, **71,** 220–222
vesicle recycling in, **71,** 204
Drosophila mitotic mutation identification and characterization
choosing mutagens
chemical
disadvantages, **61,** 321
ethyl methane sulfonate, **61,** 321
saturation screens, **61,** 321
transposon
advantages, **61,** 321
hot spots, **61,** 321–322
P element mutagen, **61,** 321
plasmid rescue, **61,** 321
cytological analysis of larval brain and imaginal discs
DAPI staining of neuroblast squashes
hypotonic swelling, **61,** 340
Mowiol preparation, **61,** 340–341
procedure, **61,** 339–340
signal to noise ratio, **61,** 339
storage, **61,** 340
dissection
distinguishing sexes, **61,** 337–338
Ephrussi–Beadle Ringer's solution rinse, **61,** 337
mutant brain morphology, **61,** 337
drug treatments of neuroblasts *in vitro*
avoiding anoxia, **61,** 343
drug carrier controls, **61,** 343
procedure, **61,** 343
drug treatments of neuroblasts *in vivo*, **61,** 344
immunofluorescence in semisquashed preparations
procedure, **61,** 342–343
uses, **61,** 342
immunofluorescence in whole mounts
procedure, **61,** 341–342
taxol use, **61,** 342
uses, **61,** 341
orcein staining of neuroblast squashes
orcein reflux, **61,** 339
procedure, **61,** 339
saving stained slides, **61,** 339
cytological analysis of syncytial mitoses
analysis of cytoskeletal elements and chromatin *in vivo*
artifact reduction, **61,** 334
fluorescent probe choice, **61,** 333–334
immunofluorescence analysis confirmation, **61,** 334
laser scanning confocal microscopy, **61,** 332
sample preparation, **61,** 332–333
clearing immunolabeled embryos
clearing solution volume, **61,** 332
phalloidin labeling, **61,** 332

Drosophila mitotic mutation identification
and characterization (*cont.*)
photobleaching, **61,** 332
procedure, **61,** 331–332
purpose, **61,** 331
formaldehyde fixation
advantages, **61,** 327–328
phalloidin staining, **61,** 327–328
procedure, **61,** 328
vitelline membrane removal, **61,** 329
formaldehyde/methanol fixation
anoxia artifacts, **61,** 327
fixation times, **61,** 327
formaldehyde concentration, **61,** 325
procedure, **61,** 325–326
rapid fixation importance, **61,** 327
storage, **61,** 327
taxol use, **61,** 325
labeling elements in fixed embryos
direct primary antibody
conjugates, **61,** 331
laser confocal analysis, **61,** 330
non-specific protein blocking, **61,** 331
Oil-green reagent, **61,** 330
primary antibody dilution
determination, **61,** 331
procedure, **61,** 330–331
methanol fixation
advantages, **61,** 330
procedure, **61,** 329–330
vitelline membrane removal, **61,** 330
life cycle
cellular blastoderm, **61,** 319
cellularization, **61,** 319
embryogenesis, **61,** 320
genetic control of mitosis
late larval development, **61,** 320
postcellularization divisions, **61,** 320
rapid syncytial divisions, **61,** 320
larvae procession, **61,** 320
oogenesis, **61,** 318
pronuclear fusion, **61,** 318
syncytial blastoderm, **61,** 319
maternal-effect mutations
basic genetic strategy
cytological screen for division
errors, **61,** 323
DNA labeling, **61,** 323
homologous balancer
chromosomes, **61,** 323
immunofluorescence analysis, **61,** 323
time-lapse mitosis analysis, **61,** 323
maternal mutation screens
advantages, **61,** 322
limitations, **61,** 322–323
web site, **61,** 323
zygotic mutations disrupting larval mitosis
advantages and limitations
chromosome architecture
examination, **61,** 335
hypotonic treatment, **61,** 335
mitotic index differential, **61,** 335
basic genetic strategy
balancer chromosomes, **61,** 336
Black Cells marker, **61,** 337
distinguishing homozygous
mutants, **61,** 336
lethal mutation defects, **61,** 336
mitotic mutations affecting target
tissues, **61,** 335–336
red mutation, **61,** 337
Tubby larval marker, **61,** 336–337
Drosophila S2 cells
apoptotic activity testing, **66,** 332–337
beta-galactosidase cell death assays, **66,** 334–335
cell death assays, **66,** 334–336
cell transfection, **66,** 333–334
green fluorescence protein cell death
assays, **66,** 335
immunoprecipitations, **66,** 336–337
Drug binding, DNA-targeting, **64,** 187–188
Drug development, cell–matrix
adhesion, **69,** 18–19
Drug dose, intensity, as determinant of
treatment efficacy, **64,** 375
Drug efflux blockers
combinations, **64,** 197
effects on drug retention, **64,** 195–196
fluorochrome retention with or
without, **64,** 201–206
Drug efflux, energy-dependent, in bacteria, **64,** 557
Drug inhibitors, cell–matrix adhesion
assays, **69,** 376–378
Drug resistance, **62,** 482, 519
Drug role in analyzing mitosis
intracellular drug concentration
determination, **61,** 268, 287–288
intracellular drug level determination
high-performance liquid chromatography
(HPLC) analysis, **61,** 288
monolayer culture directly in sterilized
scintillation vials, **61,** 287–288
monolayer or suspension culture in
flasks, **61,** 287

mechanisms of action
colchicine
apoptosis, **61,** 280
binding to tubulin and microtubules, **61,** 277–278
catastrophe and rescue frequency changes, **61,** 279
cellular uptake and loss of colchicine, **61,** 280
C-mitosis, **61,** 278
disadvantages, **61,** 280–281
drug concentration effects, **61,** 278–280
dynamic instability suppression *in vitro*, **61,** 279
effects on mitosis, **61,** 279–280
mitotic block, **61,** 280
polyploid restitution nucleus formation, **61,** 280
primate *vs.* non-primate cell line differences, **61,** 280
treadmilling suppression *in vitro*, **61,** 279
tubulin–colchicine complex, **61,** 279
nocodazole
apoptosis, **61,** 282
depolymerization half-times, **61,** 281
GTPase activity of tubulin, **61,** 281
high concentration effects, **61,** 281
inhibition reversal, **61,** 281
low concentration effects, **61,** 282
mechanism, **61,** 281
mitotic block, **61,** 282
uptake and release rates, **61,** 281
use in synchronizing cells, **61,** 282
taxol
anaphase onset delay, **61,** 285
cellular uptake and loss, **61,** 285–286
discodermolide, **61,** 283
distinct binding sites, **61,** 283
effects in human carcinoma cells, **61,** 284–285
effects on mitosis, **61,** 285
epothilones A and B, **61,** 283
high concentration effects, **61,** 283
intermediate concentration effects, **61,** 284
low concentration effects, **61,** 284, 285
microtubule bundling, **61,** 283
microtubule polymerization enhancement *in vitro*, **61,** 283
mitotic block, **61,** 284
plastic binding affinity, **61,** 282–283
protofilament structure differences, **61,** 283
solubility, **61,** 282
spindle microtubule stabilization, **61,** 282
taxotere, **61,** 283
vinblastine
direct microtubule binding, **61,** 273
drug concentrations and range of effects, **61,** 270, 274
effects on mitosis, **61,** 274–276
effects on tubulin polymer mass, **61,** 273–274
high-affinity binding sites, **61,** 273
low-affinity binding sites, **61,** 273
low drug concentration effects, **61,** 273
microtubule dynamics and mitotic block, **61,** 276–277
spindle morphology alterations, **61,** 274–275
tubulin binding, **61,** 271
tubulin conformational changes, **61,** 271–272
tubulin self-association, **61,** 271–272
vinblastine–tubulin paracrystal formation, **61,** 272
microtubule assembly dynamics
dynamic behavior parameters, **61,** 269
dynamic instability, **61,** 269
half-times of interphase and mitotic spindle microtubules, **61,** 269
microtubule composition, **61,** 268
plus and minus ends, **61,** 269
treadmilling, **61,** 269, 270
parameters and cell types, **61,** 268
use of anti-mitotic drugs
concentration dependency, **61,** 289
concentration intervals, **61,** 288
extrapolation between cell types, **61,** 289
incubation duration variations, **61,** 288
individual cells differences, **61,** 289
solubilization, **61,** 288
use of alternative anti-mitotic drugs, **61,** 289
Drug screening, expression cloning, **69,** 219–220
Drug selection
blasticidin S, **62,** 493–494, 497
cycloheximide, **62,** 491, 497, 510
paromomycin, **62,** 493–494, 497, 510, 536
table, **62,** 482
Drugs
β-lactams, inhibition of peptidoglycan layer, **64,** 558–559
fluorescent cytotoxic, cytometric assays, **64,** 477–478

Drugs, effects on protein expression, **63,** 227
Drugs, using proteins as, **71,** 325–328
Dry objective lens, **70,** 263
Dryl's medium, *see* Culture media
DSPE, *see* 1,2-Distearoyl-*sn*-glycero-3-phosphatidylethanolamine
Dual-excitation laser scanning confocal microscope, **70,** 420
Dulbecco's modified Eagle's medium (DMEM), **71,** 43, 77, 83, 92, 376
Dunn chemotaxis chamber setup, FLIM, **69,** 268–269
Duplication assays
  centrosome kinases in mammalian cells, **67,** 314–315
  for centrosomes, **67,** 289–294
Duroquinol-ubiquinone transhydrogenase, **65,** 109
Dye combinations, protocols for mitochondria analysis, **64,** 125–126
Dye efflux, bacteria, **64,** 543, 545–546
Dyes
  Drosophila embryo confocal imaging, **70,** 373
  fast membrane staining indicators, **70,** 430–431
  membrane potential second harmonic generation, **70,** 444–446
  Nernstian redistribution
    methods, **70,** 437–441
    plasma membrane potential, **70,** 434–436
    single mitochondria membrane potential, **70,** 436–437
  overview, **70,** 429–430
  slow redistribution indicators, **70,** 432–434
  suppliers for confocal microscopy, **70,** 76–77
  Tetrahymena thermophila live fluorescence microscopy, **70,** 357
DYN1–green fluorescent protein, fusion protein studies in yeast, **58,** 99
Dynactin, **61,** 137, 138
Dynamic force microscopy
  ligand-receptor recognition
    bond lifetime, **68,** 124
    kinetic rates, energies, binding pocket, **68,** 125
    unbinding force *vs.* loading rate, **68,** 124–125
  Na^+/D-glucose cotransporter
    epitope mapping, **68,** 131–133
    membrane-sidedness and functionality, **68,** 131
    overview, **68,** 130–131
  vascular endothelial cadherin
    binding strength, **68,** 128–129
    Ca^{2+} dependent, **68,** 127–128
    conformation, **68,** 126
    overview, **68,** 125–126
    thermodynamics and structural parameters, **68,** 129–130
    trans association, **68,** 127–128
Dynamic force spectroscopy
  applications, **68,** 350–351
  overview, **68,** 347–349
Dynamic range
  and logarithmic display of data, **63,** 302–304
  fluorescent particles, **63,** 318
  logarithmic amplifiers, **63,** 306
  low-light microscopy, camera characterization parameters, **56,** 32–33, 58–59
  quantitative video microscopy, resolution limits, **56,** 128–130
Dynamic reciprocity, definition, **69,** 8
Dynamic recognition force microscopy
  antibody–antigen recognition, **68,** 134–137
  recognition imaging, **68,** 134–137
Dynamitin, in pericentrin recruitment to centrosomes, **67,** 63
Dynein, **61,** 137, 142
Dynein, **62,** 54, 64
  gene disruption, **62,** 284–287
  isolation, **62,** 433–439
  vanadate-mediated photolysis (V1 reaction), **62,** 434–439
Dynein, mislocalization by pericentrin, **67,** 67
Dynein, scanning transmission electron microscopy, **53,** 159

E

E1A, immortalization of cells, **57,** 72–73
E7, immortalization of cells, **57,** 72
Early pressure screens, zebrafish
  advantages and limitations of early pressure screening, **60,** 73–75, 82
  advantages of zebrafish system for vertebrates, **60,** 71–72
  ethylnitrosourea mutagenesis of males, **60,** 76
  gynogenetic diploid embryo production, **60,** 73
  haploid *vs.* homozygous diploid embryo screening, **60,** 72–74
  immunohistochemical screening, **60,** 76–77
  maternal-effect genes, *see* Maternal-effect genes, gynogenesis-based screening in zebrafish
  motorneuron mutation screening
    comparison with mutations isolated in other screens, **60,** 79–80

design of screen, **60,** 78
phenotype in wild-type, **60,** 77–78
types of mutations, **60,** 78–79
neural crest mutation screening
comparison with mutations isolated in other screens, **60,** 81–82
design of screen, **60,** 80
normal development, **60,** 80
types of mutations, **60,** 80–81
specific phenotype screening, **60,** 76
visible phenotype screening, **60,** 82
Ease of culture, peripheral neurons, **71,** 18
EBV, *see* Epstein–Barr virus
ECM, *see* Extracellular matrix
Ecotropic packaging lines, **66,** 203
Ectopic gene expression, zebrafish, **59,** 117–125
Ectopic genes
in axonal pathfinding, **71,** 136
overexpression in grasshopper limb bud, **71,** 190–191
Ectopic vectors, in *Drosophila,* 220–222
EDTA, *see* Ethylenediamine tetraacetate
EDTA, cell–cell adhesion force measurements, **68,** 106–108
Effector caspase, **66,** 308, 454
Egg incubator, **71,** 10, 13, 42, 57
postinjection, **71,** 382
Egg, zebrafish
fertilization, *in vitro*, **60,** 15, 44, 46–47, 53, 370–372
microinjection
apparatus setup, **60,** 138
egg collection and preparation, **60,** 139–140
fish maintenance, **60,** 136–137
needle preparation, **60,** 139
plasmid preparation, **60,** 136
plate preparation, **60,** 137–138
technique, **60,** 140–141
stripping, **60,** 14
Eggs, staging for retroviral infection, **71,** 380
Eggs, *Xenopus*
and centrosome injection, temporal coupling, **67,** 215
arrested S-phase, **67,** 276–279, 281
centrosomal and noncentrosomal aster formation, **67,** 243
centrosome duplication assay logic, **67,** 289–290
centrosome-independent spindle assembly, **67,** 245–247
centrosome kinase assays, **67,** 315–316
centrosome reproduction, **67,** 270, 283
chemical compound screening, **67,** 250–251
extract preparation, **67,** 241
extract storage, **67,** 325
extraction solutions, **67,** 331
mitotic extract preparation, **67,** 170
parthenogenesis *vs.* fertilization, **67,** 212–214
permeabilized sperm nuclei, **67,** 280–281
preparation, **67,** 241
priming for extract preparation, **67,** 276
S-phase-arrested extracts, **67,** 276–279, 281
spindle assembly
assays, **67,** 243–245
associated proteins, **67,** 247–251
Eggs, *Xenopus*, confocal microscopy
antibody incubations, **70,** 391–392
bleaching, **70,** 391
borohydride reduction, **70,** 391
chromosome staining, **70,** 393–394
clearing, **70,** 394–395
handling procedures, **70,** 390–391
mounting procedures, **70,** 395–397
photobleaching, **70,** 397
secondary antibody choice, **70,** 392–393
Eggs, zebrafish, *in vitro* fertilization, **59,** 50–51
EHS components, native basement membrane proteins, **69,** 115
Elasticity, AFM of cells
cytoskeleton, **68,** 76–77
overview, **68,** 74–76
Electrochemical adsorption, DNA, **68,** 359
Electrochemical gradient, components, **63,** 468–469
Electrochromic dyes, membrane potential measurement, **70,** 444–446
Electrocompetence, of adenoviral vectors, **71,** 394
Electrofusion, cell hybridization, **57,** 116, 136–138
Electron microscope
intermediate-voltage, **65,** 253
Electron microscopy
apoptosis assay, **57,** 253
Dictyostelium centrosomes, **67,** 346–347
for centrosome reproduction assay, **67,** 300
scanning electron microscopy, *see* Scanning electron microscopy
SFM of DNA, **68,** 359
TEM
bacteriophage T_5, STM comparison, **68,** 6–7
transient fusions at depressions, **68,** 46
transmission electron microscopy, *see* Transmission electron microscopy

Electron microscopy (*cont.*)
yeast cells
embedding protocols, **67,** 34
freezing techniques, **67,** 29–34
imaging, **67,** 34–35
immunolabeling, **67,** 35–36
sectioning, **67,** 34–35
SPB 3D modeling, **67,** 37–40
Electron microscopy *in situ* hybridization (EMISH), synaptonemal complex, **53,** 253
Electron microscopy, *see* Electron microscopy *in situ* hybridization; Scanning electron microscopy; Scanning transmission electron microscopy; Transmission electron microscopy
Electron microscopy, basement membrane self-assembly, **69,** 133–135
Electron microscopy, ECM analysis, **63,** 585
Electron microscopy, *see* Microscopy, Electron
Electronic components, noise
amplifier noise, **55,** 59–60, 151–152
backreflections, **55,** 60, 152–153
beam-pointing fluctuations, **55,** 59, 62, 153
Johnson noise, **55,** 149–151
power fluctuations, **55,** 58, 152
shot noise, **55,** 58–59, 149
vibrations, **55,** 58, 154
Electronic scan zoom, confocal microscopy, **70,** 252
Electrophoresis
maize coleoptile mitochondria, **65,** 71
Electrophysiology, **62,** 363–375
Electrophysiology
cerebellum as study mutants for, **71,** 91
in *Drosophila,* **71,** 227–238
Electroporation, *see also* Transformation, **62,** 421, 485–500
Electroporation parameters
buffer composition, **71,** 349
cell density, **71,** 345–346
Free Ca^{2+} in electroporation buffer, **71,** 347–348
rest for recovery, **71,** 348
temperature of cell suspension, **71,** 348
voltage and duration of pulse, **71,** 346
Electroporation, T-cell immortalization, **57,** 77, 80–81
Electroporation technique
compared with nuclear injection, **71,** 311
compared with retroviral infection, **71,** 385
for chick primary neurons, **71,** 344–345
in PC12 cell lines, **71,** 282
transfection of primary and peripheral nervous system neurons with, **71,** 339–350
Electroretinogram, **71,** 236–237
Electrostatic adsorption, DNA
overview, **68,** 361–362
soluble cation effects, **68,** 362–363
Electrostatic cell sorters, **63,** 8
Electrostatic charge, applied to drops containing cells of interest, **63,** 45–46
Elicited mouse peritoneal macrophages
phagocytosis, **66,** 359
ELISA
TNF ligands, **66,** 505–506
tumor necrosis factor-related ligands, **66,** 516–517
ELISA, *see* Enzyme-linked immunosorbent assay
Elliptical spot, laser beam illumination, **63,** 24
Ellipticine, DNA-binding, **63,** 136
Ellis manipulator, **61,** 211
Elongation behaviors, chick forebrain and rat hippocampal neurons, **71,** 52
Elvanol mounting medium preparation, **61,** 49–50
Embedding, embryo, **69,** 345–346
Embedding, for electron microscopy, **67,** 34
Embryo
DiI labeled neurites, **69,** 355–356
slicing and embedding, **69,** 345–346
Embryo embedding
Drosophila
eye discs, **66,** 328–330
Embryo fixation
Drosophila
transmission EM, **66,** 330
Embryo imaging, morphogenetic movements, analysis, **59,** 201
Embryogenesis, *Drosophila* centrosome analysis during, **67,** 114–122
Embryonic axis formation, zebrafish, **59,** 157–172
Embryonic chick brain (cerebellum), **71,** 3
Embryonic chick ciliary ganglion, **71,** 3
Embryonic chick dorsal root ganglia, **71,** 3
Embryonic chick paravertebral chain ganglia, **71,** 3
Embryonic development, avian ganglion as subject for study of, **71,** 38–42
Embryonic hematopoiesis, gene expression, **59,** 318–320
Embryonic rat brain neurons, **71,** 3
Embryonic stem cell, *see also* Transgenic animal
applications in transgenic technology, **57,** 280–281, 292
cell line establishment, **57,** 284–285

characteristics of cell lines, **57,** 280
culture conditions, **57,** 281–282
feeder plate preparation, **57,** 282
freezing
plates, **57,** 287–288
vials, **57,** 283–284
genome alteration detection, **57,** 287
maintenance, **57,** 281–282
passaging, **57,** 283
picking colonies, **57,** 286–287
thawing
plates, **57,** 288–289
vials, **57,** 284
Embryonic vascularization, gene expression, **59,** 314–318
Embryos
cytoskeletal dynamics of zebrafish, **59,** 205–224
dejellying for blastomere injections, **71,** 133
development, long-term imaging, **70,** 142–143
Drosophila
AO staining, **66,** 325
Drosophila, CLSM
autofluorescence, **70,** 368–369
bleed-through, **70,** 369–370
collection and permeabilization, **70,** 363
fluorescent probes, **70,** 364
hard copy images, **70,** 372
image archiving, **70,** 372
image collection, **70,** 366–368
image processing, **70,** 370–372
image transfer, **70,** 370
microscope details, **70,** 365–366
mounting, **70,** 364–365
multiple labels, **70,** 372–375
overview, **70,** 361–363
staging for Tibia-1 pathway studies, **71,** 182–183
TUNEL staining
Drosophila, **66,** 326–327
Xenopus, confocal microscopy
cytoskeletal and nuclear probes, **70,** 410–413
keratin filaments, **70,** 386–388
microfilaments, **70,** 388–390
microtubule fixation, **70,** 383–386
overview, **70,** 380–382
recipes and reagents, **70,** 409–410
zebrafish, **59,** 54–55
Embryos, *Drosophila*
centrosome extract preparation, **67,** 141–142
γ-tubulin-containing complexes, **67,** 6–11
γTuRC, **67,** 3–6
γTuSC, **67,** 4–6
Embryos, laser micromanipulation, **55,** 92–94
Embryos, whole, preparation for PRINS, **64,** 58–59
EMISH, *see* Electron microscopy *in situ* hybridization
end zone, mutation effects in development, **60,** 81
Endocrine therapy, proliferating tumors, **64,** 376–377
Endocytosis, **71,** 5–8
B35 cell line analysis and, **71,** 288
circumventing degradation pathway of, **71,** 328
in *Drosophila* studies, **71,** 196
of cell adhesion molecules in B35 cells, **71,** 294–295
synaptic vesicle pathway in *Drosophila,* **71,** 204
Endocytosis, **62,** 42–43
Endogenous aleurone nuclease, **66,** 439
Endogenous c-Jun N-terminal kinase (JNK)
JNK activation
in vitro kinase assay, **66,** 190–192
Endogenous cleavage
DFF45, **66,** 301
PARP-1, **66,** 291–296
cell extracts, **66,** 291
tissue extracts, **66,** 291–292
Endogenous respiration
cyanide titration, **65,** 121
inhibition
KCN, **65,** 124
intact cells, **65,** 120–122
Endoplasmic reticulum, **62,** 40–42
Endoplasmic reticulum, green fluorescent protein targeting
chimera construction and expression, **58,** 78–79
expression in primary cultures, **58,** 80–83
fluorescence microscopy, **58,** 79–80, 83–84
multiple organelle labeling with blue-shifted protein, **58,** 81
transfection, **58,** 83
Endoplasmic reticulum, high-resolution multimode digital analysis, **56,** 186
Endoplasmic reticulum, pCa_i measurement, **70,** 461–463
Endosomes, **62,** 42
Endosperm cell division studies
culture methods, **61,** 417–418
features, **61,** 416–417
findings
actin network organization
actin microfilaments, **61,** 418
lipid coated colloidal gold particles, **61,** 418
calcium measurements

Endosperm cell division studies (*cont.*)
chlortetracycline, **61,** 419
quin–2, **61,** 419
drugs
APM, **61,** 418
colchicine, **61,** 418
cytochalasin–D, **61,** 418
oryzalin, **61,** 418
taxol, **61,** 418
laser microsurgery, **61,** 418–419
microtuble roles, **61,** 418
temperature shock, **61,** 418
ultraviolet microbeam irradiation, **61,** 418
source suitability characteristics, **61,** 417
species, **61,** 417
Endothelial cells, expressing specific markers, **63,** 575–576
Energy, ligand–receptor recognition, **68,** 125
Engelbreth–Holm–Swarm matrix, preparation, **63,** 569–570
Enucleation
for centrosome duplication assay, **67,** 291–292
MTs in cytoplasts, **67,** 47–48
Enzymatic activity, bacteria, flow cytometry, **64,** 523–524
Enzymatic quantitation
ceramide-(DAG) kinase assay, **66,** 144–147
Enzymatic treatments, in *Heliosoma* cultures, **71,** 158
Enzyme activity, identification of apoptosis based on, **63,** 532
Enzyme kinetics, time application, **63,** 180
Enzyme-linked immunosorbent assay
histone-associated DNA fragments in apoptosis, **57,** 261
hybridoma screening
human, **57,** 131–132, 138–139
mouse, **57,** 125–126
Mabs for instructive ECM identification, **69,** 44
proteoglycans, **69,** 63–65
Enzyme-linked immunosorbent assay, **63,** 290
Enzyme treatment
GAG–growth factor interaction, **69,** 92–94
heparan sulfate, **69,** 94
Enzymes
plant mitochondria, **65,** 64
Enzymes, target, availability: quality control, **64,** 180–182
Eosin-B, CSU-10 studies, **70,** 97
Ependymoma, *see* Glioma
Epi-fluorescence microscopy, *see* Fluorescence microscopy
Epi-illumination, arc, for fluorescence image analysis, **63,** 52
Epi-objective lens, characteristics and function, **70,** 266–268
Epidermal cell division studies
culture methods
onion cells, **61,** 424, 425
Tradescantia, **61,** 425
findings
Tradescantia
nuclear migration and cell division relationship, **61,** 425
nuclear response to mechanical perturbation, **61,** 425–426
onion anaphase spindle re-orientation
cytochalasin, **61,** 425
phalloidin, **61,** 425
stomatal complex formation, **61,** 424
Epifluorescence microscopy, **64,** 329
images, **64,** 332–333
Epifluorescence microscopy, development, **70,** 130
Epilepsy, cortical and hippocampal involvement in, **71,** 354
Epiplasm, **62,** 65
Episomal replication, expression cloning, **69,** 218
Epithelial cells, **66,** 223–224
kidney, $\alpha 3$ integrin null mouse, **69,** 311–313
mammary, adenovirus
generation in *Escherichia coli*, **69,** 410–413
HEK293 E1 cells, **69,** 415–416
high-titer viral stocks, **69,** 418–420
infection, **69,** 420–422
overview, **69,** 409–410
screening, **69,** 413–415
virus titration, **69,** 417–418
mammary, mouse, culturing, **69,** 409
Epithelium, nonmalignant, cell production rates, **64,** 464–468
Epitope mapping, Na^+/D-glucose cotransporter, **68,** 131–133
Epitope tag, types, **53,** 275–276
Epitope tagging, **62,** 549–557
Epitopes
cross-reacting, **63,** 287–289
effect on antibodies for immunophenotyping, **63,** 224–225
expression, fixation effect on, **63,** 227
masking, **63,** 291
Epothilones A and B, **61,** 283

Eppendorf Injectman N12 semiautomated injection system, **71,** 311
Eppendorf transjector 5246, **71,** 138
ε-(γ-glutamyl)-lysine cross link density, **66,** 115–119
  automated amino acid analysis, **66,** 118–119
  cornified envelope proteins, **66,** 117
  keratinocyte-cornified envelope isolation, **66,** 116–117
  PTU derivatization, **66,** 117–118
Epstein–Barr virus, immortalization of cells, **57,** 73
Equilibrium gel filtration, basement membrane self-assembly, **69,** 128
Equipment noise, *see* Instrumental noise
Equipment, for plant flow cytogenetics, **64,** 12
ER, *see* Endoplasmic reticulum
Ergodicity, SFM of DNA, **68,** 384–385
ERKs, **66,** 187–188
Ernst, Karl, **71,** 68
Erythrocytes
  lysing, **63,** 237–238
  lysing solution: formulation, **63,** 214
  separation from leukocytes, **63,** 210–211
Erythrocytes, lysis, and immunofluorescence staining, **64,** 345–346
Erythroid cultures, for study of erythropoiesis, **64,** 141
Erythropoiesis
  erythroid cultures for study of, **64,** 141
  fetal, **64,** 140
  sickle-cell, **64,** 148–149
ES cell, *see* Embryonic stem cell
ES cells
  integrin adhesion molecule gene mutations, **69,** 314–317
  null–α5 integrin, **69,** 320
  null–β1 integrin cells, **69,** 317–318
  null–integrin, **69,** 320
Escherichia coli, adenoviral plasmid generation, **69,** 410–413
Escherichia coli, for SPB protein *in vitro* binding, **67,** 90
Esophageal tumors, squamous, **64,** 456
EST database
  web sites, **66,** 494
EST screen
  Bcl-2 family cloning, **66,** 34
EST, *see* Expressed sequence tag
Esterification, AFM tips, **68,** 118
Estrogen receptors, application of LSCM cell position feature, **63,** 74
Estrogen receptors, LSCM analysis, **64,** 435–436
EtBr
  mtDNA, **65,** 399–400
Etching, deep, cryo-AFM, **68,** 252–253
Ethidium bromide
  bound to apoptotic subpopulations, **63,** 143
  DNA-bound, **63,** 135–136
Ethidium bromide (EtBr)
  mtDNA, **65,** 399–400
Ethidium monoazide, stock solution, **63,** 250
Ethylenediamine tetraacetate, proteoglycans, **69,** 56–57
Ethylnitrosourea mutagenesis
  initiation of positional cloning project, **60,** 262–263
  premeiotic male germ cells
    dose dependence, **60,** 24–25
    safety, **60,** 24
Etoposide, **66,** 211
  effect on topoisomerase II, **64,** 177
  spheroids repeatedly treated with, **64,** 226–227
Excitation
  multiphoton, **63,** 586
  sources, in confocal microscopy, **63,** 94–95
Excitation beam
  early flow cytometers, **63,** 6–8
  for LSCM, **63,** 59–60
Excitation energy
  in flow cytometry compared to LSCM, **63,** 83
  LSC, **63,** 52
Excitation light
  phototoxicity, **65,** 345
Excitation/emission wavelength, multicolor LSCM
  chromatic aberration, **70,** 235–237
  optical sectioning depth, **70,** 237
  resolution effects, **70,** 233–235
  spherical aberration, **70,** 235
Excitotoxicity, in hippocampal and cortical neurons, **71,** 111, 112
Exconjugant, **62,** 133, 228
  isolation, **62,** 221–222, 235–237
Exocytosis, **62,** 43–47, 189, 347–360
Exocytosis, **71,** 5–8
  in *Drosophila* studies, **71,** 196
  inhibition in PC12 cell lines, **71,** 26
  PC12 studies of, **71,** 269
Exocytosis, AFM studies
  apical plasma membrane structures, **68,** 39–40
  fusion pores, **68,** 40, 43

Exocytosis, AFM studies (*cont.*)
transient fusion at depressions, **68,** 43–47
Exogenous molecules
assisted delivery into live neurons, **71,** 326
use with *Xenopus* spinal neurons, **71,** 130
Exogenous NADH
DPI-induced apoptosis, **66,** 313
Experimental manipulations
choice of species based on, **71,** 17
motor neuron and spinal cord cultures, **71,** 80–82
Experimental needs, choosing neuronal cultures according to, **71,** 2–8
Experimental properties, of cultured neurons, **71,** 6–7
Experimental techniques, accessibility of specific neurons to, **71,** 2
Explant culture assay, zebrafish, **59,** 127–155
Explant cultures
in *Xenopus* spinal neurons, **71,** 130
preparing, of peripheral neurons, **71,** 29–30
Xenopus cocultured, **71,** 145–147
Explants
assay induction with, **59,** 150–153
isolation and culture, zebrafish, **59,** 140–150
size, zebrafish, **59,** 132
Expressed sequence tag, zebrafish
abundance in databases, **60,** 214
applications
candidate gene cloning, **60,** 232
expression profiles, **60,** 223, 225
genome mapping, **60,** 187, 190, 230–232
homolog identification, **60,** 225–226
isolation of full-length cDNA clones, **60,** 226–226
marker identification, **60,** 230
promoter isolation, **60,** 228, 230
generation, *see* cDNA clone tagging
management of tagged clones
documentation, **60,** 220–221
plasmid storage, **60,** 220
redundant clone minimization, **60,** 221–223
tissue libraries, **60,** 223
Expression cloning
candidate clones, **69,** 217
cDNA recovery, **69,** 216–217
critical controls, **69,** 219
definition, **69,** 210
disease/drug screening, **69,** 219–220
factors, **69,** 212
flow cytometry, **69,** 215–216
library quality, **69,** 213
library type, **69,** 213
lipofection, **69,** 213
materials and equipment, **69,** 214
methodology importance, **69,** 212
positive clones, **69,** 217–218
protocol variation, **69,** 218–219
screen sensitivity, **69,** 214
transfection, **69,** 215–216
Expression, levels and timing of, in biolistics, **71,** 364–365
Expression libraries, for centrosome-associated protein cDNA, **67,** 128–129
Extinction signals, flow cytometric, **63,** 112
Extracellular matrix
3-D substrates for cell culture, **63,** 563–565
AFM probe microscopy, **69,** 183–185
and ECM–cell interactions
3-D imaging, **63,** 584–587
autofluorescence, **63,** 591–593
CRM, **63,** 587–591
antigen identity, **69,** 41–42
cell adhesion and migration, **69,** 261–262
cell-assembled, *see* Cell-assembled extracellular matrix
cell-derived, **69,** 157–158
cell–matrix interactions, **69,** 9–10
cell regulation, **69,** 385–387
cell responses, **69,** 3
cell survival and function, **69,** 4
classification, **63,** 562–563
coated adhesive islands, living cell adherence, **69,** 388–390
coating micropatterned culture surface, **69,** 398
collagen, AFM imaging, **69,** 171–172
degradation, mediated by plasma membrane, **63,** 614
deposited, antigen identity, **69,** 41–42
deposited, reporter cell response, **69,** 35–36
fibronectin, **69,** 145–147
fibronectin null fibroblasts, **69,** 321
gene targeting, **69,** 310
global effects, **69,** 14
GTPases, **69,** 13–14
importance to cell biology, **63,** 561–562
instructive, *see* Instructive extracellular matrix
intact, **63,** 565–568
integrin adhesion molecule gene mutation, ES cell lines, **69,** 314–315
integrin signaling, **69,** 13
intestinal submucosa as 3-D substrate, **63,** 572–576

micropatterned culture surfaces, **69,** 390–391,393
multicomponent, *see* Multicomponent extracellular matrix
overview, **69,** 165, 195–196
protein grouping, **69,** 4
proteolyzed, internalization of, **63,** 619–623
recombinant, *see* Recombinant extracellular matrix
reconstituted, **63,** 568–572
reporter cell response, **69,** 36–37
self-assembly, **69,** 111–112, 136–137
signaling networks, **69,** 14–15
structural studies, **69,** 14
structure and properties, **69,** 3
syndecan-4 null fibroblasts, **69,** 321
tenascin, molecular pulling/force spectroscopy, **69,** 175–177
thrombospondin 2–null fibroblasts, **69,** 321–322
Extracellular matrix molecules, as culture substrate, **71,** 11–12
Extracellular matrix proteins
RGD-containing, integrin $\alpha V\beta 3$, **69,** 17
types, **69,** 8–9
Extracellular matrix proteins, in *Xenopus* cultures, **71,** 139–141
Extracellular signal-regulated kinases (ERKs), **66,** 187–188
Extract preparation
cell-free systems, **66,** 167–171
Extraction
cell membrane-anchored wild-type integrins, **69,** 225–227
integrins, **69,** 233
proteoglycans, **69,** 55–57
Extragenic suppressor screen, SPBs, **67,** 100–102
Extranuclear total value, LSCM list mode feature, **63,** 65–67
Extrinsic cues
experimental studies of, **71,** 80
in avian Purkinje neuronal cultures, **71,** 89–105
Eye discs
Drosophila
double labeling, **66,** 328
TUNEL staining, **66,** 329
Eyepiece, light microscopy, **56,** 139–140

F

F-actin, *see* Microfilaments
Förster cycle, chromophore, **58,** 10–12
F/P ratio, in protein labeling, **63,** 189–190
F(ab) fragments, use as second antibodies, **63,** 228–229
FAB classification, AML, **64,** 345, 348–351
Fab fragments, FLIM, **69,** 266–267
FACS, **65,** 88–90
FACS, *see* Fluorescence-activated cell sorting
FACS, *see* Fluorescent activated cell sorting
FACSVantage SE flow cytometer, **66,** 64
Facilities, animal, **71,** 13
Factor XIII, ECM, **69,** 146
False positivity, AML, **64,** 352
Fast axonal transport, **71,** 8
Fast Green dye, **71,** 381
Fast membrane staining potentiometric indicators, **70,** 430–431
Fc receptor binding domain, antibody, **63,** 219
Fc receptors, blocking with IgG prior to staining, **63,** 229–232
Feedback control, PFM studies, **68,** 202–203
Feeding layer, invertebrate cell culture, **57,** 193–194
Felodipine, **66,** 264
Fel cell, in Tibia-1 pathway studies, **71,** 174–176
Fenestrin, **62,** 70, 93, 96, 98, 295, 303, 442
Fermentor, cell culture scaleup, **57,** 226–227
Fertilization *vs.* parthenopgeneis, in *Xenopus*, **67,** 212–214
Fertilization, *in vitro*, **60,** 15, 44, 46–47, 53, 370–372
Fertilized chicken eggs
incubation of, **71,** 67–69
storage length *vs.* fertility, **71,** 68
Fetal bovine serum, **71,** 77, 374
for peripheral neuron cultures, **71,** 23
for spinal and motor neuron cultures, **71,** 83
Fetal calf serum, **71,** 92
FGT, *Xenopus* oocytes and embryo microtubule fixation, **70,** 383–386
Fibrillar center, *see* Nucleolus
Fibrillar collagens, AFM probe microscopy
aligning and orienting, **69,** 187–188
humidity effects, **69,** 188
Fibrillogenesis, collagen, **63,** 590
Fibrin–fibronectin matrix
advantages, **69,** 147
contraction, **69,** 150
disadvantages, **69,** 147
fibrinogen preparation, **69,** 148
immunofluorescence, **69,** 150–151
other protein inclusion, **69,** 148–150
synthesis *in vitro*, **69,** 148

Fibrin–fibronectin matrix (*cont.*)
troubleshooting, **69,** 151
Fibrinogen, scanning transmission electron microscopy, **53,** 158–159
Fibroblasts, contaminating neuronal cultures, **71,** 30
Fibroblast growth factor
HS binding probe, **69,** 98–102
HS interactions, **69,** 84–87, 89–90
regulatory HS binding probes, **69,** 95–98
Fibroblast growth factor receptor–alkaline phosphatase
fixed cell monolayers, **69,** 99
heparin binding of HS on 96-well plates, **69,** 98–99
specificity, **69,** 100–101
tissue sections, **69,** 101–102
Fibroblast growth factor receptor–HS interaction, **69,** 85–87
Fibroblasts
derivation from mutant mouse, **69,** 314
microenucleation, **65,** 409
mitochondrial respiratory complex activity, **65,** 100–101
null
α4 and α5 integrin, **69,** 320
fibronectin, **69,** 321
syndecan-4, **69,** 321
thrombospondin 2, **69,** 321–322
permeabilized human skin
ATP synthesis rates, **65,** 140
Fibronectin
as culture substrate, **71,** 12
in *Xenopus* cultures, **71,** 139, 145
Fibronectin
cell-derived ECM, **69,** 157–158
characterization, **69,** 282
definition, **69,** 7
ECM, basic function, **69,** 145–147
null fibroblasts, **69,** 321
Fibronectin–heparin–gold complexes
complex preparation, **69,** 185–186
imaging, **69,** 186–187
Fibronectin peptide ligand–integrin receptor interaction
$\alpha_5\beta_1$ immobilization, **69,** 179–181
bioartificial membrane, **69,** 177–179
force measurements, **69,** 181–182
Fibrosis, cell–matrix adhesion studies, **69,** 19
Ficoll-Hypaque density gradient separation, **63,** 213
Filament, **62,** 14-nm, 59, 96, 98
Filamentation, bacterial, drug-induced, **64,** 559
Filopoidal extensions
ability to sense guidance cues, **71,** 176–177
chick forebrain neurons, **71,** 52
extension and retraction, in Tibia-1 pathway, **71,** 172
in embryonic grasshopper limb bud, **71,** 173–176
in *Xenopus* cultures, **71,** 146–147
Filters
difference: chromosome analysis, **64,** 84–85
optical, LP and BP, **64,** 297
Filters, *see specific processes; specific types*
Filters, in wavelength selection, **63,** 115–118
Fimbrin, **62,** 57
Fin, *see* Zebrafish fin
Fine Science tools (www.finescience.com), **71,** 71
Fingerprinting, and PCR products, **64,** 407–408
Finite element modeling, AFM interfacing, **68,** 182
Finite lens systems, confocal digital imaging, **70,** 254–256
Firefly luciferase, **65,** 134
Firefly tails
Photinus pyralis
luciferase, **65,** 366
Fish liver mitochondria, **65,** 28
FISH, *see* Fluorescence *in situ* hybridization, **53, 57, 60, 64, 70,**
FISH, *see* Fluorescent *in situ* hybridization
FITC-annexin
flow cytometry assay, **66,** 342, 347
FITC-dextran, **71,** 132
monitoring untagged ciliary ganglion cultures with, **71,** 38
FITC, *see* Fluorescein isothiocyanate, **63, 64,**
Fixation
blood cells, **63,** 210
cell biopsy and, **64,** 111
cell suspensions, by acetone, **63,** 405
cells
after staining, **63,** 243
and permeabilization, **63,** 254–259
cultured cells *in vitro* and *in vivo*, **63,** 360–361
ethanol, bacterial cells, **64,** 557
in simultaneous staining procedure, **63,** 265
paraformaldehyde, **63,** 404–405
properties of cross-linking fixatives, **63,** 277–279
protocols for mitochondria analysis, **64,** 125
tested for different cell types, **63,** 408–409
thick tissue sections, **70,** 308–313, 329
Xenopus oocytes, **70,** 391

Xenopus oocytes and embryo keratin filaments, **70,** 386–388
Xenopus oocytes and embryo microtubules, **70,** 383–386
Fixation, *see* Fluorescence *in situ* hybridization; Transmission electron microscopy
Fixation, protein and gene expression, analysis, **59,** 67–68
Fixation, see Microscopy
Fixed tissue research, objective lens selection
bone, **70,** 273
brain tissue, neuroscience, and retinal imaging, **70,** 273–274
cytoskeleton, **70,** 272
genetics, **70,** 273
immunogold and silver grains, **70,** 272–273
microbiology, **70,** 275–278
ocular imagin, **70,** 274
UV imaging, **70,** 269–271
visible Ca^{2+} imaging, **70,** 271–272
Fixed yeast cells
immunofluorescent staining, **65,** 269–274
Flagellum, **62,** 90, 279
flat top, mutation effects in development, **60,** 81
flh, complementation by injection of genomic clones, **60,** 275–276
FLIM, *see* Fluorescence lifetime imaging microscopy
FLIP, *see* Fluorescence loss during photobleaching
Flooring method, **65,** 443
Florida wild-type strain, characteristics, **60,** 358
Flow cytogenetics, plant
chromosome sorting, **64,** 26–28
problems with, **64,** 4–5
Flow cytometers
addition of cell sorting, **63,** 8
analysis of plant cells: problems, **63,** 509–512
bacterial sample preparation, **64,** 513–514
BrdUrd antibodies, **63,** 11
collection of light signals from particles, **63,** 37
electronics, **63,** 123–124
extinction and absorption signals, **63,** 112
forward scatter signals, **63,** 112–113
hemopoietic cell classification, **63,** 11–12
instrument performance, verification, **64,** 293–295
instrument setup, **64,** 142
for microbes, **64,** 512–513
laser light sources, **63,** 7
light collection optics, **63,** 114–115
measurement of bacteria, **64,** 547–548
measurement sensitivity, **63,** 127–129
multiparameter, **63,** 6
optics and illumination power, **63,** 110–112
particle illumination, **63,** 20–26
performance characteristics, **63,** 302–312
time stamping of events, **63,** 171, 181
wavelength selection, **63,** 115–118
Flow cytometery, ratio imaging, **56,** 219–220
Flow cytometric analysis staining
membrane permeability, **65,** 149–150
Flow cytometry
accommodating anti-BrdUrd signal, **63,** 364
analysis of
cell–cell adhesion assays, **63,** 603
cell death, **63,** 528
cellular fluorescence, **63,** 346
PCNA expression, **63,** 423–425
and data acquisition, **63,** 370–371
and LSCM: differences, **63,** 54
annexin staining, **66,** 346–349
apoptosis
caspase 3, **66,** 63
caspases, **66,** 50
cell size, **66,** 52–57
cell sorting, **66,** 64–67
mitochondrial transmembrane potential, **66,** 81–82
monovalent intracellular ions, **66,** 57–63
vs. LSC, **66,** 100–103
apoptosis assays
annexin V/cell cycle analysis
filters, **57,** 273
performance of assay, **57,** 273–274
preparation of cells, **57,** 272
principle, **57,** 267–268
staining of cells, **57,** 272–273
TUNEL/cell cycle analysis
filters, **57,** 269
fixation of cells, **57,** 268
labeled nucleotides, **57,** 275–276
performance of assay, **57,** 269–272
principle, **57,** 259, 267
staining, **57,** 268–269, 271–272
TUNEL/surface staining, **57,** 257–259
apoptosis assays, **66,** 213–214
apoptotic cells
annexin V, **66,** 85
apoptotic *vs.* necrotic death, **66,** 71
assessment of lymphocyte activation/proliferation, **63,** 447–453
calcium imaging, **66,** 239

Flow cytometry (*cont.*)
cell cycle phase duration determination, **57,** 237
cell necrobiology, **66,** 70–72
cell preparation, **69,** 241
cell preparation and staining for, **63,** 137–138
cell sorting, *see* Fluorescence-activated cell sorting
centering particle in illuminating beam, **63,** 26–33
characterization of immortalized mononuclear cells, **57,** 81
common measurements, standardization and calibration, **63,** 312–330
data display formats, **63,** 176–178
detection, **63,** 118–123
distinguishing apoptotic cells, **63,** 492–493, 542
DNA fragmentation assay, **66,** 87–88
drawbacks, **66,** 71
expression cloning, **69,** 215–216
fluorescence compensation and logarithmic amplification, **63,** 124–127
fluorescence lifetime, **63,** 133–134
excited-state, **63,** 132–133
instrumentation, **63,** 138–141
from data to information, **63,** 41–43
from light signals to data file, **63,** 37–41
green fluorescent protein
bacteria
expression systems, **58,** 324–325
flow cytometry, **58,** 327–328, 333–335, 337–338
mutants for enhanced fluorescence, **58,** 333–335
mammalian cells
expression by retrovirus constructs, **58,** 320–322
flow cytometry, **58,** 325, 329–330, 335–336
gene therapy applications, **58,** 336
multiple reporter detection, **58,** 330
mutants for enhanced fluorescence, **58,** 328–329
transcription assays, **58,** 329
transfection, **58,** 320–321
viral infection assay, **58,** 335–336
plant nuclei, flow cytometry, **58,** 326–327, 333
plant protoplasts
flow cytometry, **58,** 326, 331–333, 337
preparation of protoplasts, **58,** 322–323
recombinant construct preparation, **58,** 323
transfection, **58,** 323–324
transgenic tobacco plant expression, **58,** 324
heterogeneous fluorescence decays, **63,** 144–145
high discrimination preparative sorting, **63,** 83
historical overview, **63,** 19–20
homodyne *vs.* heterodyne signal detection, **63,** 146
identification of caspase activation sequence, **63,** 488–490
Jurkat cells, **66,** 61–62
karyotyping, **63,** 12
labeled inhibitor analysis, **66,** 95
lasers, **58,** 316
light scatter angles, **58,** 316
MC540 staining, **66,** 355
measuring particle concentration, **63,** 329–330
poly(ADP-ribose) polymerase, **66,** 98
principle, **58,** 315–316
quantitative fluorescence, **63,** 334
removal of irrelevant cells, **63,** 80
sample mixing and delivery: time role, **63,** 171–175
signal, background, and sensitivity, **63,** 113–114
signal processing tasks, **63,** 108–110
sorting, **63,** 44–48
standardization, **63,** 300–302
terminology, **63,** 301–302
testicular germ cell, analysis of DNA, **57,** 106–107
time-resolved fluorescence measurements, **63,** 146–147, 170
TUNEL, **66,** 91
vs. LSC, **66,** 75
zero resolution imaging in, **63,** 52
Flow cytometry assay
FITC-annexin V, **66,** 342
Flow cytometry-binding
tumor necrosis factor-related ligands, **66,** 514
Flow cytometry standard
data storage format, **63,** 161
LSCM values stored as, **63,** 61–62
support, **63,** 179
Flow cytometry, *see also* Multiparameter data analysis
analysis of
drug uptake, **64,** 184
Hoechst 33342–DNA binding, **64,** 185–188
responses to antimicrobial agents, **64,** 526–529
RNA synthesis, **64,** 129–137
and sorting
application to plant chromosomes, **64,** 15–18
chromosome analysis, **64,** 22–24

theoretical flow karyotypes, **64,** 22
application to
medical microbiology, **64,** 563
microbial systems, **64,** 531–532
bacteria
basic considerations, **64,** 541–543
fluorescent dyes for, **64,** 554–555
rapidly growing, **64,** 558–562
standards and controls, **64,** 548–550
BrdUrd/DNA distribution, **64,** 132
compared with
immunohistochemistry, **64,** 315, 340
LSCM, **64,** 326–327
detection of
microbes, **64,** 514–516
MRD, **64,** 401–402
nuclear hsp70, **64,** 278
determination of Gram stain, **64,** 525–526
fluorescent detection of mitochondria, **64,** 118
fluorescent halo assay based on, **64,** 262–263
HIV disease investigation, **64,** 582–584
HIV infection monitoring, **64,** 568–582
identification of
food and drink contaminants, **64,** 529–531
specific microorganisms, **64,** 516–521
viable bacteria, with FISH, **64,** 524–525
identification of MDR, **64,** 194–196
controls, standards, and
instruments, **64,** 199–201
critical aspects, **64,** 198–199
list mode analysis, **64,** 490–491
measurement of
hemoglobin profiles, **64,** 141, 150–151
nuclear matrix protein content,
64, 272–273
nuclear protein content, **64,** 271–272
S-phase fraction, **64,** 448–449
membrane integrity analysis, **64,** 521–522
membrane potential analysis, **64,** 522–524
monitoring of HP maturation, **64,** 155–156
assessment of HP subsets, **64,** 157–167
critical aspects, **64,** 156–157
MRD analysis by
immunophenotyping, **64,** 389, 391–393
multicolor fluorescence, **64,** 578–582
mycobacteria, **64,** 562–563
nuclear matrix stability assay after heat
shock, **64,** 274–276
S-phase cell fraction, **64,** 363–364, 369–372
Flow cytometry, nuclear protein import
assay, **53,** 539
Flow karyotyping
and chromosome sorting, **64,** 19–21
detection of MRD, **64,** 401–402
instrument alignment, **64,** 15–16
univariate and bivariate analysis, **64,**
16, 22–24
Flow rate, measurements for cell-ligand
adhesion, **63,** 607–608
Fluidics, flow cytometric, **63,** 26–33
Fluid level, in biolistics, **71,** 354–367
Fluids
advanced handling, AFM studies
DNA digestion by DNase, **68,** 227
future outlook, **68,** 227
operation principles, **68,** 226–227
overview, **68,** 225–226
aqueous, *see* Aqueous fluids
molecular pulling/force spectroscopy
$\alpha_5\beta_1$ immobilization, **69,** 179–181
bioartificial membranes, **69,** 177–179
force measurements, **69,** 181–182
overview, **69,** 174
unfolding tenascin, **69,** 175–177
probe microscopy
collagen degradation, **69,** 171
collagen III assemblies, **69,** 174
laminin-1 dynamic motion, **69,** 168–171
SFM of DNA
mobility modulation and solution
dynamics, **68,** 371–374
molecular motion, **68,** 374–376
protein interactions, **68,** 376–377
water and buffers, **68,** 370–371
Fluo-3/Fura-Red ratio,
intracellular Ca^{2+}, **70,** 425
Fluorescein diacetate, **66,** 482
Fluorescein isothiocyanate
bleaching by, **63,** 99
CD15-conjugated, **63,** 80–81
coupled to gelatin, **63,** 615–625
misclassification of apoptotic cells due
to, **63,** 539
optimal labeling of proteins, **63,** 190–191
photon flux, **63,** 95
plus phycoerythrin: histograms, **64,** 490–491
scattergrams, **64,** 324–326
self-quenching, **63,** 134
Fluorescence
average, as useful metric, **63,** 177
background
and flow cytometric sensitivity, **63,** 113–114
assessment for activation
antigens, **63,** 439, 441, 443

Fluorescence (*cont.*)
principal components, **63,** 409
compensation, and logarithmic amplification, **63,** 124–127
decays, heterogeneous, **63,** 144–145
dim, in washed specimen, **63,** 234
energy transfer: problem in plant cell analysis, **63,** 512
from different types of particles, comparisons, **63,** 325–326
intensity, as indicator of substrate concentration, **63,** 180
photobleaching, **65,** 345
recovery, after photobleaching, **63,** 104–105
standardization and calibration, **63,** 312–326
Fluorescence *in situ* hybridization
diagnostic screening, **64,** 400
DNA fragment mapping, **64,** 34
Drosophila embryo confocal imaging, **70,** 374
for QDFM, **64,** 45–46
hamster/zebrafish radiation hybrid characterization
applications, **60,** 294–295
hybridization and washing, **60,** 297–298, 302
materials, reagents, and solutions, **60,** 295
metaphase spread preparation, **60,** 296–297
probe preparation, **60,** 295–296, 302
slide making, **60,** 297, 300–301
identification of viable bacteria, **64,** 524–525
in LSCM, measurement of nucleoli, **64,** 438–441
interphase nuclei, **64,** 98
karyotyping
analysis of results, **60,** 332–334
antibody detection, **60,** 332
counterstaining, **60,** 332
denaturation, **60,** 331
hybridization, **60,** 331
pretreatment of slides, **60,** 331
probe synthesis, **60,** 329–330
washes, **60,** 331–332
Ki-67 expression determined by, **64,** 434
monitoring MRD, **64,** 400–401
mouse/zebrafish somatic cell hybrid characterization
chromosome and slide preparation, **60,** 310–311
chromosome number, **60,** 313
denaturation, **60,** 311
fluorescence detection and amplification, **60,** 312
hybridization, **60,** 311
probes
Alu Type Ib, **60,** 308–309
labeling, **60,** 309
painting probe, **60,** 308
SRI, **60,** 308
washes, **60,** 311–312
PGD based on, **64,** 99
protocols, **64,** 108
relocalization for, **64,** 327, 329, 331–333, 335–336
and slide pretreatment, **64,** 112
target DNA, **64,** 398, 400
Tetrahymena thermophila rDNA
nick translation reaction, **70,** 346–347
probe synthesis, **70,** 346
sample preparation, **70,** 344–346
Fluorescence *in situ* hybridization (FISH)
antifading agents, **53,** 55
autocorrelation function of spatial arrangement, **53,** 61–63
background, lowering from secondary antibodies, **53,** 55
chromosome architecture preservation, **53,** 188–189
colocalization techniques
image overlay, **53,** 63–64
scatter plots, **53,** 66–68
visual impression, **53,** 64–66
cooled charge-coupled device camera
charge coupling, **53,** 57
comparison to other cameras
home video camera, **53,** 57–59
SIT, **53,** 57
cost, **53,** 57–58
quantum efficiency, **53,** 57
read-out rate, **53,** 57–58
counterstaining of DNA, **53,** 197
fluorescence microscopy imaging
confocal microscopy, **53,** 224
data analysis software, **53,** 225, 279–280
deconvolution microscopy, **53,** 224–225
detectors, **53,** 279
difficulty, **53,** 223–224
multiwavelength imaging, **53,** 224–225, 278–279
resolution requirements, **53,** 224–225
immunostaining of whole-mount *Drosophila* samples, **53,** 217–218
probe synthesis for whole-mount chromosomes
cloning, **53,** 191–192
fluorescence labeling approaches, **53,** 190–191, 197–198

biotinylation, **53,** 197–198, 276, 278
digoxigenin labeling, **53,** 198, 276–278
haptens, **53,** 197
nick translation, **53,** 276–277
probe selection, **53,** 196–199
fragments
endonuclease reactions, **53,** 192–193
size restriction, **53,** 190, 192, 276
RNA *vs.* DNA probes, **53,** 189–190
satellite DNA, **53,** 194–196
terminal deoxynucleotidyl transferase tailing of fragments, **53,** 193–194
sample preparation for *Drosophila* whole-mount hybridization
buffer preparation, **53,** 199–200
egg chamber
manual dissection, **53,** 207–208
mass preparation, **53,** 203–207
formaldehyde fixation
cells cultured on coverslips, **53,** 210
duration, **53,** 199
egg chambers, **53,** 203–208
embryo fixation and methanol devitellinization, **53,** 201–203
formaldehyde preparation from paraformaldehyde, **53,** 200–201
tissues on slides, **53,** 208–209
methanol fixation/devitellinization of embryos, **53,** 203
squashes, **53,** 210–211
sandwiching techniques for signal amplification, **53,** 198–199, 276
synaptonemal complex
detection systems, **53,** 277–278
fluorescence microscopy and imaging, **53,** 252, 278–280
hybridization conditions, **53,** 277
materials, **53,** 250–251
probe labeling, **53,** 276–277
proteinase K treatment, **53,** 251–252
sample preparation, **53,** 250
whole-mount *Drosophila* hybridization
equipment, **53,** 212, 215
reagents and solutions, **53,** 211–212
slide preparations, **53,** 215–217
squash preparations, **53,** 217
tube preparations, **53,** 212–215
whole-mount *Saccharomyces cerevisiae* hybridization, fixation
cell wall removal, **53,** 217, 220
difficulty, **53,** 218–219
formaldehyde fixation for rDNA probe hybridization, **53,** 219–221
immunodetection of nuclear antigens, fixation protocol, **53,** 221–223
Fluorescence *in situ* hybridization, chromosome analysis, **57,** 209, 211
Fluorescence-activated cell sorting
deflection of droplets, **58,** 316
green fluorescent protein expressed in cells and organelles
bacteria, **58,** 327–328
mammalian cells, **58,** 325, 330
nuclei from plants, **58,** 327
protoplasts, **58,** 326, 333, 337
site-directed mutants, **58,** 317–319
sensitivity, **58,** 319
Fluorescence-activated cell sorting (FACS), **61,** 131; **65,** 88–90
Fluorescence axes, calibration, problems with, **64,** 149–150
Fluorescence beads, IFI value, **64,** 88–90
Fluorescence confocal microscopy, *see* Confocal fluorescence microscopy
Fluorescence detection, PFM, **68,** 200–202
Fluorescence image acquisition
optimum conditions, **65,** 347
Fluorescence image analysis, arc epi-illumination for, **63,** 52
Fluorescence intensity
flow cytometric analysis of asthma, **64,** 498–499
integrated, *see* Integrated fluorescence intensity
Fluorescence lifetime
excited-state, **63,** 132–133
flow cytometry, **63,** 133–134
instrumentation, **63,** 138–141
measurements, **63,** 134–136
instrumentation, **63,** 140–141
phase-resolved, **63,** 136–137
Fluorescence lifetime imaging microscopy
antibody labeling, **69,** 266–267
antibody staining, **69,** 269
biophysical principles, **69,** 265–266
data analysis, **69,** 270
Dunn chemotaxis chamber setup, **69,** 268–269
Fab fragment generation, **69,** 266–267
global analysis, **69,** 275–276
image acquisition, **69,** 269
lifetime measurement methods, **69,** 262–263
live cell and confocal FLIM, **69,** 272–273
microinjection, **69,** 267–268
multiple-frequency FLIM, **69,** 273–275

Fluorescence lifetime imaging microscopy (*cont.*)
optimization and experimental design, **69,** 270–272
proteins, **69,** 262–264
transfection, **69,** 267
Fluorescence loss during photobleaching
confocal microscopy, **58,** 278, 280
controls, **58,** 280
principle, **58,** 278
Fluorescence measurements
membrane potential, **65,** 87–90
Fluorescence microscopy
BrdUrd subcellular distribution, **64,** 132–133
calcium imaging, **66,** 240–242
chromosome identification, **64,** 17
digital deconvolution, **56,** 305–315
constrained iterative deconvolution, rationale, **56,** 306–307
evaluation, **56,** 309–314
future research directions, **56,** 314–315
nearest-neighbor deconvolution
implementation, **56,** 308–309
rationale, **56,** 307–308
overview, **56,** 305–306
high-resolution multimode digital microscopy
applications, **56,** 209–214
anaphase, *in vitro*, **56,** 210–211
digitally-enhanced differential interference contrast images, **56,** 209–210
fluorescence photoactivation, **56,** 211–212
green fluorescent protein–dynein-labeled microtubules, **56,** 186, 212–214
immunofluorescent specimens, **56,** 214
migrating tissue cells, **56,** 212
computer microscope control, **56,** 207–208
design criteria, **56,** 187–190
epi-illumination optics, **56,** 194–197
image processing, **56,** 208–209
overview, **56,** 186–187
imaging system for telomeres, **64,** 73–77
low-light imaging, **56,** 19–21
membrane permeability
intermembrane mitochondrial proteins, **65,** 151
ratio imaging
detector systems, **56,** 231–233
digital image processing, **56,** 233–234
emission control, **56,** 230–231
illumination control, **56,** 230–231
instrument choice, **56,** 219–222
intracellular ion measurement, **56,** 237–251
experimental chamber, **56,** 249–250
fluorescence signal calibration, **56,** 245–247
imaging workstation components, **56,** 247–249
incorporation into intact living cells, **56,** 242–243
indicator properties, **56,** 239–244
overview, **56,** 237–238, 251
perfusion system, **56,** 249–250
pitfalls, **56,** 243–244
rationale, **56,** 238
signal-to-noise ratio, **56,** 243
spectral characteristics, **56,** 239–242
structural characteristics, **56,** 239–242
methodology
emission ratio method, **56,** 223–225
excitation ratio method, **56,** 222–223
multiple fluorophores, **56,** 225–226
optical considerations, **56,** 226–230
chromatic aberration, **56,** 227–230
spherical aberration, **56,** 226–227
overview, **56,** 217–219, 234
temporal imaging, **56,** 277–302
lifetime-resolved camera, **56,** 287–292
applications, **56,** 290–293
data acquisition, **56,** 289–291
image intensifier, **56,** 288–289
light source, **56,** 289
optics, **56,** 289
methods, **56,** 279–287
frequency domain measurement, **56,** 279–281
photobleaching effects, **56,** 190, 285–287
simultaneous multiple-lifetime component measurement, **56,** 281–285
time domain measurement, **56,** 279–281
overview, **56,** 277–279, 301–302
pump–probe microscopy, **56,** 298–301
applications, **56,** 300–301
instrumentation, **56,** 299–300
two-photon microscopy, **56,** 293–298
applications, **56,** 297
data acquisition, **56,** 296–297
detector, **56,** 296–297
instrumentation, **56,** 294–297
light source, **56,** 295
optics, **56,** 295–296
scanner, **56,** 295–296
Fluorescence microscopy, *see also* Confocal scanning laser microscopy, Fluorescence

in situ hybridization immunofluorescence microscopy
fluorophore selection
DNA-specific dyes for confocal microscopy, **53,** 54
long-wavelength fluorophores, **53,** 53–54
rhodamines *vs.* fluorescein, **53,** 53
image registration
algorithms
cross-correlation functions, **53,** 70
empirical method not requiring significant overlap, **53,** 71
simplex minimization, **53,** 70–71
bleed-through between channels for alignment, **53,** 70
fluorescent beads as alignment markers, **53,** 69–70
magnification dependence on wavelength, **53,** 68–69
spherical aberration, **53,** 71–72
imaging modalities
confocal, **53,** 50–52
digital, **53,** 51
video, **53,** 51
imaging software
customization, **53,** 72–73
vendors, **53,** 72–74
immunofluorescence, sample preparation, **53,** 50
multiply tuned dichroic mirrors
advantages, **53,** 55–56
applications with filter wheel, **53,** 56–57
spectral characteristics, **53,** 55
nuclear assembly
Drosophila system, **53,** 401, 410
mammalian systems, **53,** 362–363
sea urchin system, **53,** 439–440, 446–448
Xenopus system, **53,** 382–384
nuclear protein import assays
background in rapid screening, **53,** 535–536
imaging techniques, **53,** 538–539
indirect immunofluorescence, **53,** 537–538
out-of-focus blur removal
confocal imaging, **53,** 51–53
deconvolution, **53,** 51–53
selection of technique, **53,** 52–53
DNA replication site mapping, *see* DNA replication site, mapping *in situ*
three-dimensional data sets
display modes, **53,** 60–61
surface rendering, **53,** 59–60
volume rendering, **53,** 60
transmission electron microscopy interfacing, **53,** 101–103
Fluorescence microscopy, green fluorescent protein
cell maintenance
enclosed systems, **58,** 348
media, **58,** 347–348
perfusion chambers, **58,** 185–186, 215–216, 231, 255–256, 348–349
confocal microscopy, *see* Laser scanning confocal microscopy
data quantity, requirements for imaging, **58,** 290
deconvolution
images, **58,** 39–40
microscopy, **58,** 354–355
detectors
bioluminescence, **58,** 289
charge-coupled device camera, **58,** 36–39, 52–53, 70–71, 230–231, 256, 266, 287, 289
linearity, **58,** 37
offset, **58,** 37
photomultiplier tube, **58,** 36–37
Drosophila fusion proteins, **58,** 176–178
filters, **58,** 34–35, 109, 158–159, 177–178, 287
fluorescence resonance energy transfer, *see* Fluorescence resonance energy transfer
high-speed fluorescence microscopy
charge-coupled device camera, **58,** 230–231
deconvolution of images, **58,** 231–234
dynamic properties of centrosomes, **58,** 234–235
heated cell chamber, **58,** 231
image capture, **58,** 234
lasers, **58,** 229
overview of instrumentation, **58,** 228–229
piezoelectric focus drive, **58,** 229
imaging facility
computer systems
acquisition system, **58,** 356
archival and backup, **58,** 189–190, 357–359
client, **58,** 357
costs, **58,** 355–356
data analysis, **58,** 359–360
fileserver, **58,** 356–357
concept, **58,** 344–345
ergonomics, **58,** 347
heating/ventilation/air-conditioning and ducting, **58,** 346

Fluorescence microscopy, green fluorescent protein (*cont.*)
imaging systems, overview, **58,** 350–355
room size, **58,** 345–346
software, *see also specific programs*
data reduction and statistics, **58,** 360
drawing programs, **58,** 362–363
file format, **58,** 361–362
painting programs, **58,** 362
staff, **58,** 365–366
light path and collection parameter optimization, **58,** 289–290
light sources, **58,** 34, 109
microspectrofluorometry, **58,** 37–38, 349–350
objective lens, **58,** 34, 109
organelle-targeted protein, **58,** 79–80, 83–84
photobleaching controls, **58,** 33
printing images for publication, **58,** 363–365
single-molecule detection, *see* Total internal reflection fluorescence microscopy
time-lapse imaging
image acquisition, **58,** 94–95
image analysis, **58,** 95
instrumentation, **58,** 94
Metamorph software, **58,** 119–120
photobleaching and cell viability, **58,** 118–119
presentation of images, **58,** 95–96
time-lapse imaging of membrane fusion proteins
cameras, **58,** 266
confocal microscopy, **58,** 266–268
focal control, **58,** 265
image analysis, **58,** 268, 270
photobleaching, **58,** 265
two-photon excitation microscopy, *see* Two-photon excitation microscopy
yeast cells
budding yeast, **58,** 93–94
fission yeast
formaldehyde fixation, **58,** 135–136
methanol fixation, **58,** 135–136
mounting on agarose pad, **58,** 133–134
Fluorescence microscopy, in chick ciliary neuron cultures, **71,** 39
Fluorescence microscopy, Mabs for instructive ECM identification, **69,** 43–44
Fluorescence microscopy, *see* Microscopy, Light
Fluorescence microscopy, single fluorophore visualization
aqueous solutions, **55,** 118–123
low-background epifluorescence microscopy, **55,** 119
low-background total internal reflection fluorescence microscopy
aqueous solutions, **55,** 119–123
kinesin observation, **55,** 123, 125–127
myosin molecule ATP turnover, **55,** 123–125
Fluorescence polarization, **63,** 325–326
Fluorescence properties
calcium dyes, **66,** 236
Fluorescence quenching
and DNA-targeting drug binding, **64,** 187–188
anthracyclines, **64,** 199
Fluorescence recovery after photobleaching
confocal microscopy, **58,** 272–273, 275–276
diffusion constant calculation, **58,** 271–272, 276–278
principle, **58,** 271
qualitative analysis of lateral diffusion, **58,** 272–273
quantitative analysis of lateral diffusion, **58,** 275–276
Fluorescence resonance energy transfer (FRET) reported proteins, **71,** 132
Fluorescence resonance energy transfer
energy-level diagram, **58,** 295–296
energy transfer efficiency, **58,** 297
fluorescence microscopy imaging system
acquisition and processing of images, **58,** 306–307
deconvolution of images, **58,** 307
detectors, **58,** 305–306
filters, **58,** 305
fluorescence lifetime imaging microscopy, **58,** 309
microscope, **58,** 303–305
two-photon excitation fluorescence imaging microscopy, **58,** 309
Förster equation, **58,** 296–297
fusion proteins
applications, **58,** 103–104, 307–309
controls for microscopy, **58,** 303
design of constructs, **58,** 301
functional assays, **58,** 301–302
Pit-1 dimerization, **58,** 307–309
transient expression techniques, **58,** 302–303
green fluorescent protein
literature review, **58,** 297–298
luciferase and green fluorescent protein *in vivo*, **58,** 15

overlap integral of donor and acceptor spectra, **58,** 297
protease assays, **58,** 337
range limitations, **58,** 294, 296
requirements, **58,** 295–296
tryptophan *in vivo* in green fluorescent protein, **58,** 15–16
validation by X-ray diffraction, **58,** 298
Fluorescence resonance energy transfer, **64,** 189–190
Fluorescence resonance energy transfer, *see* Fluorescence lifetime imaging microscopy
Fluorescence spectroscopy
ceramide, **66,** 147–149
Fluorescent *in situ* hybridization (FISH)
alphoid DNA satellite probes, **61,** 66
centromere–kinetochore complex
combined DNA/protein staining
protocol, **61,** 70–72
reagents, **61,** 69–70
protocol, **61,** 68
denaturation, **61,** 67–68
detection, **61,** 68–69
hybridization, **61,** 68
probe preparation, **61,** 68
signal amplification, **61,** 69
reagents, **61,** 67
uses, **61,** 66
Fluorescent activated cell sorting, phagocytosis assay, **63,** 619–623, 625–626
Fluorescent bromodeoxyuridine (BrdU) protocol
growth assays, **66,** 221–222, 222
Fluorescent detection, mitochondria, **64,** 118
Fluorescent dextrans, **71,** 131
as single-cell injections, **71,** 138–139
nucleus injected with, **71,** 313
Fluorescent dyes
DNA staining, **63,** 9–10
for flow cytometry, **63,** 7
in analysis of cell–cell adhesion assays, **63,** 603
in flow cytometry of bacteria, **64,** 554–555
mitochondria, **66,** 379t, 380
organic, protein labeling with, **63,** 186–194
PKH26, **63,** 455
propidium iodide, **63,** 66
uptake, vital cells, **64,** 555–558
Fluorescent dyes, cell attachment assays, **69,** 287
Fluorescent fusion proteins, **71,** 131
expression of, **71,** 383
in *Xenopus* cultures, **71,** 130, 148
injecting neurons with, **71,** 314
Fluorescent halo assay, measurement of DNA damage, **64,** 258, 261–263
Fluorescent ionic dyes
monovalent intracellular ions
flow cytometry, **66,** 58
Fluorescent particle standards
commercially available, **63,** 322
making of, **63,** 320
spectrally matched and unmatched, **63,** 322–324
tertiary, **63,** 324–325
Fluorescent probes
coupling to proteins, conjugation chemistry, **63,** 186–187
for confocal microscopy, **63,** 96–99
Mito Tracker, **63,** 475–476
study of lymphocyte activation/proliferation, **63,** 453
Fluorescent proteins
pH-sensitive, **65,** 373–374
Fluorescent speckle imaging
deconvolution methods, **61,** 170
image processing
background equalization, **61,** 164
camera correction, **61,** 164
image alignment for motion analysis, **61,** 164
image stack conversion, **61,** 164–165
movie viewing, **61,** 165
unsharp mask, **61,** 164
microscopy and image acquisition
applications, **61,** 161
computer memory requirements, **61,** 163–164
differential interference contrast (DIC) microscopy, **61,** 161
imaging recording parameters, **61,** 163
MetaMorph software, **61,** 161, 164, 165, 166
microscope design, **61,** 161
microscope stability
focus drift, **61,** 162
vibration isolation table, **61,** 162
phase contrast microscopy, **61,** 161
removal of DIC imaging components, **61,** 163
resolution and light sensitivity
CCD camera, **61,** 162
detector placement, **61,** 162
lens selection, **61,** 162
quantum efficiency values, **61,** 162
total magnification equation, **61,** 162
wavelength selection
filter selection, **61,** 162–163
fixed wavelength imaging, **61,** 163
fluorescence excitation, **61,** 162
High Q filter cubes, **61,** 163

Fluorescent speckle imaging (*cont.*)
microtubule principles
assembly dynamics, **61,** 156–157
background fluorescence effect, **61,** 158
contrast equation, **61,** 158
light microscopy resolution equation, **61,** 157
resolution limit equation, **61,** 158
standard deviation equation, **61,** 158
stochastic association, **61,** 157–158
specimen methods
photobleaching prevention, **61,** 161, 163
spindles *in vitro*
fluorescent label concentrations, **61,** 160
slide-coverslip preparation, **61,** 160–161
Xenopus source, **61,** 160
tissue cells
Rose chamber method, **61,** 160
slide-coverslip chamber preparation, **61,** 160
Taricha granulosa lung cell cultures, **61,** 160
tubulin microinjection concentration, **61,** 160
tubulin purification and labeling
molar dye to protein ratio determination, **61,** 159
phosphocellulose chromatography, **61,** 159
porcine brain source, **61,** 158–159
tetramethylrhodamine labeling, **61,** 160
X-rhodamine labeling, **61,** 159
speckle motility analysis
distance calibration, **61,** 165
fluorescence photoactivation marking studies, **61,** 170
kymographs
horizontal–vertical orientation, **61,** 166
point to point advantages, **61,** 166
velocity determination, **61,** 166–167
single speckle tracking
measure pixel routine, **61,** 165
single particle tracking, **61,** 165
Fluorescent tagging
in *Xenopus* spinal neurons, **71,** 130
of cell motility, axon outgrowth, and calcium signaling, **71,** 132
of cytoskeletal proteins, **71,** 39
Fluoro-2′-deoxyuridine (FdUr), **71,** 84
Fluorochrome-conjugated probes
antibody suppliers for confocal microscopy, **70,** 76–77
thick tissue labeling, **70,** 315–321, 330–331
Fluorochromes, **64,** 441; **65,** 541
brightest, for immunophenotyping, **63,** 226–227
cellular retention, **64,** 201–206
color compensation, **63,** 318–319
cycles of excitation and fluorescence, **63,** 32
DNA-specific, for LSCM, **64,** 423–425
in detection of MRD, **64,** 391
in epitope detection, **63,** 224–225
in simultaneous multicolor applications, **63,** 201
multiple, simultaneous measurement, **64,** 580–582
nuclear DNA-encoded subunits mitochondrial DNA, **65,** 326–327
overlapping fluorescence spectra, **64,** 295–298
photobleaching, in confocal microscopy, **63,** 97, 99
photodetection, **63,** 37
phycoerythrin/Texas Red, **64,** 339
quantitation, issues regarding, **63,** 332–334
selection, **63,** 259, 261
study of drug retention and efflux, **64,** 197
tandem, **63,** 234–237
5-fluorodeoxyuridine (FDUR), in glial contamination, **71,**120–121
Fluorometric assays
caspase, **66,** 15
Fluorophore
caspase, **66,** 15
Fluorophores
blue fluorophores, **70,** 177–178
BODIPY, **70,** 171–173
cyanine fluorophores, **70,** 176
detector cross-talk in LSCM, **70,** 213, 217
fluorescein, **70,** 171
Lucifer yellow, **70,** 173
multicolor LSCM, **70,** 209–217, 221–225
multiphoton excitation imaging, **70,** 138–139
overview, **70,** 167–171
phthalocyanine derivatives, **70,** 176–177
phycobiliproteins, **70,** 174–176
red, detector cross-talk in LSCM, **70,** 213–215
rhodamine fluorophores, **70,** 173
single-wavelength excitation, LSCM, **70,** 192–201
three-color LSCM, krypton–argon ion laser, **70,** 206–207
two-color LSCM, **70,** 201–204
visualization specificity, **70,** 181
Fluorophores, conjugation, **63,** 196
Fluorophores, visualization, **55,** 118–123
low-background epifluorescence microscopy, **55,** 119
low-background total internal reflection fluorescence microscopy, **55,** 119–123

5-Fluorouracil, plus adriamycin and cyclophosphamide, **64,** 374
Fluorsave Reagent, **71,** 96
Fly heads
  isolation and extraction of, **71,** 205–206
  lipid analysis of, **71,** 214–215
  subcellular fractionation of, **71,** 206–207
Fly nervous system, *see Drosophila*
Flybase, **71,** 221
Focus assay
  transformation assays, **66,** 216–218, 217
Focus drift, in time-lapse fluorescence imagine, **71,** 319
Folding method
  recombinant caspase purification, **66,** 13–14
Folding, green fluorescent protein
  β-can fold, **58,** 3–4, 33, 299
  chromophore formation, **58,** 3–6, 21–22, 33, 299
  intermediates, **58,** 4
  thermal stability, **58,** 3–4, 26–27
  X-ray crystallography, **58,** 20–21
Food and drink, contaminants, flow cytometric analysis, **64,** 529–531
Food vacuoles, **62,** 44, 98
Force clamp experiments, noise limitations, **55,** 143–146
Force curves
  AFM, soft samples, **68,** 70
  protein unfolding, **68,** 317–318
Force–distance cycles
  dynamic force microscopy, vascular endothelial cadherin, **68,** 128
  unbinding force, ligand–receptor recognition, **68,** 121–123
Force sensor
  cell–cell adhesion force
    research results, **68,** 105–109
    single cell immobilization, **68,** 104–105
  living cell force spectroscopy
    functionalized sensors, **68,** 98
    monolayer immobilization, **68,** 99–100
    passivated force sensors, **68,** 95
    research results, **68,** 95–104
    sphere immobilization, **68,** 94
  PFM studies, calibration, **68,** 207
Force spectroscopy
  cell adhesion, *see* Cell adhesion force spectroscopy
  dynamic
    applications, **68,** 350–351
    overview, **68,** 347–349
  molecular adhesion, **68,** 14–15
  molecular interaction, **68,** 13–14
  optical and patch–clamp technique combinations, **68,** 19
  protein folding, **68,** 15–19
  protein modular structure, **68,** 15
  protein–protein interactions, **68,** 292
  protein unfolding, **68,** 317
  transverse dynamic
    application, **68,** 353–354
    overview, **68,** 351–353
Force, *see* Atomic force microscope; Gradient trap force; Optical force microscopy
Forebrain, **71,** 3
Formaldehyde
  as fixative: properties, **63,** 277–279
  fixation, *see* Fluorescence *in situ* hybridization
  plus methanol, fixation by, **63,** 280
  preparation from paraformaldehyde, **53,** 200–201
Formaldehyde fixation
  thick tissue sections, **70,** 308–311
  Xenopus oocytes and embryo keratin filaments, **70,** 386–388
  Xenopus oocytes and embryo microtubule, **70,** 383–386
  Xenopus oocytes fixation, **70,** 391
Formaldehyde, fixation for immunofluorescence microscopy, **57,** 316–317, 327–329
Formaldehyde, fixative, **64,** 8
Formate dehydrogenase, **65,** 449
Forward genetic systems, **71,** 216–228
  and *Drosophila* electrophysiology, **71,** 228–238
Forward light scatter, **63,** 34, 36
  increased in necrotic plant cells, **63,** 514
  measurements, **63,** 11
  signals, in flow cytometry, **63,** 112–113
Forward-scatter
  acquisition by correlated list mode data, **64,** 290–291
  vs. side scatter display, **64,** 309
Forward-scatter (FSC), **66,** 52–53, 80
  S49 Neo cells
    vs. PhiPhiLux contour plot, **66,** 60
    vs. SSC dot plot, **66,** 54–56
Four-dimensional reconstruction
  CSU-10 digital signal processing, **70,** 115–116
  overview, **70,** 66–68
Fourier transform, digital image processing, **56,** 105–108
Fractional DNA content
  DNA fragmentation assay, **66,** 87–90

Fractionation, and cell/culture integrity, **64,** 178–179
Fractionation, subcellular, protein kinases, **67,** 311
Frame buffers, digital image processing, **56,** 75
Frame grabbing, AFM integration, **68,** 176–177
Frame rate, low-light imaging, **56,** 33–34, 62
Frame, confocal microscopy, **70,** 54
Francisella tularensis, survival ratio, **64,** 519
FRAP, *see* Fluorescence recovery after photobleaching
Free Ca^{2+} in electroporation buffer, as electroporation parameter, **71,** 347
Freeze-slamming, **65,** 250
Freeze-substitution
  process, **67,** 29–31
  protocols, **67,** 33–34
Freezing
  animal cell culture guidelines, **57,** 16, 40–41
  embryonic stem cells, **57,** 283–284, 287–288
  freezers, **57,** 9
  lymphocytes, **57,** 129
  MEFs, **66,** 202
Freezing cells, **62,** 200–206, 211–215; **71,** 13
Freezing media, for γTuSC reconstitution, **67,** 13
FRET-based assay
  mitochondrial membrane potential, **65,** 296–306
FRET, *see* Fluorescence lifetime imaging microscopy
FRET, *see* Fluorescence resonance energy transfer
Frog neuromuscular junction neurons, endocytosis and recycling in, **71,** 5–8
Frog sensory neurons, endocytosis and recycling in, **71,** 5–8
Frozen-hydrated specimens
  TEM, **65,** 252
Frozen sections, centromere–kinetochore complex
  confocal microscopy, **61,** 77
  CREST immunofluorescence staining, **61,** 77
  cryomicrotome, **61,** 77
  three-dimensional projection software, **61,** 77
Full-length LIGHT, **66,** 504
Function-based screens
  genes, **66,** 455
Function-blocking antibodies, **71,** 130, 148
Functional analysis techniques, in *Drosophila* nervous system, **71,** 224–228
Functional assay, verification of antibody specificity, **63,** 292
Functional assays, for γTuSC and γTuRC, **67,** 18–23
Functional proteins (peptides), **71,** 131
Functional redundancy studies, **71,** 200–201
Functional viruses
  growing, **71,** 369–375
  producing, **71,** 375–378
Fungal Genetics Stock Center, **61,** 353–354
Fura-2, **66,** 235
Fura 2-acetoxymethyl, intracellular ion ratio imaging, properties, **56,** 239–244
  incorporation into intact living cells, **56,** 242–243
  pitfalls, **56,** 243–244
  signal-to-noise ratio, **56,** 243
  spectral characteristics, **56,** 239–242
  structural characteristics, **56,** 239–242
Fura Red, pCa_i, **70,** 456
fused somites, mutation effects in development, **60,** 31–33
Fusion markers
  mitochondrial GFP, **65,** 278–281
Fusion pores, pancreatiaacinar cells, AFM studies, **68,** 40, 43
Fusion protein expression, anti-zKLP antibodies, **59,** 238–240
Fusion protein intermediates, **65,** 208
Fusion protein, *see also specific proteins*
  budding yeast
    expression, **58,** 91–92, 110, 116–117
    vectors, **58,** 88–89
  characterization
    fluorescence intensity, **58,** 91
    subcellular localization, **58,** 92–93, 110–114
  Drosophila constructs
    β-galactosidase fusion protein, **58,** 170
    GAL4 system in targeting, **58,** 167–168, 174
    maternal and early embryonic expression of fusion proteins, **58,** 174
    tau fusion proteins, **58,** 170–173
    UAS vector and green fluorescent protein fusions, **58,** 168–170
  fission yeast plasmids
    behavior of proteins *in vivo*, **58,** 124, 128, 130–131
    copy number, **58,** 124–126
    integration into genome, **58,** 126
    promoters, **58,** 127, 129–130
    screening of expression, **58,** 131–133

types, **58,** 124, 128
vectors, **58,** 130
fluorescence recovery after photobleaching
confocal microscopy, **58,** 272–273, 275–276
diffusion constant calculation, **58,** 271–272, 276–278
principle, **58,** 271
qualitative analysis of lateral diffusion, **58,** 272–273
quantitative analysis of lateral diffusion, **58,** 275–276
fluorescence resonance energy transfer applications, **58,** 103–104
functional assays, **58,** 263–264
linkers, **58,** 90–91
membrane proteins
cell preparation, **58,** 264–265
expression levels, **58,** 263
fluorescence loss during photobleaching
confocal microscopy, **58,** 278, 280
controls, **58,** 280
principle, **58,** 278
organelle-targeting chimeras, construction and expression, **58,** 76, 78
pharmaceutical applications, **58,** 103
quantitative analysis, **58,** 270–271
targeting assays, **58,** 263
terminus selection for fusion, **58,** 89–90
time-lapse imaging
cameras, **58,** 266
confocal microscopy, **58,** 266–268
focal control, **58,** 265
image analysis, **58,** 268, 270
photobleaching, **58,** 265
Fusion proteins
for centrosome-associated proteins, **67,** 131–132
for SPB protein *in vitro* binding, **67,** 90
GST, binding to glutathione sepharose, **67,** 91
protein A, and SPB precipitation, **67,** 87–88
two-hybrid vector test, **67,** 78–79
Fusion proteins, screening in PC12 cell lines, **71,** 283
Fusion, hybridomas, Mab identification of instructive ECM, **69,** 42–43

G

G-innexin-1, **71,** 182
GABA-gated channels, **71,** 213
and B35 cell lines, **71,** 288
GABAergic and glycinergic inhibitory interneurons, **71,** 213
β-Galactosidax, two-hybrid assay, **67,** 81–82
GAG chains, proteoglycans, ELISAs, **69,** 63
GAG–growth factor interaction, enzyme treatment
chondroitin sulfate, **69,** 93
heparan sulfate, **69,** 93–94
heparin, **69,** 93–94
overview, **69,** 92–93
Gain
basic camera adjustments, **56,** 5–9
low-light imaging, **56,** 33
GAL 1–10 promoter, **61,** 372
GAL1 promoter, **66,** 456
GAL10 promoter, **66,** 456
GAL4 system, green fluorescent protein targeting in *Drosophila*, **58,** 167–168, 174, 284–285
GAL4-VP16, cell type marking in *Arabidopsis thaliana*, **58,** 146–147
Galactose
induction, **65,** 282
Galactose-induced mitochondrial green fluorescent protein markers
induction, **65,** 279–281
Galactose-inducible GFP markers, **65,** 278–279
Gallus gallus (common domestic chick), **71,** 68
Gamma-ray mutagenesis
blastula stage embryos, **60,** 60–62
characterization of gene mutations, **60,** 64–66
sperm, **60,** 62–63
spermatogonia, **60,** 63–64
translocations, **60,** 66–67
Gamma-tubulin homologs, **61,** 25–26
Gamma, basic camera adjustments, **56,** 6
Gaps, in axonal neurofilament array, **71,** 315
Gastric adenocarcinomas, potential doubling time, **64,** 462
Gating
CD45 as reagent, **64,** 573
CD45 scattergram, **64,** 325–326
CD45/SSC, **64,** 345, 348–356
color, **64,** 311
and DNA histograms, **64,** 142
for immunophenotyping, **64,** 321–322
light scatter lymphocyte, **64,** 571
lymphocyte isolation by, **63,** 75
multiparameter data analytic approach, **64,** 308
on scatter characteristics, **63,** 43
procedures for lymphocytes, **63,** 439

Gating (*cont.*)
scattergram, **63,** 55
Gating, OHC transduction channel, **68,** 161–165
GD25 cells, β1 integrin null cells, **69,** 318
GE11 cells, β1 integrin null cells, **69,** 318
Gel electrophoresis
DNA fragmentation, **66,** 438–448
Gel electrophoresis, *Dictyostelium* centrosomes, **67,** 345–346
Gel electrophoresis, DNA ladder detection with, **63,** 494–495, 499
Gel electrophoresis, multicomponent ECM matrix proteins, **69,** 202–203
Gel filtration
basement membrane self-assembly, **69,** 128
proteoglycans, **69,** 60–61
Gel-filtration analysis
TNF ligands, **66,** 511
Gel-filtration chromatography
Superose 12 Column
Apaf-1 purification, **66,** 22–23
Gel mobility shift assay, base-unpairing region binding protein, **53,** 338–339, 348
Gel overlay, **62,** 461–464
Gel types
BN-PAGE, **65,** 233
Gelatin, coupled to FITC, **63,** 615–625
Gels, thick collagen, **69,** 406–407
Gene confirmation
nonimmortal cells, **66,** 210
Gene delivery, immortalization of cells, **57,** 77–78, 80–81, 95–98, 105–106
Gene expression, and cell shape, as mechanism for contact effect, **64,** 221–222
Gene expression, vascular and blood, **59,** 313–333
conclusions and future directions, **59,** 332–333
during development, **59,** 320–327
gene expression in embryonic hematopoiesis, **59,** 323–327
genes in dorsoventral specification, **59,** 320–321
embryonic hematopoiesis, **59,** 318–320
blastula and gastrula periods, **59,** 318
hemangioblast, **59,** 319–320
pharyngula and adult period, **59,** 319
segmentation period, **59,** 318–319
embryonic vascularization, **59,** 314–318
blastula and gastrula periods, **59,** 315
introduction, **59,** 314–315
pharyngula period, **59,** 315–316
segmentation period, **59,** 315
vessel structure, **59,** 316–318
introduction, **59,** 314
mutations affecting vascularization and hematopoiesis, **59,** 327–332
blood cell progenitors, **59,** 330
blood differentiation and maintenance, **59,** 330–331
blood specifications, **59,** 329–330
endothelial cell differentiation, **59,** 327–328
hypochromic blood, **59,** 331–332
introduction, **59,** 327
photosensitive mutations, effect on blood, **59,** 332
vascular formation and patterning, **59,** 328–329
vascular integrity, **59,** 329
Gene fusions, for SPB precipitation, **67,** 82–84
Gene gain-of-function experiments, **71,** 370
Gene knockout, *see* Transformation
Gene mapping, by hybridization onto genomic DNA fibers, **64,** 48, 50
Gene of interest
generating mutations in, **71,** 201–202
inserting into a shuttle vector, **71,** 393
Gene rearrangements, and PCR, **64,** 404–407
Gene replacement
mtDNA
Saccharomyces cerevisiae
mitochondria, **65,** 388–392
Gene rescue, **66,** 487–490
Gene targeting, ECM, **69,** 310
Gene targeting methods, **71,** 216
Gene therapy, HIV infection, **64,** 584
Gene transfer
retroviral-mediated, **66,** 202–205
Gene translocations, **71,** 202
GenePix 4000A, **65,** 441
Genes
colors
intensity ratio, **65,** 440–441
differentiation-induced, **66,** 36
function-based screens, **66,** 455
internally redundant
yeast arrays, **65,** 447
into two-hybrid vectors, **67,** 75–76
plant γ-tubulin, **67,** 196–200
proapoptotic, **66,** 204–205
Genes, *see also* Mutant strains
essential and nonessential, **62,** 146–147, 172
nomenclature, **62,** 561–562
Genes, integrin adhesion molecule, **69,** 314–317
Gene targeting ECM, **69,** 310
Genetic analyses

complementation tests, **62,** 237–338
crosses, **62,** 217–238
isolation of assortants, **62,** 125–126
mapping, **62,** 159–163, 263–277
mating types, **62,** 128–130
methods, **62,** 227–238
mutant analysis, **62,** 251–261
three-way cross, **62,** 223
Genetic analysis, SPBs
dosage-dependent suppressor screen, **67,** 97–100
extragenic suppressor screen, **67,** 100–102
synthetic dosage lethality, **67,** 108
synthetic lethal screen, **67,** 103–107
Genetic manipulation
Dictyostelium discoideum, **66,** 485–491
Genetically engineered [Ca^{2+}] indicators, pCa_i
calibration, **70,** 468–469
expression, **70,** 464–465
measurements, **70,** 467–468
overview, **70,** 460–461
pitfalls, **70,** 469–470
targeting, **70,** 461–464
Genetics, confocal objective lens selection, **70,** 273
Genetics, laser manipulation, **55,** 81–83
Genetics, zebrafish system, **59,** 7–8
Genome
genetic code, **62,** 150–151, 565–566
mitochondrial
size, **65,** 420
size, **62,** 134
Genome analysis, *see* Amplified fragment length polymorphisms; Random amplified polymorphic DNAs; Simple sequence length polymorphisms; Single strand conformation polymorphisms; Zebrafish
Genome sequencing, *Drosophila,* 198–202
Genomic DNA preparation, *see* Zebrafish
Genomic exclusion, **62,** 153–155, 234–235, 527
Genomic library, *see* Bacterial artificial chromosome; cDNA clone tagging; P_1-derived artificial chromosome; Yeast artificial chromosome
Genomic vectors, in *Drosophila,* 220–222
Genomics, in centrosome kinase isolation, **67,** 311–312
Genomics, zebrafish system, **59,** 7–8
Geometric distortion, low-light microscopy, **56,** 60
Geometrical optics, *see* Ray optics regime
Germ cell, *see* Testicular germ cell
Germinal vesicle breakdown, **61,** 41
Germline, *see* Micronucleus
GFAP, fluorescence confocal microscopy applications, **70,** 326
GFP
Aequorea victoria, **65,** 373
cell death assays
Drosophila S2 cells, **66,** 335
cytochrome *c*, **66,** 373–375
fusion markers, **65,** 278–281
visualization, **65,** 376
GFP-alpha5 integrin receptors, **71,** 152
GFP expression, **71,** 134
in B35 neuroblastoma cell lines, **71,** 291
GFP, *see* Green fluorescence protein
GFP, *see* Green fluorescent protein
GFP, *see* Green fluorescent proteins
gfp, introns, **58,** 141
GFPmut1, **65,** 374
Giant fiber system *(Drosophila),* 234
Giant multiprotein complexes
bovine heart mitochondria, **65,** 240
Giant protein complexes, **65,** 239
Giemsa staining, **62,** 319–321
Giemsa staining, chromosome analysis, **57,** 209
Gilbert-Developmental Biology (www.devbio.com), **71,** 85
Glass, as culture substrate, **71,** 12
Glass coverslips
for peripheral neurons, **71,** 19
for spinal and motor neuron cultures, **71,** 80
preparation of,
Glass-bottom dishes
for *Heliosoma* cultures, **71,** 169
for time-lapse imaging, **71,** 187
Glass surface, preparation, **69,** 327
Glassware, animal cell culture, **57,** 13, 54–55
Glia contamination, **71,** 30
in cortical neuronal cultures, **71,** 121
in hippocampal neuron studies, **71,** 112
Glia stocks, for long-term hippocampal and cortical neuronal cultures, **71,** 121
Glia-conditioned medium
and speed of axonal development in cortical cultures, **71,** 120
in long-term hippocampal cultures, **71,** 117–121
Glial coculture, chick forebrain neurons, **71,** 63
Glial fibrillary acidic protein (GFAP) immunoreactivity, **71,** 119
Glioma
cell isolation from tumors
enzymatic digestion, **57,** 148

Glioma (*cont.*)
mechanical dispersion, **57,** 149–150
sampling site, **57,** 148
cell line establishment
anaplastic astrocytoma, **57,** 159
efficiency, **57,** 151–152
ependymoma, **57,** 159
heterogeneity of cell lines, **57,** 162
maintenance, **57,** 152–154
medulloblastoma, **57,** 160–161
nestin-positive cell line, **57,** 160
oligodendroglioma, **57,** 159
progenitor cells, **57,** 161
stabilization, **57,** 152
characterization of cell lines
immunostaining markers for lineage and phenotype, **57,** 155–158
neoplasticity assays
genetic analysis, **57,** 158
xenotransplantation, **57,** 158–159
heterogeneity of disease, **57,** 147, 162
primary culture
bovine corneal endothelial cell basement membrane
biological effects, **57,** 161–162
cell attachment to culture dishes, **57,** 151, 161
goals, **57,** 150
Gln^{133}
sequencing chromatogram, **66,** 127
Glu-222, proton transfer in green fluorescent protein, **58,** 12–13, 26
Glutamate
as extrinsic factor in Purkinje neuronal cultures, **71,** 90
gated channels in *Drosophila* studies, **71,** 213
neuronal sensitivity to, **71,** 91
uptake studies of, **71,** 91, 97–100
Glutamate transporters, **71,** 91
in developing Purkinje neurons, **71,** 102–105
Glutamatergic excitatory interneurons, **71,** 140
Glutamine, in chick ciliary neuron cultures, **71,** 43
Glutaraldehyde fixation, **62,** 325
thick tissue sections, **70,** 311, 313
Xenopus oocytes, **70,** 391
Xenopus oocytes and embryo keratin filaments, **70,** 386–388
Xenopus oocytes and embryo microtubule, **70,** 383–386
Glutaraldehyde, cell fixation
immunofluorescence microscopy, **57,** 316–317, 328–329
transmission electron microscopy, **57,** 298, 300
Glutathione *S*-transferase
fusion protein, **67,** 91
Glutathione sepharose, **67,** 91
Glycerol-3-phosphate CoQ reductase
hamster brown adipose tissue mitochondria, **65,** 24
Glycerol phosphate dehydrogenase (GPD)
schematic map, **65,** 357
Glycocalix, elasticity, AFM studies, **68,** 75
Glycoproteins, basement membrane proteins, **69,** 118–120
Glycosaminoglycanase, proteoglycans, **69,** 61
Glycosaminoglycans, **69; 71,** 12
cell adhesion assays, **69,** 298–302
cell invasion assays, **69,** 298–299, 302–307
insoluble molecule interactions, **69,** 298
molecule bindings, **69,** 298
ubiquity, **69,** 297–298
Glycosyl-phosphatidylinositol, **62,** 37, 48
Glyoxylate cycle, **62,** 51
GMPCPP polarity-marked microtubule preparation, **61,** 150
gobbler, **60,** 81
Gold-coated surfaces
fabrication, **69,** 393
micropatterned culture surface, **69,** 397
Gold coating procedure, in biolistics, **71,** 356
Gold particles, use in biolistics, **71,** 355
Gold, *see* Scanning transmission electron microscopy, Transmission electron microscopy
Gold, membrane protein labeling, **55,** 176–178
golden, **60,** 61–62, 64
Golgi apparatus, **62,** 33, 40–42
Golgi complex, pCa_i measurement, **70,** 464
Golgi vesicles
in mitochondrial preparations, **65,** 4
Golgi, green fluorescent protein targeting
chimera construction and expression, **58,** 79
photobleaching studies, **58,** 280
Government agricultural agents, as sources for chick neuronal cultures, **71,** 57
gp210, nuclear pore complex structure, **53,** 5–6, 24
GPD
schematic map, **65,** 357
GPI-anchor, **62,** 37, 48
G-proteins, heterotrimeric, **62,** 38
Gradient centrifugation

Dictyostelium discoideum, **67,** 340–342
Spisula MT nucleation, **67,** 157–158
Gradient gel preparation, **65,** 234
Gradient trap force
dielectric sphere force calculations, single-beam gradient laser trap, **55,** 1–25
arbitrary trap location, **55,** 11–16, 23–25
gradient trap force, **55,** 9–16
index of refraction effects, **55,** 19–20
light forces, **55,** 4–8
mode profile effects, **55,** 16–20
overview, **55,** 1–4, 20–21
ray force, **55,** 21–22
transmission electron microscope mode profile effects, **55,** 16–19
Y axis trap focus, **55,** 10–11, 22–23
Z axis trap focus, **55,** 9–10
multiple traps, trapping configurations, **55,** 206–208
object holding, **55,** 206
object manipulation, **55,** 206–207
static force exertion, **55,** 207–208
Gram stain, bacteria, **64,** 525–526
Granular component, *see* Nucleolus granule cells
effects on Purkinje neurons, **71,** 101
in Purkinje neuronal cultures, **71,** 91–92
Granule lattice proteins, **62,** 46
Granulocytes, peripheral blood, apoptosis in, **63,** 480
Granulocytic/monocytic pathway, **64,** 160, 167
Granzyme B, **66,** 4–5
Grasshopper *(Schistocerca)* limb bud, **71,** 3
advantages for axonal guidance studies, **71,** 172–173
culturing, **71,** 183
developmental properties of, **71,** 5
ectopic expression in embryo, **71,** 185–186
filleting to access the Tibia-1 pathway, **71,** 188–189
study of extracellular guidance cues in, **71,** 172
Tibia-1 pathway study of, **71,** 173–187
Grasshopper spermatocytes, *see* Micromanipulation of chromosomes and spindles
Gray value, digital image processing
contrast enhancement, **56,** 101–104
contributing factors, **56,** 84–85, 89
intensity quantification, **56,** 94–95
Green fluorescence protein (GFP), **66,** 204
cell death assays
Drosophila S2 cells, **66,** 335
cytochrome *c*, **66,** 373–375
Green fluorescence protein, immunofluorescence microscopy, **57,** 322–323
Green fluorescent protein
green reporter gene identification of transgenic founder fish, **60,** 143
promoter analysis in living embryos
fluorescence imaging, **60,** 146
polymerase chain reaction-based mutagenesis of constructs, **60,** 145–146
promoter/enhancer construct generation, **60,** 144–147
green fluorophore, detector cross-tack in LSCM, **66,** 213–217
Green fluorescent protein, **62,** 169, 550
Green fluorescent protein (GFP)
Aequorea victoria, **65,** 373
cell death assays Drosophila S2 cells, **66,** 335
cytochrome *C*, **66,** 373–375
fluorescence properties
aequorin, **61,** 114
fluorochrome structure, **61,** 114
GFP choice
confocal microscopy, **61,** 117
dual-receptor experiments, **61,** 117–118
microtubule analysis, high-resolution multimode digital microscopy, **56,** 186, 212–214
single-probe experiments, **61,** 117
photostability, **61,** 114
spectral mutations
blue-emitting mutant, **61,** 115
fluorescence resonance energy transfer, **61,** 115
GFP-UV variant, **61,** 115
multicolor fluorescence abilities, **61,** 115
S65T bright green mutant properties, **61,** 115
yellow mutant, **61,** 115
spectrum, **61,** 114–115
thermostability and folding mutations
posttranslational chromophore formation rates, **61,** 117
wild-type properties, **61,** 115, 117
transcript mutations, WT-GFP, **61,** 117
fusion markers, **65,** 278–281
fusion protein construction strategies
cloning PCR fragments
nucleotide buffers, **61,** 120–121
TA cloning, **61,** 121
cloning vectors

Green fluorescent protein (GFP) (*cont.*)
CMV immediate early promoter, **61,** 123
tetracycline-regulated vector system, **61,** 123
DNA polymerase
commercial sources, **61,** 120
fidelity *in vitro*, **61,** 120
mutational frequency, **61,** 120
DNA sequence fidelity assessment, **61,** 122
linker amino acids, **61,** 118
PCR protocol
agarose gel electrophoresis analysis and purification, **61,** 121
TA vector ligation, **61,** 121
primers
organization, **61,** 119–120
restriction site, **61,** 118–119
spacer codons, **61,** 118
target priming sequence, **61,** 118
recombinant PCR
principle, **61,** 122
requirements, **61,** 122
uses, **61,** 122
target protein properties, **61,** 118
template
concentration, **61,** 120
plasmids, **61,** 120
mammalian cell expression
stable transfection
FACS (fluorescence-activated cell sorting), **61,** 131
picking colonies, **61,** 130–131
selection, **61,** 129–130
transformation, **61,** 129
titering transfection reagents
protocol, **61,** 128
variables, **61,** 126–127
transient transfection
calcium phosphate transfection, **61,** 124
cationic lipid-mediated transfection, **61,** 124
cell plating densities, **61,** 124–125
detection times, **61,** 126
efficiency variables, **61,** 124
glass coverslip preparation, **61,** 123–124
lipofectamine transfection protocol, **61,** 125–126
verifying GFP fusion proteins
activity and localization, **61,** 131–132
biofluorescent dominant-negative mutant proteins, **61,** 133
phenotypic analysis, **61,** 133
visualization, **65,** 376
yeast microtubule and spindle dynamics quantitation
applications *in vivo*
dynamic behavior of microtubule, **61,**
cytoskeleton and spindle poles, **61,** 379
molecular level kinetic analysis, **61,** 379
protein function of spindle pole, **61,** 379–380
spindle pole organization, **61,** 380–381
astral microtubule assembly dynamic considerations, **61,** 370
C-terminal dynein-GFP probe, **61,** 370
difficulties, **61,** 369–370
image acquisition and processing, **61,** 378–379
imaging difficulties and solutions
acceptable signal to noise ratio, **61,** 378
confocal microscopy, **61,** 378
noise sources and reduction, **61,** 376–377
resolution and magnification, **61,** 377–378
imaging system
CCD camera, **61,** 375, 378
DIC and fluorescence modes filter wheel, **61,** 375
excitation filter wheel, **61,** 375
MetaMorph software, **61,** 376
multiple bandpass dichromatic mirror, **61,** 375
inducing production of GFP fusion protein
cell midlogarithmic growth phase, **61,** 372
induction times, **61,** 372
role of galactose, **61,** 372
protein–GFP fusion and promoter selection
choice of vector cassette, **61,** 371
endogenous promoters, **61,** 371
GAL1–10 promoter characteristics, **61,** 372
HIS3 promoter characteristics, **61,** 371, 372
induction levels and strategies, **61,** 372
integrative vector advantages, **61,** 371
low-level constituitive promoters, **61,** 371
MET promoter characteristics, **61,** 371–372
promoter choice, **61,** 371
saturation, **61,** 371
quantifying fluorescence
autofluorescence curves, **61,** 373
distribution curves, **61,** 373

laser scanning confocal microscopy, **61,** 373
optimum GFP fusion protein amount, **61,** 375
phenotypes, **61,** 375
promoter strength, **61,** 373
real-time quantitation, **61,** 373
single cell protein level measurements, **61,** 374–375
Green fluorescent protein-fusion constructs
primer design, **65,** 260–261
Green fluorescent protein-fusion proteins
expression, **65,** 281
localization, **65,** 281
Green fluorescent protein-Isp6p
fluorescence patterns, **65,** 280
Green fluorescent protein-labeled mitochondria
representative paths, **65,** 348–349
Green fluorescent protein markers
cells expressing
fixation, **65,** 281
Green fluorescent protein, microtubule analysis, high-resolution multimode digital microscopy, **56,** 186, 212–214
Green fluorescent protein, subcellularly targeted, **64,** 121
Green fluorescent proteins
Drosophila embryo confocal imaging, **70,** 375
pCa_i measurement, **70,** 460–461, 464
Green fluorophore, detector cross-talk in LSCM, **70,** 213, 217
Green helium neon laser, **71,** 151
Growth assays
fluorescent BrdU protocol, **66,** 221–222, 222
growth curves, **66,** 219–220
nonimmortal cells, **66,** 219–222
[^{3}H]thymidine incorporation, **66,** 220–221
Growth chamber, **65,** 343–344
Growth conditions, **71,** 6–7
Growth cone
formation, *Heliosoma* advantageous for study of, **71,** 158
motile *vs.* nonmotile, **71,** 159
motility
correlating with cell biological mechanisms, **71,** 157
in *Xenopus* cultures, **71,** 130
ultraviolet photolysis study of, **71,** 152–154
protrusive activity of, **71,** 3
rate of advance, as criterion for choice of neuronal culture, **71,** 2
responses, experimental studies of, **71,** 80
size, **71,** 8–9
as criterion for choice of neuronal culture, **71,** 2
Growth control, zebrafish fin, **59,** 301–303
Growth curves
growth assays, **66,** 219–220
Growth factor signaling, HS role, **69,** 84–85
Growth media
for γTuSC reconstitution from Sf9 cells, **67,** 13
for two-hybrid assay, **67,** 79
Growth medium, in dissociating chick ciliary ganglia, **71,** 46–48
Growth rate, neural-enriched cultures for study of, **71,** 143
Growth suppression assay, **66,** 464–465
GST, *see* Glutathione *S*-transferase
GTPases
cell-matrix adhesion assays, **69,** 378
extracellular matrix, **69,** 13–14
Guanylate cyclase, **62,** 38
Guidance cues, repulsive, **71,** 177
guidance cues, **71,** 2
exemplified in Tibia-1 pathway, **71,** 172
Guidance molecules
annulin, **71,** 182
G-innexin-1, **71,** 182–183
laminin, **71,** 181–182
REGA-1, **71,** 182–183
semaphorins, **71,** 171–181
GVBD, *see* Germinal vesicle breakdown
Gynogenesis, *see* Early pressure screens, zebrafish; Maternal-effect genes, gynogenesis-based screening in zebrafish

H

H148N green fluorescent protein, proton transfer, **58,** 13
P-Hydroxybenzylideneimidazolidinone, *see* chromophore, green fluorescent protein
HA-tagged proteins, **67,** 86–87
Haemanthus katherinae, see Spindle birefringence
Hair cells
AFM studies
lateral link effects, **68,** 155–159
measurements, **68,** 148–150
inner, organ of Corti, **68,** 144, 146
outer, *see* Outer hair cells
Hairy root cultures, **64,** 26
Halo–comet method, measurement of DNA damage, **64,** 263–266
Halogenated pyrimidines

Halogenated pyrimidines (*cont.*)
in cell proliferation research, **64,** 449–452
in vivo labeling of brain tumors, **64,** 462
Hamburger, Viktor, **71,** 68
Hamilton syringe, **61,** 226
Hamster brown adipose tissue mitochondria, **65,** 23–24
Hamster/zebrafish radiation hybrid
applications, **60,** 288, 298
construction
cell line preparation, **60,** 289–290
colony picking and hybrid growth, **60,** 291–292, 300–301
donor cell irradiation, **60,** 290
fusion of cells, **60,** 290–291
materials, **60,** 289
overview, **60,** 288
troubleshooting, **60,** 299–300
DNA extraction, **60,** 292
fluorescence *in situ* hybridization characterization
applications, **60,** 294–295
hybridization and washing, **60,** 297–298, 302
materials, reagents, and solutions, **60,** 295
metaphase spread preparation, **60,** 296–297
probe preparation, **60,** 295–296, 302
slide making, **60,** 297, 300–301
freezing, **60,** 292
genome mapping, **60,** 298–299
polymerase chain reaction characterization
agarose gel electrophoresis of products, **60,** 294, 301
amplification reactions, **60,** 293–294
DNA stock plate preparation, **60,** 293, 301
materials, **60,** 292–293
scoring of results, **60,** 294, 300
Handedness, *see* Patterning
Handling properties, **71,** 11–12
Hank's balanced salt solution, **71,** 71
in spinal and motor neuron cultures, **71,** 82
Haploid screen, zebrafish
behavioral screens, **60,** 54–55
embryo characteristics, **60,** 48, 51, 73–74
embryo production
fertilization, *in vitro*, **60,** 44, 46–47, 53
ultraviolet irradiation, **60,** 47–48
gamma-ray mutagenesis
blastula stage embryos, **60,** 60–62
characterization of gene mutations, **60,** 64–66
sperm, **60,** 62–63
spermatogonia, **60,** 63–64
translocations, **60,** 66–67
in situ hybridization screens, **60,** 55–56, 61, 68
limitations of haploids, **60,** 57, 59, 72–74
line selection
*AB-derived wild type, **60,** 52–53
AB wild type, **60,** 52, 62
clonal lines, **60,** 51–52
morphology screen, **60,** 53–54
mosaicism in F_1 screens, **60,** 59–60
overview, **60,** 44
phenotype compared to diploid, **60,** 56–57
polymerase chain reaction screen, **60,** 56, 68
Harvey, William, **71,** 68
HCMV
tumor necrosis factor-related ligands, **66,** 524
Heart development, cell line tracing, **59,** 359–364
analysis, **59,** 362–364
in vivo and *in situ* analyses, **59,** 362–363
field regulation, **59,** 363–364
introduction, **59,** 359–360
methodology, **59,** 360–362
cell cluster photolabeling, **59,** 361–362
dye filling of embryos, **59,** 360
Heat shock
altered DNA replication patterns after, **64,** 276–278
hsp70 localization after, **64,** 283
nuclear matrix stability assay after, **64,** 274–276
oxidative potential of intracellular milieu after, **64,** 280–281
Heat shock, gynogenesis induction, **60,** 5–6, 15
Heat treatment, for bacterial staining, **64,** 558
Heating assay, laser tweezers, **55,** 43–45
HEC cells, adhesion force measurements, **68,** 100–103
HEC/RL cells, culture on coverslips, **68,** 109
α-Hemolysin
AFM studies, **68,** 270
Staphylococcal, AFM studies, **68,** 270
HeK293 cells
maintaining, **71,** 397–398
producing adenovirus in, **71,** 397–400
transfection with adenovirus DNA, **71,** 398–399
HEK 293 cells
transfection, **66,** 504–505
HEK 293T cells
transfection, **66,** 504–505
HEK293 E1 cells, virus production, **69,** 415–416
HeLa cells
MIMS–EYFP, **65,** 375
HeLa cells, Ki-67 expression, **63,** 411–414
Heliosoma buccal ganglion neurons, **71,** 165
large size of, **71,** 8

Heliosoma saline, **71,** 160
Heliosoma trivolvis, 3
advantages of culturing, **71,** 159
conditioned medium cultures, **71,** 168–169
culturing neurons of, **71,** 157–169
developmental properties of, **71,** 5
dissociation of whole ganglia, **71,** 166
extracting single neurons from ganglia of, **71,** 164–167
freshwater aquaria required for, **71,** 13
initial dissection of, **71,** 160–163
maintaining animals, **71,** 159
nerve cell culture protocols, **71,** 163–169
polylysine-attached cultures, **71,** 168
speed of culture adhesion, **71,** 169
time-consuming nature of initial dissection, **71,** 163
Helium–cadmium lasers, **70,** 189
Helium neon laser, LSC with, **64,** 337
Helium–neon lasers
multicolor laser scanning confocal microscopy, **70,** 189
two-color LSCM, **70,** 204
helix, mutation effects in development, **60,** 81
Hemagglutinating virus of Japan, cell hybridization, **57,** 112, 114–115
Hematological malignancy, BrdUrd infusions, **64,** 464
Hematopoiesis, integrin-null ES cells, **69,** 320
Hemoglobin
accumulation pattern, **64,** 140
bivariate profiles
effect of TGF-β, **64,** 145–147
flow data, **64,** 143–144
isolation of fetal nucleated red cells, **64,** 144–145
problems and limitations, **64,** 149–151
study of sickle-cell erythropoiesis, **64,** 148–149
Hemopoietic progenitors
circulating, **64,** 155
committed, **64,** 154
maturation, flow cytometric monitoring, **64,** 155–156
progenitor subsets, assessment, **64,** 157–167
Hemostasis, zebrafish, analysis, **59,** 337–355
Heparan sulfate
enzyme treatment protocols, **69,** 94
FGF receptor interaction, **69,** 85–87
GAG-growth factor interaction, **69,** 93–94
growth factor signaling mechanism, **69,** 84–85
polypeptide growth factor regulation, **69,** 83–84
regulatory, *see* Regulatory heparan sulfate
structure, **69,** 102–103
sulfation, chlorate treatment, **69,** 90–91
Heparan sulfate-deficient cells, **69,** 94–95
Heparan sulfate-growth factor
chlorate treatment, **69,** 89–90
HS-deficient cells, **69,** 94–95
Heparin
GAG-growth factor interaction, **69,** 93–94
HS binding, **69,** 98–99
Heparin-binding growth factors, **69,** 84
Hepatocytes, micropatterned culture surface studies, **69,** 390
HEPES, as media buffer, **71,** 22
Heptane glue, **67,** 115–116
Herpesvirus saimiri, immortalization of cells, **57,** 74
Heterobifunctional reagents, in protein labeling, **63,** 195–201
Heterochromatin regions, DNA replication in, **64,** 282
Heterokaryon, **62,** 155–156, 190–191, 513–514
construction, **62,** 237, 258–261
strains (tables), **62,** 219–220, 270
Heterokaryon formation, **61,** 363, 366–367
Heteroplasmy, **65,** 313, 406, 409–410
Hexacyanoferrate, **65,** 112
Hexadecanethiol, **69,** 396–397
Hexagonally packed intermediate layer, AFM studies
bacterial pore single protomer, **68,** 287
native membrane protein conformation, **68,** 280–282
Hexidium iodide, gram-positive bacterial stain, **64,** 525–526
HFKs, *see* Human foreskin keratinocytes
High molecular weight DNA
isolation, **66,** 440
High-performance liquid chromatography analysis (HPLC), **61,** 288
High-power fields, and mitotic index, **64,** 362
High-pressure freezing, **65,** 250–252; **67,** 32
High Q filter cubes, **61,** 163
High-resolution atomic force microscopy
contact mode imaging, **68,** 260–261
tapping mode imaging, **68,** 261–263
High-resolution microscopy
multimode digital system, **56,** 185–214
applications, **56,** 209–214
anaphase, *in vitro*, **56,** 210–211

High-resolution microscopy (*cont.*)
digitally-enhanced differential interference contrast images, **56,** 209–210
fluorescence photoactivation, **56,** 211–212
green fluorescent protein–dynein-labeled microtubules, **56,** 186, 212–214
immunofluorescent specimens, **56,** 214
migrating tissue cells, **56,** 212
cooled charge-coupled device camera, **56,** 199–207
background reference image subtraction, **56,** 205–206
camera features, **56,** 199–200
cooling advantages, **56,** 203–205
dark reference image subtraction, **56,** 205–206
linearity, **56,** 203–205
noise, **56,** 203–205
quantum efficiency, **56,** 200–202
readout, **56,** 202–203
shading, **56,** 206
signal-to-noise ratio, **56,** 206–207
design criteria, **56,** 187–190
digital imaging system
computer control, **56,** 207–208
image analysis, **56,** 208–209
image processing, **56,** 208–209
software, **56,** 209
image formation, **56,** 155–161
bias retardation, **56,** 160–161
image features, **56,** 158–160
microscope design, **56,** 155–158
microscope components, **56,** 191–199
epi-illumination optics, **56,** 194–197
focus control, **56,** 198–199
imaging optics, **56,** 191–194
trans-illumination optics, **56,** 197–198
vibration isolation, **56,** 199
microscope selection, **56,** 168–179
analyzer, **56,** 171–173
beam-splitting prisms, **56,** 174
camera detector, **56,** 175–176
compensation methods, **56,** 174
condenser, **56,** 173–174
considerations, **56,** 168–169
filters, **56,** 171
illumination, **56,** 169–171
magnification, **56,** 175–176
objective, **56,** 173–174
polarizer, **56,** 171–173
specimen chambers, **56,** 175
specimen stage, **56,** 174
vibration isolation, **56,** 176
overview, **56,** 153–155, 161–167, 186–187
analog contrast enhancement, **56,** 162–165
digital contrast enhancement, **56,** 165–167
video basics, **56,** 161–162
performance test specimens, **56,** 179–183
diatoms, **56,** 179–180
embedded skeletal muscle, **56,** 181–183
squamous cheek cells, **56,** 180–181
perfusion chamber studies, **56,** 253–271
assembly, **56,** 270–271
chamber design, **56,** 256–260, 265–266
construction, **56,** 260–265, 267–269
historical perspectives, **56,** 256
overview, **56,** 253–256
temperature regulation, **56,** 269–270
video components
analog processor, **56,** 176–177
camera, **56,** 176–177
considerations, **56,** 168–169
digital image processor, **56,** 177–178
digital storage, **56,** 179
peripherals, **56,** 178–179
time–date generator, **56,** 178
video-enhanced differential interference contrast light microscopy, **56,** 153–183
High-spatial resolution
in *Heliosoma* studies, **71,** 157
in live-cell imaging, **71,** 151
High-temporal resolution, in live-cell imaging, **71,** 151
High-voltage electron microscope (HVEM), **65,** 252–253
Hippocampal neurons, **71,** 3
acquisition of, **71,** 113–117
advantages of, **71,** 18
culture characteristics, **71,** 119–121
culturing, **71,** 111–126
long-term culture methods, **71,** 121–126
organelle traffic in dendrites of, **71,** 8
oxidative stress on, **71,** 12
short-term culture methods, **71,** 117–121
Hirt supernatant, cDNA recovery, **69,** 216–217
HIS3 promoter, **61,** 371, 372
Histochemistry, **65,** 313–317
COX, **65,** 314–317, 326
methods, **65,** 326
mtDNA
COX 1 gene, **65,** 325
SDH, **65,** 314, 326

Histograms
bivariate, **64,** 301–303
bivariate DNA *vs.* BrdUrd, **63,** 364–366, 371
CD45/SSC, **64,** 347
describing intensity distributions, **63,** 41–43
DNA
and gating, **64,** 142
in bivariate profiles of hemoglobin, **64,** 144
DNA frequency, **63,** 534
FITC/PE, **64,** 490–491
fluorescence, **64,** 291–292
single- and two-parameter, **63,** 164–166
single-antibody histogram analysis, **64,** 289–290
tail moment, **64,** 246
three-dimensional plots and, **63,** 166
univariate, **63,** 245, 247
Histone acetyltransferases, **62,** 379–392
Histone H3
phosphorylation
during cell cycle, **63,** 343–344
monoclonal antibody, **63,** 350–353
plus antibody, as marker of mitotic cells, **63,** 344–345
plus cyclin A or B1, immunocytochemical detection, **63,** 345–346
Histone, associated DNA fragments in apoptosis, **57,** 261
Histones, **62,** 144–148, 379–381
acetylation, **62,** 148, 379–392
antibodies to, **62,** 380
gel analyses, **62,** 388–390
isolation, **62,** 382–385, 390–392
HIV infection
CLASSIF1 analysis, **64,** 503–506
data management and verification, **64,** 574–578
immunophenotyping
multicolor fluorescence, **64,** 578–582
T cell, **64,** 568–574
multiparameter flow cytometric analysis, **64,** 490
research applications
cell death, **64,** 583
immunophenotypic, **64,** 582–583
monitoring gene therapy, **64,** 584
HL-60 cells
confocal image, **66,** 122
HLA-DR/CD8, HIV immunophenotype, **64,** 503–504
HLA-DR, HP stained with, **64,** 158–160
Hoaglund's nutrient solution, **64,** 7–8
Hoechst 33342
cytotoxicity, **64,** 215, 217
DNA binding, flow cytometric analysis, **64,** 185–188
Hoechst dye, *Drosophila* embryo confocal imaging, **70,** 373
Holes
invadopodia assay, **63,** 623–625
produced by invadopodia, **63,** 619–620
Holey study
mitochondria, **66,** 365–391
Hollow fiber, cell culture scaleup, **57,** 223–224
Homemade microdissection device, **65,** 324
Homogeneity
desirable in PC12 cell lines, **71,** 282
of chick ciliary ganglion neurons, **71,** 38
Homogenization
mitochondria, **66,** 369–370
Homologous balancer chromosomes, **61,** 323, 336
Homologous double crossovers
mtDNA sequences, **65,** 388–389
Homologous recombination, **66,** 490–491
Homologous recombination
in BJ5183 *E. coli,* 394–397
in human kidney, **71,** 389
Homoplasmic-mutated clones, **65,** 410
Hood, animal cell culture
horizontal flow hoods, **57,** 4–5
laminar flow hoods, **57,** 5–6
placement in laboratory, **57,** 10
sterilizing, **57,** 6, 13–14
Horse serum, **71,** 77, 83, 122
Horseradish peroxidase, in Purkinje neuronal cultures, **71,** 96
Host range, **66,** 203
hox genes, comparative genome mapping and duplication, **60,** 152–153, 156–157
HP, *see* Hemopoietic progenitors
HPI layer, *see* Hexagonally packed intermediate layer
HPLC
ceramide, **66,** 147–149
HPLC elution profile
tryptic involucrin peptides, **66,** 128
HPLC, *see* High-performance liquid chromatography analysis
HPV, *see* Papillomavirus
Hrk
domains, **66,** 33
yeast two-hybrid, **66,** 36
HS *rho* petites, **65,** 423
HS, *see* Heparan sulfate

HSA, *see* Human serum albumin
Hsp, **62,** 61, 441, 450
Hsp72, **71,** 91
hsp70
  localization after heat shock, **64,** 283
  nuclear
    flow cytometric detection, **64,** 278
    immunofluorescence
      localization, **64,** 279–280
HT1080, recombinant laminin
    generation, **69,** 122–123
HT29 cells
  LIGHT-mediated killing, **66,** 502
[^{3}H]thymidine incorporation
  growth assays, **66,** 220–221
HTLV-1, *see* Human T-cell leukemia virus-1
Human cytomegalovirus (HCMV)
  tumor necrosis factor-related ligands, **66,** 524
Human disease, zebrafish system, **59,** 6–7
Human foreskin keratinocytes, **69,** 30–32
Human growth hormone, secretion assays
    using, **71,** 278–283
Human mitochondrial genome
  description, **65,** 312
Human mtDNA molecule
  description, **65,** 414
Human mtRNA
  core, **65,** 420
Human papillomavirus, *see* Papillomavirus
Human platelet mitochondria, **65,** 25–28
  biochemical parameters, **65,** 26
  coupled mitochondrial particles, **65,** 26–27
  crude mitochondrial membranes, **65,** 25–26
Human serum albumin, lateral force
    mapping, **68,** 133–134
Human T-cell leukemia virus-1, immortalization
    of cells, **57,** 73–74
Humatrix, preparation from myoepithelial
    tumors, **63,** 570–571
Humidity, fibrillar collagen AFM probe
    microscopy, **69,** 188
Humidity, SFM, DNA, organic
    solvents, **68,** 366–367
HVEM, **65,** 252–253
Hybridization
  Bcl-2 family cloning, **66,** 33–34
Hybridization, cells
  applications, **57,** 117–118
  history of techniques
    hemagglutinating virus of Japan, **57,** 112
    humanization of antibodies, **57,** 112–113, 127
    polyethylene glycol, **57,** 112
  hybridomas, *see* Hybridoma
  invertebrate cells, **57,** 197
  principles of techniques
    electrofusion, **57,** 116
    hemagglutinating virus of Japan, **57,**
      114–115
    polyethylene glycol, **57,** 115–116, 124
  selection of hybrid cells, **57,** 116
Hybridization master mix, **64,** 101
Hybridization, protein and gene expression,
    analysis, **59,** 68–71
Hybridoma
  applications of monoclonal antibodies
    human
      cancer diagnosis, **57,** 140–141
      light chain modification, **57,** 141–142
      multifunctional antibodies, **57,** 141
    mouse, **57,** 126–127
  definition, **57,** 118
  human–human hybridoma preparation
    cell fusion
      A_4H_{12} cells, **57,** 136
      efficiency, **57,** 132, 134–135
      electrofusion, **57,** 136–138
      HO-323 cells, **57,** 135
      NAT-30 cells, **57,** 135
      RF-S1 cells, **57,** 136
    immunization of lymphocytes *in vitro*
      cancer cell antigens, **57,** 130–132
      rationale, **57,** 129–130
      soluble antigens, **57,** 132–134
    immunization of lymphocytes using SCID
        mice, **57,** 134
    lymphocyte preparation
      cryopreservation, **57,** 129
      lymph node, **57,** 128–129
      peripheral blood, **57,** 128
    screening with enzyme-linked
        immunosorbent assay, **57,** 131–132,
        138–139
  mouse–mouse hybridoma preparation
    apparatus, **57,** 119–120
    fusion of lymphocytes and myeloma
        cells, **57,** 124–125
    immunization of mouse
      emulsion preparation, **57,** 121–123
      injection, **57,** 123
      overview, **57,** 121
    lymphocyte preparation, **57,** 123–124
    myeloma cell preparation, **57,** 124
    parent cell lines, **57,** 119
    reagents and medium, **57,** 120–121

screening with enzyme-linked immunosorbent assay, **57,** 125–126
Hybridomas
fusion and preparation, **69,** 42–43
Mab against integrin-associated proteins, **69,** 235–237
Hybridoma cells, **67,** 351–352
Hydrazine-linked beads
apoptosis
antidinitrophenyl antibodies, **66,** 131–132
Hydrocephalus, **71,** 291
Hydrodynamic drag, micromechanical measurement, **55,** 139–140
Hydrogen peroxide
submitochondrial particles, **65,** 82
Hydrolysis, and protein concentration, in protein labeling, **63,** 189–190
N-Hydroxysuccinimide ester
Plus maleimide, **63,** 197–198, 200
protein labeling, **63,** 192
Hydrophobic interaction chromatography, proteoglycans, **69,** 59–60
Hydroxyquinone analogues, **65,** 106
Hydroxyurea-arrested CHO cells, **67,** 260–261
Hydroxyurea, cell cycle synchronization, **57,** 240–241
Hygromycin, **66,** 207
Hygromycin B, **66,** 424, 426, 431
Hyperspectral imaging
plant senescence, **66,** 449
Hypersuppressive (HS) *rho* petites, **65,** 423
Hyperthermia, effects on cellular organelles, **64,** 270
Hypomorphic phenotypes, **62,** 545
Hypoxanthine guanine phosphorybosyl transferase, **65,** 400–401
Hypoxia, **66,** 211

I

I167T green fluorescent protein, excitation peaks, **58,** 142
IFI, *see* Integrated fluorescence intensity
IgG, *see* Immunoglobulin G
IgM, *see* Immunoglobin M
IHC, *see* Inner hair cells
IL-3 withdrawal assay
Bcl-2 family, **66,** 38–39
Illumination
arc epi-illumination, **63,** 52
basic intensity adjustments, **56,** 9–10
flow cytometric, **63,** 110, 112
fluorescence lifetime imaging, **56,** 289, 295
high-resolution multimode digital microscopy
epi-illumination optics, **56,** 194–197
Köehler illumination, *see* Köehler illumination
trans-illumination optics, **56,** 194–197
in microscopy, **63,** 90
light microscopy, **56,** 136, 138
particle, by flow cytometer, **63,** 20–26
power, flow cytometric, **63,** 114
pump–probe microscopy, **56,** 299–300
ratio imaging control, **56,** 230–231
source, for fluorescence microscopy, **64,** 74
stability, **64,** 75–76
video-enhanced differential interference contrast light microscopy, **56,** 169–171
Image acquisition
equipment, **65,** 344–345
microarray research, **65,** 441–442
Image analysis, invadopodia holes assay, **63,** 625
Image analysis software
Website, **65,** 442
Image-based cytometry, for fluorescent halo assay, **64,** 258, 261–262
Image formation, in confocal microscopy, **63,** 92–96
Image processing
microarray research, **65,** 442–445
Image processing, *see* Digital image processing; Video microscopy
Image transfer function, contrast enhancement, **56,** 101–104
Images, comet, analysis, **64,** 240–242
Imaginal discs
AO staining
Drosophila, **66,** 325–326
Imaging
3-D, ECM and ECM–cell interactions, **63,** 584–587
cytoplast MTs, **67,** 48–51
for electron microscopy, **67,** 34–35
multiphoton autofluorescence, **63,** 592
reflected light, **63,** 587–591
Imaging, advantageous in large neurons of grasshopper limb bud, **71,** 173
Imaging chambers, in *Xenopus* live cell cultures, **71,** 148–150
Imaging system
mitochondrial motility
schematic diagram, **65,** 345
Imaging, *see also* Confocal imaging

Imaging, *see also* Confocal imaging (*cont.*)
procedures, morphogenetic movements, **59,** 193–195
Imaging, *see* Confocal scanning laser microscopy; Fluorescence *in situ* hybridization; Fluorescence microscopy; Transmission electron microscopy
Imaging, *see* Fluorescence microscopy, Laser scanning confocal microscopy
IMM, **65,** 357–358
separation
potato mitochondria, **65,** 67
Immobilization antigens, **62,** 37
Immobilized centrosome recovery assay, for *Spisula* MT nucleation, **67,** 153–155
Immortalization, cell lines
gene delivery, **57,** 77–78
genes for transfection
adenovirus, **57,** 72–73
Epstein–Barr virus, **57,** 73
herpesvirus saimiri, **57,** 74
human T-cell leukemia virus-1, **57,** 73–74
oncogenes, **57,** 74–76
p53, **57,** 76–77
papillomavirus, **57,** 71–72
simian virus, **57,** 40, 70–71
invertebrate cells, transgenesis, **57,** 198–199
spontaneous immortalization, **57,** 70
T-cell immortalization by oncogene transfection
bone marrow extract preparation, **57,** 79
characterization of cells, **57,** 81–83
efficiency, **57,** 81
electroporation, **57,** 80–81
oncogene expression plasmid construction, **57,** 79
peripheral blood mononuclear cell preparation, **57,** 79–80
yolk lipoprotein preparation, **57,** 79
telomere length in immortalized cells, **57,** 84
testicular cells, *see* Testicular cell
tumorigenicity, **57,** 84
Immunization, *Dictyostelium* centrosomes, **67,** 350
Immunization, hybridoma production, *see* Hybridoma
Immunization, Mab identification of instructive ECM, **69,** 41–42
Immunization, with cognate antigen, **63,** 396
Immunoaffinity chromatography
integrin-associated proteins, **69,** 237–239
proteoglycans, **69,** 61–63
Immunoassay, **71,** 403
Immunoblotting, **71,** 11
Immunochemical detection, PCNA, **63,** 421–423
Immunocytochemical techniques, signaling molecules, **69,** 262
Immunocytochemistry
cytochrome *c*, **66,** 373
detecting migration and expression in B35 cells, **71,** 301–302
detection of cytosolic proteins in PC12 cell lines, **71,** 283
in intact and permeable PC12 cell lines, **71,** 283–285
in Tibia-1 pathway studies, **71,** 187
limitations in *Heliosoma* cultures, **71,** 169
with protein expression studies in *Drosophila,* 216
Immunodepletion, **61,** 401–402
Immunodepletion, spindle assembly proteins, **67,** 248–250
Immunoelectron microscopy, green fluorescent protein fusion proteins in fission yeast, **58,** 136–137
Immunofluorescence
analysis
membrane permeability, **65,** 152–154
cell-assembled ECM staining, **69,** 153
identification of microorganisms, **64,** 517–520
in situ, localization of hsp70, **64,** 279–280
indirect
Saccharomyces cerevisiae, **65,** 268, 272–274
intracellular, fixative solution affecting, **63,** 280–281
localization
mitochondrial DNA, **65,** 327
quantification of intracellular antigens, **63,** 12
synthetic matrix, **69,** 150–151
Immunofluorescence microscopy
antibodies
double labeling, **57,** 321
handling, **57,** 319
incubation protocol, **57,** 329–330
primary antibodies, **57,** 319–320
secondary antibodies, **57,** 320–321
titration, **57,** 319
cameras, **57,** 314, 323–324
cell culture, **57,** 314–315
cell line authentication
glioma markers for lineage and phenotype, **57,** 155–158

species-specific
immunofluorescence, **57,** 206–207
centrosome proteins, **67,** 317–318
counterstaining, **57,** 324, 330
Dictyostelium centrosomes, **67,** 342–343
direct *vs.* indirect approach, **57,** 314
fluorophores
commonly used probes, **57,** 321–322
filter sets, **57,** 321–322
green fluorescence protein, **57,** 322–323
focal planes, **57,** 325
γ-tubulin antibody, **67,** 182–183, 185
γ-tubulin effects on MT nucleation, **67,** 187
microscope adjustments, **57,** 323
morphology of cells, **57,** 325
mounting cells, **57,** 324–325, 331
MT nucleation, **67,** 171
pericentrin staining, **67,** 55–56
plant γ-tubulin, **67,** 206–207
protein kinases, **67,** 306–308
sample preparation
antigen preservation, **57,** 315–317
epitope shielding, **57,** 318–319
fixation
buffers, **57,** 326
fixative selection, **57,** 316–318
formaldehyde, **57,** 316–317, 327–328
glutaraldehyde, **57,** 316–317, 328–329
methanol, **57,** 316–317, 327
stock solutions, **57,** 327
permeabilization of cells, **57,** 318, 327–329
simian virus 40 large T antigen, **57,** 99–100
Immunofluorescence microscopy, mitotic
cells, **63,** 347–349
Immunofluorescence microscopy, *see*
Fluorescence microscopy
Immunofluorescence, *see* Microscopy, Light
Immunofluorescent staining
fixed yeast cells, **65,** 269–274
Saccharomyces cerevisiae, **65,** 265–266
Immunofluorescence staining techniques
background staining, **70,** 184–185
fluorophore visualization specificity, **70,** 181
overview, **70,** 178–180
primary antibody specificity, **70,** 183–184
secondary antibody specificity, **70,** 181–183
Immunoglobulin
IgG
blocking of Fc receptors prior to
staining, **63,** 229–232
structure, **63,** 273
removal by washing, **63,** 233
Immunoglobulin G, **70,** 181–183
Immunoglobin M, cryo-AFM
imaging, **68,** 249–250
Immunogold, **70,** 272–273
Immunogold EM, *in situ* localization of proteins
protocol, **61,** 65–66
reagents, **61,** 63–65
Immunogold, *see* Microscopy, Electron
Immunohistochemistry
compared with flow cytometry, **64,** 315, 340
methods, **65,** 326–328
single antigen assay, **64,** 314
Immunohistochemistry, Purkinje neuronal
cultures, **71,** 97–98
Immunolabeling, yeast cells, **67,** 35–36
Immunolocalization
COX
KSS, **65,** 319
mitochondrial DNA
respiratory chain, **65,** 317–318
mitochondrial proteins
ABC method, **65,** 327–328
paraffin-embedded samples, **65,** 319–321
nuclear DNA
respiratory chain, **65,** 317–318
Immunolocalization of probes, protein and
gene expression, analysis, **59,** 71–73
Immunolocalization, phosphotyrosine, **63,** 617
Immunoperoxidase, **65,** 319
Immunoperoxidase labeling, paraffin-embedded
tissue sections, **67,** 56–57
Immunophenotyping
by LSCM, **64,** 316–327
detection of MRD, **64,** 388–397
leukemia and lymphoma, **64,** 313–316
malignant cells, **63,** 224–225
multicolor fluorescence, **64,** 578–582
problems during, **63,** 438
T lymphocytes, for HIV, **64,** 568–574
with multiple antibodies, **63,** 218
with single color, **63,** 228
Immunoprecipitation, **62,** 449–453
cell-matrix adhesion assays, **69,** 370–372,
375–376
Cox II
in organello translation, **65,** 220–222
DNA-binding proteins, **53,** 343, 345
for pericentrin enrichment, **67,** 64
instructive ECM, **69,** 47
integrin-associated
proteins, **69,** 234–235, 239–241
RNA-associated proteins, **53,** 574–575

Immunoprecipitation (*cont.*)
SPB components
antibody couplings, **67,** 84
cell lysis, **67,** 85–86
controls, **67,** 89
gene fusion construction, **67,** 82–84
HA- and Myc-tagged proteins, **67,** 86–87
protein detection, **67,** 88
with polyclonal antibodies, **67,** 87
Immunoprecipitation, probing antibody specificity, **63,** 290–292
Immunoprecipitations
Drosophila S2 cells, **66,** 336–337
TGase-interacting proteins, **66,** 119
tumor necrosis factor-related ligands, **66,** 515–516
Immunoscreening, for centrosome-associated protein cDNA, **67,** 128–129
Immunostaining
olivary nucleus
MERRF, **65,** 320
yeast mitochondria, **65,** 44
Import machinery component
radiolabeled precursor protein
cross-linking, **65,** 227–228
Importin
nuclear protein import roles, **53,** 518–519
purification of recombinant proteins
importin α, **53,** 530–531
importin β, **53,** 531–532
types, **53,** 518
IMS
reporter targeting, **65,** 355–356
In organello translation
cox II
immunoprecipitation, **65,** 220–222
products
plant mitochondria, **65,** 70–71
In situ hybridization (ISH)
to mtDNA, **65,** 321–322, 328–329
MERFF, **65,** 323
In situ end labeling (ISEL), **66,** 394, 402–406
materials, **66,** 403–404
protocol, **66,** 404–405
In situ hybridization
double-fluorescent, **59,** 77–78
to RNA, protein and gene expression, analysis, **59,** 63–64
two-color, protein and gene expression, analysis, **59,** 73–77
In situ hybridization, *see also* Fluorescence *in situ* hybridization
applications, **57,** 334–335
autoradiography, **57,** 339–340, 342
combining with *in situ* ligand binding, **57,** 348–349
hybridization conditions, **57,** 338–339, 341–342
posthybridization treatment, **57,** 339, 342
prehybridization of cells, **57,** 337–338, 341
principle, **57,** 334
probes, types and labeling, **57,** 338, 341
reagent and buffer preparation, **57,** 340
RNase inhibition, **57,** 335
tissue preparation and sectioning, **57,** 335, 337, 340–341
In situ hybridization, fluorescence, **63,** 67–68, 70, 78
In situ ligand binding
applications, **57,** 335, 342–343
autoradiography, **57,** 345, 347
combining with *in situ* hybridization, **57,** 348–349
hybridization conditions, **57,** 343, 345, 347
interpretation of results, **57,** 347–348
posthybridization, **57,** 345, 347
prehybridization, **57,** 347
principle, **57,** 334–335
probe labeling, **57,** 343, 346
reagent and buffer preparation, **57,** 345–346
sample processing, **57,** 343
tissue sectioning, **57,** 346
In situ ligation of labeled DNA fragments (ISLG-LDG), **66,** 406–407
In vitro transcription
preprotein synthesis, **65,** 193–194
In vitro translation
preprotein synthesis, **65,** 194–196
ammonium sulfate precipitation, **65,** 195–196
in vitro apoptosis
assays, **66,** 171–181
cell line models, **66,** 417–432
in vitro cleavage
DFF45, **66,** 301–302
PARP-1, **66,** 296–299
apoptotic cell extracts, **66,** 297–298
automodification reaction, **66,** 297
DEVDase detection, **66,** 298–299
inhibition assay, **66,** 298
purified caspases, **66,** 298
in vitro kinase assay
JNK activation, **66,** 190–192
In vitro neuron morphology, **71,** 97
In vivo neuron morphology, **71,** 97

In vivo protein dynamics, *Drosophila,* 227–228
In vivo therapies, potential applications for, **71,** 337
Inclusion bodies
caspase purification, **66,** 12–14
Inclusion bodies, centrosome-associated proteins, **67,** 132
Incubation
of fertilized chicken eggs, **71,** 69–71
postinjection with retroviral vectors, **71,** 382
Incubation, cell attachment assays, **69,** 287
Incubation, with γ-tubulin antibody *in vitro*, **67,** 181–183
Incubator, animal cell culture, **57,** 6–7, 52
Index of refraction, dielectric sphere force effects, single-beam gradient laser trap, **55,** 19–20
India strain, gene mapping, **60,** 358
Indirect immunofluorescence
mammalian centromere–kinetochore complex
CREST anti-sera
collection, **61,** 59
titering, **61,** 60
cytocentrifugation, **61,** 61
double- and triple-staining techniques
antibody cross-reaction avoidance, **61,** 63
image processing software, **61,** 63
mitotic shake-off, **61,** 60
staining of cell monolayers
protocol, **61,** 62–63
reagents, **61,** 61–62
plant centromere–kinetochore complexes, **61,** 77–78
Saccharomyces cerevisiae, **65,** 268, 272–274
Indirect immunofluorescence microscopy
Dictyostelium centrosomes, **67,** 342–343
protein kinases, **67,** 306–308
Indirect immunofluorescence, *Tetrahymena thermophila*
overview, **70,** 338
reagents, **70,** 340–342
sample preparation, **70,** 342–344
Indium-tin oxide
AFM of protein adsorption, **68,** 15
bacteriophage STM, **68,** 5
Indo-1, **66,** 235
Indo-1, measurement of intracellular Ca, **63,** 520, 522
Induced proximity model, **66,** 454
Infection methods, *vs.* transfection methods, **71,** 370
Infection protocol
MEFs, **66,** 205–210
Infinity lens systems, confocal digital imaging, **70,** 254–256
Infinity space, light microscopy, **56,** 139
Inhibitors
acid sphingomyelinase, **66,** 152
aspartic proteinase, **66,** 251
cell-permeable, **66,** 256t–258
ceramide synthase, **66,** 157
cysteine proteinase, **66,** 251
metallo-proteinase, **66,** 252
neutral sphingomyelinase, **66,** 154–155
protein kinase, **66,** 368
protein synthesis, **66,** 368
proteinase, **66,** 249–259
serine proteinase, **66,** 250
sphingomyelin synthase, **66,** 158–159
topoisomerase, **66,** 368
Inhibition, of high-affinity glutamate transport, **71,** 102–105
Initialization, instrument, for lifetime measurements, **63,** 141
Initiator caspases, **66,** 4, 308, 454
Injection
centrosome, and egg activation, **67,** 215
for *Drosophila* centrosome analysis, **67,** 118–120
microinjection
γ-tubulin antibody, **67,** 184–185
MTs in cytoplasts, **67,** 47–48
Injection, cells into center of sheath stream, **63,** 28–30
Injection sample, preparation of, in *Xenopus* cultures, **71,** 131–133
Injection volume, calibrating, **71,** 135
Injury models, chick motor neurons as basis for, **71,** 69
Inner hair cells, organ of Corti, **68,** 144, 146
Inner mitochondrial membrane (IMM), **65,** 357–358
separation
potato mitochondria, **65,** 67
Inner ring, Aufderheide rotocompressor device, **70,** 350
Inositol 1,4,5-trisphosphate, **62,** 38
Inositol phospholipids, **62,** 37
Insect cells
baculovirus
tumor necrosis factor-related ligands, **66,** 503–504
Insect cells, recombinant proteins, pCOCO characterization, **69,** 77

Insect cells, recombinant proteins, pCOCO (*cont.*)
expression and purification, **69,** 76–77
protein production, **69,** 74–75
recombinant baculovirus, **69,** 75–76
Insect flight muscle mitochondria, **65,** 24
Insect Sf9 cells
recombinant baculovirus expression, **66,** 221
Insert sizes, retroviral vectors, **71,** 371–378
Insertional mutagenesis, **66,** 487–490, 489
Instructive extracellular matrix
definition, **69,** 28
Mab approach, **69,** 29–32
Mab identification
antigens and immunization, **69,** 41–42
cloning, **69,** 47
fusion and hybridoma preparation, **69,** 42–43
initial characterization, **69,** 47
primary screens, **69,** 43–44
purification, **69,** 47–48
secondary screens, **69,** 44–47
purification problems, **69,** 28–29
reporter cell responses
adhesion assays, **69,** 37–40
characterization, **69,** 48–49
coculturing, **69,** 37
deposited ECM, **69,** 35–36
ECM testing, **69,** 34–35
overview, **69,** 33–34
soluble ECM, **69,** 36–37
Instrumental noise, **55,** 147–152
amplifier noise, **55,** 59–60, 151–152
backreflections, **55,** 60, 152–153
beam-pointing fluctuations, **55,** 59, 62, 153
Johnson noise, **55,** 149–151
power fluctuations, **55,** 58, 152
shot noise, **55,** 58–59, 149
vibrations, **55,** 58, 154
Instrumentation
Boyden chamber, **63,** 554–555
comet assay (Olive methods), **64,** 257–258
flow cytometer
improvements, **64,** 512
sensitivity for bacteria, **64,** 547–548
verification, **64,** 293–295
flow cytometry-based fluorescent halo assay, **64,** 262–263
fluorescence lifetime flow cytometry, **63,** 138–141
for chromosome translocation study, **64,** 102
for PRINS, **64,** 57–58
for QDFM, **64,** 37–38
historical overview, **63,** 4–9
image-based cytometry, **64,** 261
immunophenotyping of malignancy, **64,** 314
nuclear matrix stability assay, **64,** 275–276
performance in detecting low levels of fluorescence, **63,** 316–317
Intact cells
superoxide anion production assay, **66,** 311
Intact cells, binding force measurement, **68,** 178–179
Intact tissues, fluorescence microscopy
problems, **70,** 303–304
solutions, **70,** 304–308
Integrated cytochrome *c* oxidase
KCN titration, **65,** 123–124
Integrated fluorescence intensity, telomeres, **64,** 73
quantification algorithm, **64,** 88–91
segmentation algorithm, **64,** 80–82
validation of method, **64,** 92
Integrators, timing and performance, **63,** 153–154
Integrin adhesion molecule, gene mutations
cell derivation, **69,** 315–317
overview, **69,** 314–315
Integrin-associated proteins
cell labeling, **69,** 239
cell preparation for flow cytometry, **69,** 241
immunoaffinity purification, **69,** 237–239
immunoprecipitation, **69,** 234–235, 239–241
integrin role, **69,** 360–361
monoclonal antibodies, **69,** 235–237
Integrin binding assay, microtiter-based
buffers, **69,** 252
data, **69,** 253–254
general considerations, **69,** 250–251
method, **69,** 252–253
Integrin ligand binding pocket, affinity cross-linking, **69,** 248–250
Integrin-mediated adhesion assay, signal transduction studies
adhesion assay, **69,** 364–365
cell lysates, **69,** 366–370
cell preparation, **69,** 363–364
coimmunoprecipitation, **69,** 375–376
controls, **69,** 365–366
drug inhibitors, **69,** 376–378
GTPases, **69,** 378
immunoprecipitation, **69,** 370–372
kinase assays, **69,** 375
matrix-coated plates, **69,** 362–363

molecular genetics, **69,** 379–382
phospho-specific antibodies, **69,** 374–375
Western immunoblotting, **69,** 372–374
Integrin receptor–fibronectin peptide ligand interaction
$\alpha_5\beta_1$ immobilization, **69,** 179–181
bioartificial membrane, **69,** 177–179
force measurements, **69,** 181–182
Integrins
affinity chromatography, lectin column, **69,** 234
cell functions, **69,** 359–360
cell–matrix adhesion assays, **69,** 361
cell membrane-anchored wild-type, **69,** 225–227
characterization, **69,** 282
coregulators, **69,** 360–361
extraction from tissues and cells, **69,** 233
ion-exchange chromatography, **69,** 232–233
ligand binding, SPR, **69,** 254–256
null ES cells, **69,** 320
recombinant soluble, production, **69,** 227–228
redox status analysis, **69,** 256–258
signaling, **69,** 13, 360
soluble, affinity chromatography, **69,** 228–233
Tag sequences, **69,** 232
VLA-4, cell-matrix adhesion, **69,** 18
$\alpha_v\beta_3$ Integrin, adhersion force measurement, **68,** 182–184
αVβ3 Integrin
cell–matrix adhersion, **69,** 17–18
RGD-containing ECM proteins, **69,** 17
α3 Integrin-null mouse, **69,** 311–313
$\alpha3\beta1$ Integrin-null cells, **69,** 319–320
α4 Integrin-null fibroblasts, **69,** 320
α5 Integrin-null ES cells, **69,** 320
β1 Integrin-null ES cells, **69,** 317–318
β1 Integrin-null somatic cell lines, **69,** 318–319
Intensified cameras
features, **56,** 37–39
camera types, **56,** 39
intensifier formats, **56,** 38–39
low-light microscopy
advanced technology, **56,** 62
dynamic range, **56,** 58–59
geometric distortion, **56,** 60
overview, **56,** 46–47
sensitivity, **56,** 47–57
charge-coupled cameras compared, **56,** 56–57
definition, **56,** 47–48
high sensitivity enhancement, **56,** 49–54
signal-to-noise ratio, **56,** 47–49, 54–56
shading, **56,** 60
spatial resolution, **56,** 59
temporal resolution, **56,** 60
usability, **56,** 61
Intensity
CFSE staining, appropriate levels, **63,** 379–380
distributions, histograms describing, **63,** 41–43
fluorescence
as indicator of substrate concentration, **63,** 180
standardization, **63,** 319
laser beam illumination, **63,** 23–24
Interference contrast microscopy, *see* Video-enhanced differential interference contrast light microscopy
Interference filters, short- and long-pass, **63,** 117–118
Interference reflection contrast imaging, **70,** 281
Interference reflection microscopy, **63,** 607
Interference, optical, **63,** 116
Interleukin-3 (IL-3) withdrawal assay
Bcl-2 family, **66,** 38–39
Intermediate filaments, **62,** 443
Intermediate filaments, scanning transmission electron microscopy, **53,** 159
Intermediate-voltage electron microscope (IVEM), **65,** 253
Intermembrane mitochondrial proteins
fluorescence microscopy
membrane permeability, **65,** 151
Intermembrane space (IMS)
reporter targeting, **65,** 355–356
Intermolecular forces, *see also* Gradient trap force, Optical force microscopy
measurements, **55,** 101–104
Internal eliminated sequence (IES), **62,** 134–138
Internal granule cell layer, Purkinje neuronal cultures, **71,** 90
Internally redundant genes
yeast arrays, **65,** 447
International Society for Analytical Cytology, **70,** 485
Internet resources, *see* World Wide Web resources, Zebrafish
Interphase spindle pole bodies, **67,** 171–173
Interstitial connective tissues
ECM, **63,** 563
reconstituted ECM, preparation, **63,** 571–572
Interval analysis, estimating number of cells per division, **63,** 388

Intestinal submucosa
as 3-D ECM substrate, **63,** 572–576
intact, preparation, **63,** 567–568
reconstituted interstitial ECM prepared from, **63,** 571–572
Intracellular calcium
genetically engineered [Ca^{2+}] indicators
calibration, **70,** 468–469
expression, **70,** 464–465
measurements, **70,** 467–468
overview, **70,** 460–461
pitfalls, **70,** 469–470
targeting, **70,** 461–464
living cells, confocal microscopy, **70,** 417–418, 426
overview, **70,** 453–455
perfused afferent arteriole, **70,** 423–425
singe-cell measurement
prostate adenocarcinoma, **66,** 241–242
visible-light [Ca^{2+}] indicators, **70,** 455–460
Intracellular events, as criteria for neuronal culture choice, **71,** 5–8
Intracellular membranes, **62,** 347
Intracellular pH, **65,** 372
Intracellular pH, confocal microscopy
BCECF measurement, **70,** 418–420
cortical collecting tubule, **70,** 422–423
future development, **70,** 426
overview, **70,** 417–418
Intracellular pools
calcium flux, **66,** 234
Intracellular protein localization, neural-enriched cultures for study of, **71,** 143
Intracellular signalling, in B35 cells, **71,** 291–292
Intramolecular forces, force spectroscopy
protein folding, **68,** 15–19
protein modular structure, **68,** 15
Intrasegmental epithelium, in Tibia-1 pathway studies, **71,** 174
Introns, **62,** 150–151
Invadopodia
activity, as measurement for ECM degradation, **63,** 615–619
holes assay, **63,** 623–625
holes produced by, **63,** 619–620
Invasion assay
collagen, **69,** 203–204
multicomponent ECM, **69,** 198, 201
Invertebrate cell culture
buffering system, **57,** 194
feeding layers, **57,** 193–194
feeding schedules, **57,** 195
gas phase, **57,** 194
hybridization of cells, **57,** 197
importance, **57,** 188–189
intact organisms, **57,** 197
media
additives, **57,** 191–192, 196–198
selection, **57,** 190–191
nonadherent cell and fragment rejection, **57,** 193
passaging
cell harvest, **57,** 196–197
passaging without splitting, **57,** 197
proliferation encouragement, **57,** 196
scoring of culture conditions, **57,** 195
single-cell *vs.* tissue fragment preparation, **57,** 193
species
confirmation of origin, **57,** 198
selection, **57,** 190
success by species, **57,** 188–189
sterile culture cleanup, **57,** 192
temperature optimization, **57,** 194
transgenesis for immortalization, **57,** 198–199
trypsinization in culture maintenance, **57,** 197
Inverted microscope, required for transfection, **71,** 311
Involucrin peaks
TGase 1 enzyme
HPLC separation, **66,** 126
Iododeoxyuridine, **64,** 449–450
Iodoodeoxyuridine, *see* DNA replication site, Mapping *in situ*
Ion channel function, in *Drosophila* studies, **71,** 196
Ion channels, **62,** 363–375
Ion channels, voltage-sensitive, AFM–path–clamp combination, **68,** 12–13
Ion-exchange chromatography
Apaf-1 purification
Mono Q Column, **66,** 22
Ion-exchange chromatography, integrins, **69,** 232–233
Ionizing radiation
contact effect, **64,** 217–218
inhibition of DNA loop rewinding, **64,** 261–262
producing strand breaks, **64,** 245
Ionizing radiation, mutagenesis via, **71,** 198
Iontophoresis, **61,** 433
IPTG, **66,** 343
Irradiation source, zebrafish eggs, **59,** 49–50

ISAC, *see* International Society for
Analytical Cytology
ISEL, **66,** 394, 402–406
materials, **66,** 403–404
protocol, **66,** 404–405
ISEL, apoptosis assay, **57,** 255–256
ISH
to mtDNA, **65,** 321–322, 328–329
MERFF, **65,** 323
ISH, *see In situ* hybridization
ISLG-LDG, **66,** 406–407
Isoenzyme phenotyping, cell lines, **57,** 207–209
Isolated cytochrome *c* oxidase
KCN titration, **65,** 125–126
Isolated mitochondria
cytochrome *c* release, **66,** 376–377
Isolated neurons, culturing with
Xenopus, 140
Isolation
fetal nucleated red cells: bivariate
profiles, **64,** 144–145
nuclear matrix protein, **64,** 273
nuclear protein, **64,** 271–272
Isometric plot, two-parameter
histograms, **63,** 164–166
Isopropylthiogalactoside (IPTG), **66,** 343
Isopycnic gradient banding, **61,** 75
Isotype control, in Ki-67 protein assay, **63,** 410
Isozymes, **62,** 15
ITO, *see* Indium-tin oxide
IVEM, **65,** 253
IVEM tomography, for *Spisula* MT
nucleation, **67,** 155–156

J

J774A.1 macrophage
phagocytosis, **66,** 359
janus, developmental role, **60,** 3
JAR cells
adhesion force measurements, **68,** 100–104
culture on cantilever, **68,** 109–110
JC–1, **65,** 88
apoptosis
LSC, **66,** 82
LSC
apoptosis, **66,** 82
mitochondrial membrane potential,
65, 292–293
JC-1, mitochondrial dye, **64,** 119–120
JC-1 probe
applications, **63,** 478–479
detection of variations in $\Delta\Psi$, **63,** 469
disadvantages of, **63,** 474
green and orange fluorescence, **63,** 476–477
staining with, **63,** 472, 477–478
JNKs. *see* C-Jun N-terminal kinases
Johnson noise, **55,** 149–151
Jun, immortalization of cells, **57,** 74–75
Jurkat cell line, **66,** 418, 420t–423t,
424–426, 425
flow cytometry, **66,** 61f–62
transfection, **66,** 424, 426

K

K-antigens, **62,** 442
Kalman averaging, **70,** 54
Karlp, **67,** 89–92
Karyonides, **62,** 9, 134, 228
Karyoskeletal protein-enriched fraction
biochemical composition of *Drosophila*
embryo fraction, **53,** 29–30
heat stabilization, **53,** 30–31
morphology, **53,** 3, 24, 29
nucleic acid interactions during interphase
and mitosis/meiosis, **53,** 413
preparation
nuclease treatment, **53,** 15–16, 27
purification of nuclei
chicken liver, **53,** 13–14
Drosophila, **53,** 25–27
frog liver, **53,** 13–15
guinea pig liver, **53,** 12–13
rat liver, **53,** 6–12
subfractionation, **53,** 16–18, 27–30
Karyotyping, flow cytometric, **63,** 12
KCN
endogenous respiration inhibition, **65,** 124
KCN titration
COX activity
intact cells, **65,** 122–123
integrated COX activity, **65,** 123–124
isolated COX activity, **65,** 125–126
Kearns-Sayre syndrome (KSS), **65,** 312, 316
COX
immunolocalization, **65,** 319
Keratin filaments
Xenopus oocytes, **70,** 412
Xenopus oocytes and embryos, **70,**
386–388
Keratinocyte-cornified envelope isolation
ε-(γ-glutamyl)-lysine cross link
density, **66,** 116–117

Keratinocytes
immortalized, derivation, **69,** 313–314
reporter, instructive ECM studies with Mab, **69,** 29–32
Kernel, CSU-10 neighborhood-based processing, **70,** 107
KFs, *see* Keratin filaments
Ki-67
analysis in
fresh cytological specimens, **64,** 431–432
histological tissue sections, **64,** 432
antibody, observed in nucleus, **64,** 433–434
proliferation marker, **64,** 364
Ki-67 protein
and cell surface antigen, simultaneous staining, **63,** 407–408
expression in HeLa cells, **63,** 410–414
expression regulation, **63,** 401
MAbs against, **63,** 402
marker in neoplasms, **63,** 400
Kidney cells, monkey, infected, AFM studies, **68,** 8–9
Kidney epithelial cell lines, derivation, **69,** 311–313
Kinase assay, cell–matrix adhesion assays, **69,** 375
Kinesin, **62,** 65, 495
Kinesin, low-background total internal reflection fluorescence microscopy
ATP turnover, **55,** 125–127
movement observation, **55,** 123
Kinesin, single-molecule measurements with green fluorescent protein
advantages, **58,** 64–65
assays, **58,** 65–67
overview, **58,** 49–51
processivity, **58,** 65
Kinesins, PFM studies
basic structure, **68,** 24–25
mechanical behavior, **68,** 26
mechanical properties, **68,** 25–27
Kinetics
L-aminopeptidase, **63,** 77–78
antibody–antigen reaction
in permeabilized cells, **63,** 282–287
in solution, **63,** 281–282
antibody binding, **63,** 220–224
cell cycle, estimation by BrdUrd incorporation, **63,** 357–372
enzyme, time application, **63,** 180
parameter estimation, **63,** 366–369
Kinetics, ligand–receptor recognition, dynamic force microscopy, **68,** 125
Kinetics, tumor cell, **64,** 361
relation to tumor biology, **64,** 365–366
Kinetids, **62,** 52
Kinetochore activity assays *in vitro*, in *S. cerevisiae*
band shift assay
band shift gels, **61,** 152
CEN 3 DNA labeling by T4 DNA kinase, **61,** 152
CEN 3 probe isolation, **61,** 151–152
kinetochore-binding reaction
reduction of nonspecific protein binding, **61,** 153
salt concentration sensitivity, **61,** 152
tracking dye, **61,** 153
mechanism, **61,** 146
microtubule-binding assays
GDP *vs.* GTP binding, **61,** 151
microtubule end *vs* lattice binding, **61,** 151
microtubule preparation
capped microtubule preparation, **61,** 149–150
GMPCPP polarity-marked microtubule preparation, **61,** 150
polarity-marked microtubule preparation, **61,** 149
preparation of carrier DNA, **61,** 148
preparation of CEN 3 beads
coupling of CEN DNA to beads, **61,** 147–148
DNA manufacture protocol, **61,** 146–147
yeast extract preparation, **61,** 150–151
Kinetodesmal fiber, **62,** 53, 57
KNAT3–green fluorescent protein, fusion protein studies in *Arabidopsis thaliana*, **58,** 148
Köehler illumination
alignment, **56,** 142–143
benefits, **56,** 143
contrast, **56,** 143–146
high-resolution multimode digital microscopy, **56,** 194, 198
overview, **56,** 140–142
resolution, **56,** 143–146
Krypton–argon ion lasers
air-cooled examples, **70,** 228–231
double-labeled specimen imaging, **70,** 225–228
fluorophore visualization specificity, **70,** 221–225
multicolor LSCM, **70,** 188
three-color LSCM, **70,** 206–207

two-color LSCM, **70,** 204–206, 225
Krypton ion lasers, **70,** 188
KSS, **65,** 312, 316
COX
immunolocalization, **65,** 319
Kymographs, **61,** 166–167

L

L-15-based media, **71,** 27, 34
for chick forebrain cultures, **71,** 52, 61
for peripheral neurons, **71,** 22–23
in *Heliosoma* cultures, **71,** 161
in studies of spinal and motor neuron cultures, **71,** 82
Labeled inhibitor analysis
caspases, **66,** 93–97
data analysis, **66,** 96–97
flow cytometry, **66,** 95
LSC, **66,** 95–96
materials, **66,** 95
Labeling
antibodies, FLIM, **69,** 266–267
blue fluorophores, **70,** 177–178
BODIPY, **70,** 171–173
cell, integrin-associated proteins, **69,** 239
cells *in vitro* and *in vivo*
critical aspects, **63,** 369
methods, **63,** 360–361
cyanine fluorphores, **70,** 176
fluorescein, **70,** 171
γTuSC and γTuRC, **67,** 21–23
immunoperoxidase, **67,** 56–57
Lucifer yellow, **70,** 173
overview, **70,** 167–171
phthalocyanine derivatives, **70,** 176–177
phycobiliproteins, **70,** 174–176
radiolabeling, proteoglycans, **69,** 57
rhodamine fluorophores, **70,** 173
target proteins, **69,** 262
thick tissues, **70,** 315–321, 330–331
with CFSE, **63,** 378–379
yeast cells, **67,** 35–36
Labeling index, **63,** 355–356
^{3}H-thymidine, *see* ^{3}H-Thymidine labeling index
as surrogate measure of dynamic proliferation, **64,** 473
in clinical studies of tumor proliferation, **64,** 456, 459, 462–463
in vivo halogenated pyrimidine, **64,** 450–451
proliferation in human tissues and tumors, **64,** 470–471
quantification of biomarker expression, **64,** 449
Labeling index, cell cycle phase duration determination, **57,** 234–235
Labeling, antibody, membrane protein AFM imaging, **68,** 264–265
Laboratory design, animal cell culture, **57,** 10–11, 52–53
lac repressor, visualization of large-scale chromatin organization
applicability to different species, **58,** 220
binding to operator sequence, **58,** 204
chromonema fiber visualization, **58,** 219
dynamics of amplified region, **58,** 218
gene amplification and cell cloning, **58,** 212–214, 220
green fluorescent protein–repressor–nuclear localization sequence fusion protein
construction, **58,** 211–212
perfusion chamber for microscopy, **58,** 215–216
phototoxic effects, **58,** 216–218
operator introduction into Chinese hamster ovary cells
operator repeat
construction, **58,** 206–208
manipulation for stability, **58,** 208–210
overview, **58,** 205
repressor–nuclear localization sequence fusion protein
construction, **58,** 210–211
immunostaining in operator repeat detection, **58,** 214–215
Lactacystin, **66,** 255
Lactate dehydrogenase, apoptosis assay, **57,** 261
L-glutamate transport (GluT), **71,** 92, 102–103
Lambda phage, DNA, B–S transition, **68,** 332
Lamellipodial extensions
chick forebrain neurons, **71,** 52
extension and retraction of, in Tibia-1 pathway, **71,** 173
in embryonic grasshopper limb bud, **71,** 175–176
in *Xenopus* cultures, **71,** 146–147
Lamin
antibody microinjection
antibody production and purification, **53,** 592–593, 595
apparatus, **53,** 593–594
cell cycle phase, dependence of effects, **53,** 594–595

Lamin (*cont.*)
electron microscopy monitoring, **53,** 594
immunofluorescence microscopy monitoring, **53,** 594
assembly in cell-free systems
antibodies for detection, **53,** 438–439
disassembly assays, **53,** 412–413
electron microscopy, **53,** 437–438
immunodepletion studies, **53,** 441–443
immunofluorescence microscopy, **53,** 439–440
induction, **53,** 436–438
interphase *vs.* mitotic/meiotic isoform assembly, **53,** 405–407
monitoring of aggregation and polymerization, **53,** 405
regions in head-to-tail binding, **53,** 406
regulation by lamin isoforms, **53,** 407–408
temporal relationship to nuclear envelope formation, **53,** 420
Western blotting of lamins, **53,** 440–441
fluorescence image microscopy, **53,** 62–63, 66–68, 102
localization, **53,** 5
microinjection into living cells
assembly kinetics analysis, **53,** 599
controls, **53,** 598–599
fluorescence labeling of lamins, **53,** 597–598
interspecies studies, **53,** 600
microinjection technique, **53,** 598
photobleaching studies, **53,** 599–600
nucleic acid interactions during interphase and mitosis/meiosis, **53,** 413
phosphorylation, **53,** 368
posttranslational isoprenylation, **53,** 592
purification of *Drosophila* proteins
histidine-tagged recombinant proteins, **53,** 404
immunoaffinity chromatography of native proteins, **53,** 402–404
purification of recombinant *Xenopus* lamins
cation-exchange chromatography, **53,** 597
chloroform/methanol precipitation, **53,** 597
expression vectors, **53,** 595–596
inclusion body solubilization, **53,** 596
transmission electron microscopy, **53,** 102
types
Drosophila, **53,** 24
major isotypes, **53,** 591
rat, **53,** 5
sea urchin, **53,** 436
Xenopus laevis, **53,** 5
Laminar flow hood, in chick embryo preparation, **71,** 71
Laminin, **71,** 21, 92–93, 181–182
as chick ciliary neuron substrate, **71,** 37
as culture substrate, **71,** 12
for peripheral neurons, **71,** 19
basement membrane formation, **69,** 112–114
characterization, **69,** 282–283
in *Xenopus* cultures, **71,** 140, 145
inducing rapid outgrowth of axons with, **71,** 33–34
recombinant heterotrimeric, **69,** 120–123
Laminin-1, dynamic motion, **69,** 168–171
Laminin-5
instructive ECM studies with Mab, **69,** 29–32
preparation, **69,** 123–124
Laminins, AFM imaging, **68,** 218–219
Lamps, *see* Illumination
Langmuir trough, AFM studies
small Teflon wells, **68,** 235–236
trough only, **68,** 234–235
Laplacian filters, convolution, **56,** 113–114
Large amplitude swelling
membranes, **65,** 154–155
Large invertebrate neurons, **71,** 3
Laser light scattering, **66,** 79–80
Laser light sources
confocal microscopy, **63,** 90–91, 94–95
introduction of, **63,** 7
ion, **63,** 20–21
Laser scanning confocal microscopy
Arabidopsis thaliana green fluorescent protein expression
FM 1-43 counterstaining of plasma membranes, **58,** 146
growth in sterile culture, **58,** 144–145
instrumentation, **58,** 143–145
multichannel microscopy
cyan and yellow-shifted mutants, **58,** 148
double labeling with green fluorescent protein variants, **58,** 149
KNAT3–green fluorescent protein studies, **58,** 148
rationale, **58,** 147–148
principle, **58,** 143–144
propidium iodide counterstaining of cell walls, **58,** 145–146
CENP-B–green fluorescent protein, fusion protein studies of centromere dynamics
cell culture in perfusion chamber, **58,** 186–187

data storage and transport, **58,** 189–190
focal planes, **58,** 185–186
image collection, **58,** 188–189
morphometric analysis, **58,** 193–194
motion analysis, **58,** 192–193
optics, **58,** 187–188
quantitative analysis, **58,** 191–192
time-lapse imaging, making and playing movies, **58,** 190–191
commercial systems, **58,** 351–352
Drosophila
fusion proteins in nervous system development, **58,** 178–180
sample preparation, **58,** 288
fluorescein controls, **58,** 42
gain/offset settings, **58,** 43
lasers, **58,** 353
membrane fusion protein photobleaching studies
fluorescence loss during photobleaching
controls, **58,** 280
microscopy, **58,** 278, 280
principle, **58,** 278
fluorescence recovery after photobleaching
diffusion constant calculation, **58,** 271–272, 276–278
microscopy, **58,** 272–273, 275–276
principle, **58,** 271
qualitative analysis of lateral diffusion, **58,** 272–273
quantitative analysis of lateral diffusion, **58,** 275–276
multiphoton microscopy, **58,** 353–354
Ncd–green fluorescent protein in *Drosophila*
embryo preparation, **58,** 156–158
filters, **58,** 158–159
mutant analysis, **58,** 160
oocyte preparation, **58,** 156–157
photoactivation applications, **58,** 160–161
subcellular localization, **58,** 159–160
time-lapse imaging, **58,** 158
organelle-targeted protein, **58,** 84
photodamage, **58,** 353
point-spread functions, **58,** 351–352
principle, **58,** 40
setup of microscope, **58,** 43–44
standards, preparation and imaging, **58,** 44–45
time-lapse imaging, yeast, **58,** 45–46
Laser scanning cytometer
analysis of apoptosis, **63,** 541–543; **64,** 425–430
cell attributes measured by, **64,** 422–423
estrogen receptor analysis, **64,** 435–436
FISH protocol, measurement of nucleoli, **64,** 438–441
in MRD detection, **64,** 389–390
microscope plus PC, **63,** 55–57
proliferation-associated antigen analysis, **64,** 430–434
scan mirror and stepped stage, **63,** 57
transcription factor analysis, **64,** 437–438
Laser-scanning cytometer (LSC), **66,** 71–72
apoptosis
JC-1, **66,** 82
mitochondrial transmembrane potential, **66,** 82
uniqueness, **66,** 100–102
vs. flow cytometry, **66,** 100–103
apoptotic cells
annexin V, **66,** 85
red maximal pixel, **66,** 78, 79
cell preparation, **66,** 75–77
chromatin condensation, **66,** 77–89
components, **66,** 73
cytocentrifugation, **66,** 76
vs. touch preparation, **66,** 75
DNA fragmentation assay, **66,** 88
labeled inhibitor analysis, **66,** 95–96
microscope slides, **66,** 76–77
nuclear fluorescence, **66,** 74
poly(ADP-ribose) polymerase, **66,** 98–99
principles, **66,** 72–75
TUNEL, **66,** 91–92
vs. flow cytometry, **66,** 75
Laser scanning cytometry
and flow cytometry: differences, **63,** 54
area feature, **63,** 67–71
assays
in cancer therapy, **64,** 475–476
on fluorescent cytotoxic drugs, **64,** 477–478
cell position feature, **63,** 71–76
cellular DNA content analysis, **64,** 423–425
compared with flow cytometry, **64,** 326–327
description, **64,** 316
extranuclear total value feature, **63,** 65–67
future directions, **63,** 82–84
immunophenotyping by, **64,** 316–327
in clinical studies, **64,** 474–475
laser excitation energy, **63,** 52
maximum pixel value feature, **63,** 64–65
multiple cell exclusion feature, **63,** 79–80
probe spot counting feature, **63,** 78
solid phase, for cell preparation, **63,** 80–82
time feature, **63,** 77–78

Laser scanning cytometry (*cont.*)
total value feature, **63**, 62–63
Laser scissors
applications
cell fusion, **55**, 85–87
chromosome surgery, **55**, 81–83
genetics, **55**, 81–83
laser-assisted hatching, **55**, 92–94
laser zona drilling, **55**, 88–92
membrane optoporation, **55**, 85–87
mitosis, **55**, 83–85
motility, **55**, 83–85
reproduction, **55**, 87–94
oocyte manipulation, **55**, 88–92
preembryo manipulation, **55**, 92–94
sperm manipulation, **55**, 87–88
interaction mechanisms, **55**, 74–78
mechanical limitations, *see* Micromechanical limitations
overview, **55**, 71–73
Laser trap, optics, PFM, **68**, 200–203
Laser tweezers
accessories, **55**, 39–40
alignment, **55**, 37–38, 65–67
analysis, **55**, 38–39, 67–68
applications
cell fusion, **55**, 85–87
chromosome surgery, **55**, 81–83
genetics, **55**, 81–83
kinesin molecule ATP turnover manipulation, **55**, 125–127
laser-assisted hatching, **55**, 92–94
laser zona drilling, **55**, 88–92
membrane optoporation, **55**, 85–87
mitosis, **55**, 83–85
motility, **55**, 83–85
reproduction, **55**, 87–94
oocyte manipulation, **55**, 88–92
preembryo manipulation, **55**, 92–94
sperm manipulation, **55**, 87–88
calibration, **55**, 37–38, 65, 162, 179
cell membrane mechanics
bound *vs.* unbound receptor molecules, **55**, 189–192
fence effects, **55**, 188–192
laser calibration, **55**, 179–181
maximal force calibration, **55**, 161–162
membrane skeleton protein transport, **55**, 188–189
overview, **55**, 160–161, 174–176
receptor molecule dragging obstacles, **55**, 181–185
representative experiments, **55**, 181–183
trajectory analysis, **55**, 183–185
single-particle tracking
elastic intercompartmental barriers, **55**, 185–188
future research directions, **55**, 192
gold labeling particle preparation, **55**, 176–178
protein motion, **55**, 176
video microscopy, **55**, 178–179
tether effects, **55**, 188–192
trap calibration, **55**, 162–165
feedback, **55**, 62–65
high-resolution position measurement, **55**, 56–57
imaging, **55**, 35–36, 53–56
interaction mechanisms, **55**, 78–81
laser choice, **55**, 33–35, 48–50
mechanical limitations, *see* Micromechanical limitations
microscope choice, **55**, 30–33
multiple trap synthesis, **55**, 205–215
acousto-optic deflectors, **55**, 208–209
computerized position control, **55**, 212–213
mechanical deflectors, **55**, 208
noise, **55**, 210–212
overview, **55**, 205–206, 213–215
position, **55**, 210–212
trapping configurations, **55**, 206–208
object holding, **55**, 206
object manipulation, **55**, 206–207
static force exertion, **55**, 207–208
noise sources, **55**, 58–62, 210–212
optical layout, **55**, 35–36, 50–53
organelle manipulation, *in vivo*, **55**, 195–202
Spirogyra cytoplasmic streaming, **55**, 199–202
optical tweezer setup, **55**, 196–197
overview, **55**, 195–196
reticulomyxa, **55**, 197–199
artificial membrane tubes, **55**, 198–199
cell preparation, **55**, 197
organelle movement, **55**, 197–198
overview, **55**, 29–30, 47, 68, 71
parts list, **55**, 40
ray optics regime single-beam gradient laser trap, dielectric sphere force calculations, **55**, 1–25
arbitrary trap location force, **55**, 11–16, 23–25
index of refraction effects, **55**, 19–20
light forces, **55**, 4–8

mode profile effects, **55,** 16–20
overview, **55,** 1–4, 20–21
ray force, **55,** 21–22
Y axis trap focus force, **55,** 10–11, 22–23
Z axis trap focus force, **55,** 9–10
setup, **55,** 36–37
translation, **55,** 38
trap forces, *see* Gradient trap force
video recording, **55,** 38–39, 178–179
Lasers
air-cooled argon ion laser, **70,** 209–221
air-cooled krypton–argon ion laser, **70,** 221–231
argon ion lasers, **70,** 187–188, 193–195, 201–204, 209–221
helium–cadmium lasers, **70,** 189
helium–neon lasers, **70,** 189, 204
krypton–argon ion lasers, **70,** 188, 204–207, 221–231
krypton ion lasers, **70,** 188
light source overview, **70,** 185–187
tunable dye lasers, **70,** 189
Lasers, laser tweezer system, **55,** 33–35, 48–50
Lat-A, **65,** 335
Lateral force mapping
human serum albumin, **68,** 133–134
recognition imaging, **68,** 133–134
Lateral resolution
chromatic aberration corrections, **70,** 279–280
wide-field confocal microscopy, **70,** 249–250
laughing man, mutation effects in development, **60,** 81
LB01 lysis buffer, **64,** 10
Lead, toxicological effects on nervous system, **71,** 290
Lectin column, integrin affinity chromatography, **69,** 231, 234
Leica systems
objective lenses, **70,** 286
slow-scan CLSM, **70,** 31–33
Leigh's syndrome, **65,** 136
Leishmania infantum, **65,** 29
Lens selection
bone and dental cell biology, **70,** 273
brain tissue, neuroscience, and retinal imaging, **70,** 273–274
confocal microscopy, **70,** 155–156
cytoskeleton, **70,** 272
genetics, **70,** 273
immunogold and silver grains, **70,** 272–273
microbiology, **70,** 275–278
ocular imaging, **70,** 274
UV imaging, **70,** 269–271
visible Ca^{2+} imaging, **70,** 271–272
Lenses
behavior, **70,** 158
infinity lens systems, **70,** 254–256
laser beam-focusing, **63,** 23–24
maximum resolution parameters, **70,** 324–326
objective, *see* Objective lenses
positioning for collection of light signals, **63,** 33–36
Leukemia
acute
detection of MRD, **64,** 388
follow-up and prognosis, **64,** 356
acute lymphoblastic, *see* Acute lymphoblastic leukemia
acute myeloid, *see* Acute myeloid leukemia
chronic lymphocytic, detection of MRD, **64,** 395–396
immunophenotyping, **64,** 313–316
results, **64,** 325–326
PCR analysis, **64,** 412–413
Leukemia cells, rat basophilic, AFM studies, **68,** 38
Leukocyte common antigen, *see* CD45
Leukocytes
clinical chemistry parameter in asthma, **64,** 497
erythrocyte separation from, **63,** 210–211
identification, cell preparation for, **63,** 217–250
three scan pass system, **64,** 338–339
Levels of expression, in biolistics, **71,** 364
Leydig cell, establishment of cell lines
cell preparation
fractionation over discontinuous Percoll gradient, **57,** 96–97
isolation, **57,** 96
clone selection, **57,** 98–99
histology of immortalized cells, **57,** 100–101
immunodetection of simian virus 40 large T antigen, **57,** 99–100
medium for immortalized cell culture, **57,** 99
overview, **57,** 94–95
passaging, **57,** 99
seminiferous cord formation *in vitro* with other testicular cells, **57,** 101–103, 105
transfection
calcium phosphate transfection with simian virus 40 large T antigen gene, **57,** 97–98
principles, **57,** 95–96
Lifetime imaging, *see* Temporal imaging
Lifetime in culture, as criterion for choice of neuronal culture, **71,** 2

Ligand affinity chromatography, soluble
integrin, **69,** 233
Ligand binding, integrins, SPR, **69,** 254–256
Ligand–receptor recognition
AFM tips
cleaning, **68,** 117–118
crosslinker binding, **68,** 119
esterification, **68,** 118
ligand binding, **68,** 119
ligand density and functionality, **68,** 119–120
probe surfaces, **68,** 120–121
applications, **68,** 116
dynamic force microscopy
bond lifetime, **68,** 124
kinetic rates, energies, binding
pocket, **68,** 125
unbinding force *vs.* loading rate, **68,** 124–125
unbinding force
distribution, **68,** 123
force–distance cycle, **68,** 121–123
Ligands, DNA interactions, SFM
studies, **68,** 368–369
Ligation-based cloning methods, **71,** 413–414
Light
absorbance, phycobiliproteins, **63,** 194–195
collection optics, **63,** 114–115
effect on migration assays, **63,** 551
fluorescent, collection, **63,** 32
full-length, **66,** 504
interaction with cells, in flow
cytometry, **63,** 110–118
oscillation between front and rear
mirrors, **63,** 20–21
Light emission
aequorin, **65,** 360–361
Light forces, ray optics regime, single-beam
gradient laser trap, dielectric sphere force
calculations, **55,** 4–8
LIGHT-mediated killing
HT29 cells, **66,** 502
Light microscope studies, **71,** 3
Light microscopy
alignment, **56,** 135–146
components, **56,** 136–140
condensers, **56,** 138
diffusers, **56,** 138
eyepiece, **56,** 139–140
filters, **56,** 138
infinity space, **56,** 139
lamp collector, **56,** 138
light source, **56,** 136
nosepiece, **56,** 139
objective, **56,** 139
specimen stage, **56,** 139
substage, **56,** 138–139
video adapters, **56,** 140
Köehler illumination, **56,** 140–146
alignment, **56,** 142–143
benefits, **56,** 143
contrast, **56,** 143–146
overview, **56,** 140–142
resolution, **56,** 143–146
overview, **56,** 135
apoptosis assay, **57,** 253
bone marrow examination, **64,** 347–348
cell lineages, identification, **64,** 348
digital video microscopy compared, **56,** 65–66
microscopes for animal cell culture, **57,** 7–8
multimode microscopy, *see* Multimode
digital microscopy
relocalization for, **64,** 327, 329, 331–333,
335–336
vertebrate cell monolayer study, perfusion
chamber, **56,** 253–271
assembly, **56,** 270–271
chamber design, **56,** 256–260, 265–266
construction, **56,** 260–265, 267–269
historical perspectives, **56,** 256
overview, **56,** 253–256
temperature regulation, **56,** 269–270
video-enhance contrast microscopy, *see*
Video-enhanced differential interference
contrast light microscopy
Light microscopy/electron microscopy
correlation
flat embedding
cleaning, **61,** 305
Coplin jars, **61,** 304
dehydration steps, **61,** 304
ethanol clearing, **61,** 304
infiltration, **61,** 304–305
Petri dish, **61,** 304
indirect immunofluorescence
caged spindle proteins, **61,** 298
GFP tags, **61,** 298
light microscopy (LM)
differential interference contrast (DIC)
LM uses, **61,** 301
dynamic freeze
media perfusion intervals, **61,** 302
perfusion chamber, **61,** 302
framing rate, **61,** 301
objective marker, **61,** 302
observable kinetochore dynamics, **61,** 299

phase contrast LM uses, **61,** 301
polarization LM uses, **61,** 301
video-enhanced LM advantages, **61,** 299–301
obtaining required ultrastructural information
from sequential sections
application examples, **61,** 308
Sterecon system, **61,** 308
stereo viewing, **61,** 309
thick section advantages, **61,** 308–309
with 3D display, **61,** 308
without 3D display, **61,** 308
z-axis resolution, **61,** 308
from thick section tomography
applications, **61,** 309–311
limitations, **61,** 311
z-axis resolution, **61,** 309
sectioning preparation
acid removal wash, **61,** 306
block trimming, **61,** 307
cell block excision, **61,** 307
cell relocation determination, **61,** 306–307
coverslip removal, **61,** 305–306
etching, **61,** 306
plastic scoring, **61,** 306
spatial resolution limits, **61,** 298
Light microscopy, confocal microscopy comparison, **70,** 7–9
Light microscopy, deconvolution, **63,** 584–585
Light pulse, processing, **63,** 31–32
Light scatter
as contouring parameter in LSCM, **64,** 319
changes in apoptotic plant cells, **63,** 512, 514
elimination by dissolving particles, **63,** 321
flow cytometric analysis of bacteria, **64,** 512–513
flow cytometric measurement, **63,** 493
forward, *see* Forward light scatter
gating on, **63,** 43
lymphocyte gate, **64,** 571, 577
plus nucleic acid dyes: detection of microbes, **64,** 514–516
profiles, microorganisms, **64,** 516
standardization and calibration, **63,** 327–328
studied with multiangle scatter sensor, **63,** 7
Light signals
continuum to flow cytometric data file, **63,** 37–41
from particles, collection, **63,** 33–37
Light source, confocal microscopy, **70,** 16–19
Lillie, Frank Rattray, **71,** 68
Lilium longiflorum, see Spindle birefringence
Limitations, *see* Micromechanical limitations
Limiting dilution method, PCR, **64,** 409–411
Line scanners, confocal microscopy, **63,** 95–96
Lineage commitment, analysis in zebrafish, **59,** 128–130
Linear sucrose gradient centrifugation, *Spisula* MT nucleation, **67,** 157–158
Linearity
and data accuracy, **63,** 304–309
charge-coupled device cameras
basic adjustments, **56,** 6
high-resolution multimode digital microscopy, **56,** 203–205
low-light conditions, **56,** 31–32
fluorescent particles, **63,** 318
quantitative video microscopy, resolution limit corrections, **56,** 130
LINEs, zebrafish, **60,** 108
Linked assays, **65,** 112–114
limitations, **65,** 114
succinate-cytochrome c reductase, **65,** 113–114
succinate-hexacyanoferrate reductase, **65,** 112–113
Linker histones, **62,** 145–146
Lipid asymmetry, **66,** 340–342
annexin assays, **66,** 342
Lipid bilayers, AFM studies
charged bilayers, **68,** 238–240
ligand-linked bilayers, **68,** 237–238
membrane elasticity, **68,** 75–76
supported, *see* Supported lipid bilayers
Lipid biochemistry
analysis in *Drosophila,* **71,** 214–215
in *Drosophila,* **71,** 204
Lipid distribution
assays, **66,** 341
Lipid extraction
ceramide-(DAG) kinase assay, **66,** 144–145
Lipid isolation
metabolic labeling, **66,** 138–139
Lipid-symmetric erythrocytes
phagocytosis, **66,** 360
Lipid visualization
metabolic labeling, **66,** 140–141
Lipids, **62,** 36–37
Lipofection, expression cloning, **69,** 213
Lipofection, with *Xenopus* cultures, **71,** 136–138
Lipophilic cationic dyes
membrane permeability, **65,** 149
Liposomes
annexin V purification, **66,** 344–345
Liquid chromatography, **71,** 199
Liquid crystals, STM studies, **68,** 6

Liquid N_2, **71,** 13
Liquid nitrogen storage tank, animal cell culture, **57,** 9
List mode analysis
  exhaustive, asthma, **64,** 499–500, 502
  flow cytometry, **64,** 490–491
Listerine, use in *Heliosoma* cultures, **71,** 162
Live cell fluorescence lifetime imaging microscopy, **69,** 272–273
Live cell imaging
  fluorescent, **71,** 316–320
  in superior cervical ganglia neurons, **71,** 306–321
  in *Xenopus* cultures, **71,** 148–154
Live cell research, *see* Intracellular calcium, Intracellular pH
Live cells
  adhesion force measurements
    monolayer immobilization, **68,** 99–100
    research results, **68,** 100–104
  AFM and confocal imaging, **68,** 184
  AFM–optical microscopy combination
    infected monkey kidney cells, **68,** 8–9
    instrument configuration, **68,** 7–8
  AFM studies
    future studies, **68,** 47–48
    overview, **68,** 37–38
  cell–cell adhesion force
    research results, **68,** 105–109
    single cell immobilization, **68,** 104–105
  cell–surface adhesion force measurements
    functionalized sensors, **68,** 98
    passivated force sensors, **68,** 95
    research results, **68,** 95–99
    sphere immobilization, **68,** 94
Living cells
  reactive oxygen species, **65,** 91–92
Loading rate, ligand–receptor recognition, **68,** 124–125
Local probes
  AFM instrumentation, **68,** 2–3
  definition, **68,** 1–2
  STM instrumentation, **68,** 2
Localization
  mitochondrial DNA
    immunofluorescence, **65,** 327
Logarithmic amplifier, **63,** 38–40, 42, 124–127
  and data accuracy, **63,** 304–309
  conventional flow cytometry, **63,** 139
Logarithmic scale, display of data, **63,** 302–304
Look-up tables, digital image processing
  contrast enhancement, **56,** 78–79, 104
  image data conversion, **56,** 75–76
loose ends, mutation effects in development, **60,** 78
Long-term cultures
  advantageous in study of Tibia-1 pathway, **71,** 172–173
  characteristics of hippocampal and neuronal, **71,** 124–126
  media for, **71,** 23
  reducing nonneuronal contamination in, **71,** 30
Low-background epifluorescence microscopy, single fluorophore visualization, aqueous solutions, **55,** 119
Low-background total internal reflection fluorescence microscopy
  kinesin molecule
    ATP turnover, **55,** 125–127
    movement observation, **55,** 123
  myosin molecule ATP turnover, **55,** 123–125
  single fluorophore visualization, aqueous solutions, **55,** 119–123
Low-density cultures, chick forebrain neurons, **71,** 53–54
Low-light microscopy, **56,** 19–43
  advanced technology, **56,** 62
  camera characterization parameters, **56,** 22–35
    camera noise, **56,** 26–28
    dynamic range, **56,** 32–33, 58–59
    fill factor, **56,** 24–26
    frame rate, **56,** 33–34, 62
    gain, **56,** 33
    imaging rate, **56,** 34–35
    linearity, **56,** 31–32
    quantum efficiency, **56,** 23–26
    sensitivity
      definition, **56,** 47–48
      enhancement, **56,** 49–54
      intensified cameras compared, **56,** 56–57
      parameters, **56,** 34–35
      signal-to-noise ratio, **56,** 26–28, 47, 54–56
    spatial resolution, **56,** 28–31, 34, 59–60
    spectral respones, **56,** 23
    time response, **56,** 32
    uniformity, **56,** 31–32
  charge-coupled device cameras, *see* Charge-coupled device cameras
  detector features, **56,** 35–42
    color video cameras, **56,** 39–40
    complementary metal-oxide semiconductor imagers, **56,** 41–42
    fast-scan cameras, **56,** 37
    silicon intensified target cameras, **56,** 40–41
    slow-scan cameras, **56,** 36–37

geometric distortion, **56,** 60
intensified cameras, *see* Intensified cameras
overview, **56,** 19–22, 42, 46–47
applications, **56,** 19–20
detection methods, **56,** 21–22
image quality, **56,** 20–21
shading, **56,** 60
temporal resolution, **56,** 60
usability, **56,** 61–62
Low-temperature intermediates, **65,** 208
LSC, *see* Laser scanning cytometer, **66,** 71–72
LSCM, *see* Laser scanning confocal microscopy
LSCM, *see* Laser scanning cytometry
LSCM, *see* Multicolor laser scanning confocal microscopy
Lucifer yellow, **70,** 173
Luciferase, **65,** 137
ATP, **65,** 367–371
ATP-dependent luminescence, **65,** 366
calibration, **65,** 368
firefly, **65,** 134
fluorescence resonance energy transfer with green fluorescent protein, **58,** 15
future, **65,** 371–372
metabolite interference, **65,** 370–371
mitochondrial targeting, **65,** 368–371
oxygen requirement for bioluminescence, **58,** 285–286
Photinus pyralis firefly tails, **65,** 366
recombinant targeted immunolocalization, **65,** 369
transcription monitoring in *Drosophila*, **58,** 284–286, 290
Luciferase-luciferin, **65,** 134–136
Luciferin, **65,** 137
Luciferin-luciferase system, **65,** 134–136
Lucigenin-based assay
superoxide anion concentration, **66,** 312–313
Lumbosacral dorsal root ganglia (DRGs), dissecting, **71,** 25
Luminescence measurements
calcium imaging, **66,** 239–240
Luminometer, **65,** 362
Lung resistance related protein, **64,** 194
Lymphadenopathy, immunophenotyping results, **64,** 324–325
Lymphocytes
activated/proliferating
detection, **63,** 437–439
flow cytometry
advantages, **63,** 447–449
shortcomings, **63,** 449–453
activation, multiparametric analysis, **63,** 434
and blast cells: intermediate cell population between, **64,** 353
application of LSCM cell position feature, **63,** 74–75
B and T
CD45 positive, **64,** 333
scattergrams, **64,** 322
B-lineage ALL, **64,** 392–393
B, pattern of antigenicity, **64,** 331
gating procedures, **63,** 439
in vivo responses, monitoring, **63,** 392–395
intravenous transfer, **63,** 393–394
metaphase spreads, **64,** 102
percentage of positive cells, **63,** 445
subpopulations, reference ranges, **64,** 575
subsets, bulk culture, **63,** 448–449
T, immunophenotyping, **64,** 568–574
Lymphoid tissue, enzymatic cell isolation from, **63,** 407
Lymphoma, immunophenotyping, **64,** 313–316
Lymphoproliferative disorders, detection of MRD, **64,** 395–396
Lymphoreticular cells, LSC analysis, **64,** 319–320
Lysing reagent, prepared daily, **63,** 249–250
Lysolecithin, sperm permeabilization, **53,** 373, 421
Lysosomes, **62,** 43–45
in mitochondrial preparations, **65,** 4
Lysozyme receptor, **62,** 366–367

M

MAb, *see* Monoclonal antibody
Macromolecular assembly, time application, **63,** 180
Macromolecules, AFM studies, **68,** 290
Macronuclear assortment, **62,** 15
Macronuclear DNA
isolation, **62,** 239–250
Macronucleus, **62,** 32, 69, 71
ARP, **62,** 138, 263
development, **62,** 130–133
division, **62,** 72–74, 128
DNA synthesis, **62,** 71–73
transformation, *see also* Transformation, **62,** 163–172
Macrostome, **62,** 90
Madin–Darby kidney cells, AFM studies, **68,** 37–38
MAG, *see* Myelin-associated glycoprotein
Magnesium sulfate, stock solution, **64,** 11

Magnification, electronic, in confocal microscopy, **63,** 92–94
Maintenance, of peripheral neuron cultures, **71,** 28
Maize coleoptile mitochondria electrophoresis, **65,** 71
Maize leaf, **66,** 448
Major histocompatibility complex, class II, expression, **63,** 451–452
Malate, **65,** 137
Malphigi, Marcello, **71,** 68
Maltose-binding protein, anti-zKLP antibodies, **59,** 237–238
Mammalian cells
centrosome-associated protein expression, **67,** 132–133
centrosome duplication assay, **67,** 314–315
centrosome overproduction, **67,** 257–260
γ-tubulin
in ectopic MT nucleation, **67,** 185–189
in MT nucleation, **67,** 179–185
in centrosome protein analysis, **67,** 261–264
tumor necrosis factor-related ligands, **66,** 504–505
Mammary glands, epithelial cells
adenovirus
generation in *Escherichia coli,* **69,** 410–413
HEK293 E1 cells, **69,** 415–416
high-titer viral stocks, **69,** 418–420
infection, **69,** 420–422
overview, **69,** 409–410
screening, **69,** 413–415
virus titration, **69,** 417–418
mouse cell cultures, **69,** 409
MAPKs, **66,** 187–188
Mapping
AFM analysis, **68,** 179–180
human serum albumin, **68,** 133–134
Na^+/D-glucose cotransporter, **68,** 131–133
recognition imaging, **68,** 133–134
Mapping, *see also* Quantitative DNA fiber mapping
cDNA, **64,** 51
cloned probes onto DNA molecules, **64,** 50–51
MAR, *see* Matrix attachment region
Marker antigens
mitochondria
yeast, **65,** 265
Marker molecules
cell surface staining: CFSE labeling method, **63,** 381–382
empirical, PCNA as, **63,** 419–420
for LSCM total value, **63,** 63
histone H_3 plus antibody, **63,** 344–345
Ki-67 protein, **63,** 400
mitotic cells, **63,** 350–353
specific, endothelial cells expressing, **63,** 575–576
Marker rescue
nonrespiring cytoductants, **65,** 390–392
Masking, incorporated BrdUrd, **64,** 133
Mass spectrometry, complementation with scanning transmission electron microscopy, **53,** 143
Maternal-effect genes, gynogenesis-based screening in zebrafish
applicability to various traits, **60,** 17
egg
fertilization, *in vitro*, **60,** 15
stripping, **60,** 14
gynogenesis induction
early pressure, **60,** 5–6, 15–16
election of method, **60,** 5–6
heat shock, **60,** 5–6, 15
janus, **60,** 3
materials, **60,** 13
mutagenesis dose, **60,** 8, 10
mutation recovery
F_1 females, **60,** 11
F_2 siblings, **60,** 11–12
F_3 survivors, **60,** 12
pilot screen, **60,** 10–11
rationale, **60,** 3–4
selection of lines, factors
gynogenote yield, **60,** 6–7
hybrids *vs.* inbreds, **60,** 8
lethal/sterile-free genetic background, **60,** 7
sex ratio, **60,** 7–8
solution preparation, **60,** 12–13
somitabun, **60,** 3
sperm
collection, **60,** 13–14
ultraviolet inactivation, **60,** 14
testosterone treatment, **60,** 16
yobo, **60,** 3
Mating, **62,** 218–220
Mating types, *see also* Genetic analyses; *see also* Star strains, **62,** 12, 128–130, 223
Mating, yeast cells, **67,** 80–81
Matrigel, cell culturing, **69,** 408
Matrigel invasion assays, cell migration, **63,** 555–556
Matrigel matrix (BD Biosciences), **71,** 34, 46, 47, 92

as culture substrate, **71,** 11–12
in chick ciliary neuron cultures, **71,** 42
Matrix-adherent cells, cell–matrix adhesion assays, **69,** 368
Matrix attachment region (MAR), *see also* Base-unpairing region
definition, **53,** 324
structure, **53,** 324
Matrix-coated plates, signal transduction studies, **69,** 362–363
Matrix dynamics, atomic force microscopy, **69,** 164–165
Matrix metalloproteinase 2, TSP2, 9
Matrix proteins
conjugation to polyacrylamide substratum, **69,** 329–330
multicomponent ECM, **69,** 201–204
Matrix, proteoglycan anion-exchange chromatography, **69,** 58–59
Matrix-targeted recombinant probe
generic map, **65,** 356
Maximum pixel value, LSCM list mode feature, **63,** 64–65, 68
Maximum projection, CSU-10 digital signal processing, **70,** 115
MC5, in simultaneous staining, **63,** 264–266
MC540, **66,** 341
MC540 staining, **66,** 354–355
flow cytometry, **66,** 355
protocol, **66,** 354
MCF7 cells, **66,** 355
MCF-7 cell line, **66,** 419, 420–423, 425, 428–429
transfection, **66,** 428
Mcl-1, **66,** 36
MDCK cells, *see* Madin–Darby kidney cells
MDR, *see* Multiple drug resistance
Mean generation time, determination, **57,** 232
Mechanical limitations, *see* Micromechanical limitations
Mechanical stimulation, polyacrylamide substratum, **69,** 336–337
Media
buffers, **57,** 24–25
chemical contamination, **57,** 54
components, **57,** 20–21
filtration, **57,** 8
for chick forebrain neurons, **71,** 59–62
for long-term cultures, **71,** 23
for peripheral neurons, **71,** 22–25
for spinal and motor neuron cultures, **71,** 80–85
growth and freezing, for γTuSC reconstitution, **67,** 13
growth, for two-hybrid assay, **67,** 79
invertebrate cell culture
additives, **57,** 191–192, 196–198
selection, **57,** 190–191
optimization, **57,** 27–29
pH control, **57,** 22
preparation, **57,** 24–26
screening for biological activity, **57,** 23
selection, **57,** 22–23
storage of liquid medium, **57,** 12, 24
supplementation, **57,** 12–13, 20–21, 27
troubleshooting, **57,** 26–27
types, **57,** 21, 23
water purification, **57,** 8, 25–26
Media, *see* Culture media
Media, embryo, preparation, **69,** 343
Media supplements, for spinal and motor neuron cultures, **71,** 80–85
Median filters, spatial noise reduction, **56,** 100–101
Medulloblastoma, *see* Glioma
MEFs, *see* Mouse embryo fibroblasts
Meiosis, **62,** 96–97, 130–132, 295
Meiosis, *see also* Synaptonemal complex
chromosome behavior, **53,** 257–258
essential gene identification in *Saccharomyces cerevisiae*
expression cloning, **53,** 263–264
mutant screening
dityrosine deposition, **53,** 260
ether sensitivity, **53,** 259, 261
heterozygote selection, **53,** 261
microscopic examination of sporulation, **53,** 259–260
mutagenesis, **53,** 259
principle, **53,** 258–259
recombination between ectopic sites, **53,** 261
recombination-deficient mutants, **53,** 261–263
transposon tagging, **53,** 264
pulsed-field gel electrophoresis analysis of recombination intermediates in *Saccharomyces cerevisiae*
contour-clamped homogeneous electric field electrophoresis, **53,** 267–269
DNA preparation, **53,** 268
double-strand break detection, **53,** 266–268
principle, **53,** 267
Southern blot analysis of recombination intermediates in *Saccharomyces cerevisiae*

Meiosis, *see also* Synaptonemal complex (*cont.*)
cell growth and sporulation, **53,** 265–266
DNA preparation, **53,** 266
double-strand break detection, **53,** 264–265
Meiotic recombination, **62,** 163
Melanoma
data pattern classification, **64,** 493–494
malignant, proliferation, **64,** 463
multiparameter flow cytometric analysis, **64,** 489–490
Melanosomes
in mitochondrial preparations, **65,** 4
MELAS, **65,** 312, 316
Melittin, **66,** 264
Melting temperature, calculation for oligonucleotide probes, **57,** 338–339
Membrane
channels, **62,** 363–375
during conjugation, **62,** 92–95
lipids, **62,** 36–37
potential, **62,** 37, 369–372
proteins, **62,** 37–38, 366
skeleton, **62,** 65–67
traffic, **62,** 42–47, 347–350
Membrane bound
mitochondrial protein, **65,** 47
Membrane-bound ligands
tumor necrosis factor-related ligands, **66,** 519–520
Membrane disruption
skeletal muscle mitochondria complex I, **65,** 102
Membrane dynamics
B35 cell ines, **71,** 288
monitoring with stryl dyes, **71,** 226–227
Membrane fraction, in protein function studies, **71,** 206–213
Membrane integrity, bacteria, **64,** 521–522
Membrane invasion assays, cell migration, **63,** 556
Membrane marker proteins
spheroplasts, **65,** 171
Membrane permeability, **65,** 147–157
cytometric analysis, **65,** 149
detection
intact cells, **65,** 151
flow cytometric analysis staining, **65,** 149–150
fluorescence microscopy
intermembrane mitochondrial proteins, **65,** 151
immunofluorescence analysis, **65,** 152–154
isolated mitochondrial changes, **65,** 154–155
lipophilic cationic dyes, **65,** 149
mitochondrial alteration
intact cells, **65,** 148–155
Membrane potential
fast membrane staining indicators, **70,** 430–431
measurements
mitochondria, **65,** 84–85
reactive oxygen species assays, **65,** 83–85
submitochondrial particles, **65,** 85
mitochondria **66,** 379–380
fluorescence measurements, **65,** 87–90
Nernstian redistribution of permeant cationic dyes
methods, **70,** 437–441
plasma membranes, **70,** 434–436
single mitochondria membrane potential, **70,** 436–437
nonlinear optical imaging
methods, **70,** 446–448
second harmonic generation, **70,** 441–444
second harmonic generation enhancement, **70,** 444–446
two-photon excited fluorescence, **70,** 441–444
probes, **65,** 88
slow redistribution indicators, **70,** 432–434
Membrane potential analysis, bacteria, **64,** 522–523
Membrane potential, neural-enriched cultures for study of, **71,** 143–144
Membrane potential-sending dyes, **65,** 339–342
DASPMI, **65,** 340
$DIOC_6(3)$, **65,** 340
mitoTracker dyes, **65,** 341–342
rhodamine, **65,** 123, 341
rhodamine B hexyl ester, **65,** 341
Membrane proteins, AFM studies
antibody labeling, **68,** 264–265
assembly, **68,** 285, 291–292
force-induced conformational changes, **68,** 272–274
native
HPI layer changes, **68,** 280–282
OmpF porin conformation, **68,** 282–285
surface structure changes, **68,** 278–280
oligomerization, **68,** 270–272
polypeptide end removal, **68,** 267–270
polypeptide loop removal, **68,** 267
polypeptide loop replacement, **68,** 265–267
Membrane protein solubilization
BN-PAGE, **65,** 234
Membrane recovery, speed of, **71,** 348
Membrane separation
plant mitochondria, **65,** 64–69

Membrane vesicle, *see* Vesicle
Membranes, **62,** 35–52
AFM imaging, **68,** 63
AFM–path–clamp combination, **68,** 12
AFM tip, probe surfaces, **68,** 121
brush border vesicles, **68,** 131
Na^+/D-glucose cotransporter, **68,** 131
plasma
apical structures, **68,** 39–40
PFM imaging, **68,** 194–196
Membranes, *see* Cell membrane mechanics
Membranes, bioartificial, preparation and characterization, **69,** 177–179
(1-Mercaptoundec-11-yl)tri(ethylene glycol), **69,** 396
Memory
regulated in *Drosophila,* **71,** 237–238
storage in hippocampus and cortex, **71,** 112
Mercoyanine 540 (MC540), **66,** 341
Mercury arc light, **63,** 6
Merops
web site, **66,** 249
MERRF, **65,** 312, 316
ISH to mtDNA, **65,** 323
olivary nucleus
immunostaining, **65,** 320
MESF, *see* Molecules of equivalent soluble fluorochrome
Mesocricetus auratus, **65,** 23–24
Messenger RNA localization, *see In situ* hybridization
MET promoter, **61,** 371–372
Metabolic labeling, **66,** 136–142; **71,** 11–12
alkaline methanolysis, **66,** 139
bacterial sphingomyelinase assay, **66,** 142
cell culture, **66,** 138
lipid isolation, **66,** 138–139
lipid visualization, **66,** 140–141
protein measurement, **66,** 141
thin-layer chromatography, **66,** 139–140
Metal-oxide semiconductor imagers, features, **56,** 41–42
Metallo-proteinase inhibitors, **66,** 252
MetaMorph software, **61,** 161, 164–166, 376
Metamorph software, **71,** 319
Metamorph software, time-lapse imaging, **58,** 95, 119–120
Metaphase
chromosomes, preparation for PRINS, **64,** 58, 66
lymphocyte, spreads, **64,** 102
Metaphase accumulation, plant chromosomes, **64,** 19, 21
Metaphase spread, preparation from hamster/zebrafish radiation hybrids, **60,** 296–297
Methanol
fixation of cell suspensions, **63,** 406
plus formaldehyde, fixation by, **63,** 280
Methanol, fixation for immunofluorescence microscopy, **57,** 316–317, 327
Methanol, thick tissue sections fixation, **70,** 313
Methocel, **71,** 308
Methods
culturing of chick ciliary neurons, **71,** 46–47
dissection of chick ciliary ganglion, **71,** 43–44
dissociation of chick ciliary ganglion, **71,** 46
introducing reagents to neuronal cultures, **71,** 331–336
Methylcellulose, for peripheral neuron cultures, **71,** 24–25
Metrizamide, **71,** 85
MFP, *see* Molecular force probing
{1-[(Methylcarbonyl)thio]undec-11yl}tri (ethylene glycol), **69,** 396
Mica, AFM imaging, **69,** 166
Mica, DNA
multivalent cation activation, **68,** 361–362
silane treatment, **68,** 360–361
solution dehydration, **68,** 360
thermodynamic equilibrium, **68,** 363–364
Microarray research
image processing, **65,** 442–445
quantitative methods, **65,** 440–446
image acquisition, **65,** 441–442
ratiometric method, **65,** 440–441
Microarrays
data
normalization, **65,** 444–445
variability, **65,** 444–446
image analysis
ratio computing, **65,** 442
low value rejection, **65,** 443–444
perturbed mitochondrial function
cellular responses, **65,** 446–451
Microbial systems, application of flow cytometry, **64,** 531–532
Microbiology, medical, flow cytometric applications, **64,** 563
Microbodies, *see* Peroxisomes
Microcarrier
cell growth, **57,** 222–223
types, **57,** 221–222
Microcell hybridization system, **65,** 407–409

Microcell-mediated chromosome
transmitochondrial hybrid cells, **65,** 407–409
Micrococcal nuclease, chromatin structure analysis, **53,** 508–509
Microdissection device
homemade, **65,** 324
Microexudates, **71,** 12
Microfilaments, *Xenopus* oocytes and embryos, **70,** 388–390
Microinjection
affecting choice of neuronal culture, **71,** 2
axonal outgrowth, **69,** 355
FLIM, **69,** 267–268
MTs in cytoplasts, **67,** 47–48
of adherent cells, **71,** 312
of Tibia-1 pathway neurons, **71,** 185–186, 190
preparation of glass-bottom culture dishes for, **71,** 187–188
somitic migration, **69,** 345
studies of, in chick ciliary cultures, **71,** 41
suitability of peripheral neurons to, **71,** 139
survival of, **71,** 18
γ-tubulin antibody, **67,** 184–185
Microinjection, *see also* Transformation, **62,** 469–482
Microinjection of mitotic cells
advantages
kinetochore fiber disassembly, **61,** 220
timing precision, **61,** 220–221
transfection, **61,** 221
variable injectate concentration, **61,** 221
applications, **61,** 219–220, 228–229
cell choice
cell lines, **61,** 221–223
key characteristics, **61,** 221
plating density, **61,** 222
synchronization, **61,** 222
difference between micromanipulation, **61,** 221
injection timing
nuclear envelope breakdown, **61,** 223
role of calcium, **61,** 223
stage determination, **61,** 223
procedure
method
cell volume measurement, **61,** 228
hydraulic micromanipulator, **61,** 226
injectate volume measurement, **61,** 228
injector pressure settings, **61,** 226
interphase injection, **61,** 227
media evaporation prevention, **61,** 227–228
mitotic injection, **61,** 227
needle centering, **61,** 227
phase condenser, **61,** 226, 227
phase objective, **61,** 226, 227
vibration dampening, **61,** 227
microinjection chambers
coverslipped bottom culture dishes, **61,** 225
modified Rose chamber, **61,** 223–224
Petri dish, **61,** 224
micropipettes
diameters, **61,** 225
Hamilton syringe, **61,** 226
inert fluorescent molecules, **61,** 226
prevention of chromatography effect, **61,** 226
puller, **61,** 225
Microinjection of plant cells
animal cell differences, **61,** 428
methods
injection procedure
fluorescent tracer, **61,** 432
injection site choice, **61,** 432
iontophoresis, **61,** 433
large-bore needle problems, **61,** 432–433
needle clogging, **61,** 433
protein aggregation/clumping, **61,** 433
microinjector
pressure considerations, **61,** 432
water-based system, **61,** 431
microscope slide chamber
coverslip preparation, **61,** 428
glass etching, **61,** 428
needles
dimensions, **61,** 429, 431
loading, **61,** 431
puller, **61,** 429
pull times, **61,** 429, 431
stamen hair cells preparation, **61,** 429
Microinjection, DNA in transgenic zebrafish generation
apparatus setup, **60,** 138
egg collection and preparation, **60,** 139–140
fish maintenance, **60,** 136–137
needle preparation, **60,** 139
plasmid preparation, **60,** 136
plate preparation, **60,** 137–138
technique, **60,** 140–141
Microinjection, *see* Lamin, *Xenopus*
Micromanipulation of chromosomes and spindles
applications, **61,** 209
grasshopper spermatocytes

cell preparation, **61,** 214
preparing follicles, **61,** 215
proper tonicity, **61,** 215
sources, **61,** 214
testis dissection, **61,** 214
equipment
condenser numerical aperture, **61,** 211
Ellis manipulator properties, **61,** 211
image processor, **61,** 211
image storage, **61,** 211
microforge, **61,** 212
mounting considerations, **61,** 211
objective, **61,** 211
phase contrast microscopy, **61,** 210–211
polarization microscopy, **61,** 211
vibration absorption, **61,** 212
voltage-driven crystal bending, **61,** 211
favorable properties, **61,** 210
microneedle fabrication
final tip, **61,** 213–214
first joint, **61,** 212–213
microneedle cleaning, **61,** 214
needle shaft shaping, **61,** 213
second joint, **61,** 213
microneedle positioning, **61,** 215–216
manipulating cell components
centrosome manipulation, **61,** 217
chromosome manipulation, **61,** 216–217
focusing considerations, **61,** 216
microtubule manipulation, **61,** 217
Micromanipulation, *see* Laser scissors, Laser tweezers
Micromechanical limitations
Brownian motion
hydrodynamic drag, **55,** 139–140
power spectrum, **55,** 136–138
trap calibration, **55,** 138–139
noise
instrumental noise sources, **55,** 147–154
amplifier noise, **55,** 59–60, 151–152
backreflections, **55,** 60, 152–153
beam-pointing, **55,** 59, 62, 153
dynamic probe interactions, **55,** 146–147
electronics, **55,** 58–62, 147–152
force clamp experiments, **55,** 143–146
Johnson noise, **55,** 149–151
position clamp experiments, **55,** 140–143
power fluctuations, **55,** 58, 152
shot noise, **55,** 58–59, 149
vibration, **55,** 58, 154
overview, **55,** 129–132, 154
spectral data analysis
calculation, **55,** 134–135
interpretation, **55,** 132–134
Micronuclear DNA
isolation, **62,** 239–250
Micronucleus, **62,** 32, 71, 74–75
amicronucleate cells, **62,** 7, 75, 148–149
chromosomes, **62,** 134
division, **62,** 74–76
transcription, **62,** 148–149
transformation, **62,** 172–173, 501–511
Micropatterned culture surfaces
alkanethiolate SAMs, **69,** 393, 395–397
cell plating, **69,** 393, 398
ECM, **69,** 390–391
ECM cell regulation, **69,** 386–387
ECM coating, **69,** 393, 398
gold-coated surface fabrication, **69,** 393, 397
hepatocyte studies, **69,** 390
pattern stamping, **69,** 393, 397–398
PDMS stamps, **69,** 392–393, 395
photolithographic master creation, **69,** 392–395
photomask creation, **69,** 392–394
pitfalls, **69,** 399–400
ultraviolet light, **69,** 391–392
Microprojectile bombardment, **65,** 392
Microsatellite repeat sequences, zebrafish, **60,** 104–105
Microscope
fluorescence, **65,** 344
Microscope slides
LSC, **66,** 76–77
Microscope, AFM interface, **68,** 174–176
Microscopes
developmental history, **63,** 89–90
historical overview, **63,** 4–5
Microscopes, *see also specific types*
laser tweezer system, **55,** 30–33, 53–56
Microscopes, *see specific types*
Microscopy
apoptosis, **66,** 104
atomic force, *see* Atomic force microscopy
basement membrane self-assembly, **69,** 133–135
confocal fluorescence, **69,** 262, 272–273
Dictyostelium centrosomes, **67,** 342–343
digital fluorescence, *see* Digital fluorescence microscopy
dish preparation, **69,** 346
electron, *see* Electron microscopy
fluorescence lifetime imaging, *see* Fluorescence lifetime imaging microscopy
for centrosome reproduction assay, **67,** 271–276, 300

Microscopy (*cont.*)
  for *Drosophila* centrosome analysis, **67,** 121
  immunofluorescence, *see* Immunofluorescence microscopy
  live cell fluorescence lifetime imaging, **69,** 272–273
  Mabs for instructive ECM identification, **69,** 43–44
  multiple-frequency fluorescence lifetime imaging, **69,** 273–275
  probe, *see* Probe microscopy
  protein kinases, **67,** 306–308
  tagged protein kinases, **67,** 308–310
Microscopy, electron, **62,** 323–330
  double labeling, **62,** 338–339
  embedding and sectioning, **62,** 328–330
  fixation, **62,** 324–327
  immuno-e.m., **62,** 333–342
  SEM, **62,** 327–328
Microscopy, light
  brightfield, **62,** 313–321
  confocal fluorescence microscopy, **62,** 305–310
  DAPI staining of nuclei, **62,** 291–310
  fluorescence microscopy, **62,** 291–310, 459–461
  Giemsa staining, **62,** 319–321
  Protargol staining, **62,** 68–69, 317–319
  silver staining, **62,** 314–317
  SYTOX staining of nuclei, **62,** 297–310
Microscopy, *see* Electron microscopy, Immunofluorescence microscopy, Light microscopy
Microscopy, *see specific technique*
Microscopy, *see specific types*
Microstome, **62,** 90
Microtiter plates, bound HS, heparin binding, **69,** 98–99
Microtubule motors, kinesin-like, **59,** 227–248
  anti-zKLP antibodies, production, **59,** 233–242
    anticipated results, **59,** 241–242
    fusion protein expression, **59,** 238–240
    maltose-binding protein, **59,** 237–238
  cloning strategy, zebrafish genes, **59,** 229–234
    isolation and reverse transcriptase, **59,** 230–231
    isolation of cDNAs, **59,** 233
    materials, **59,** 229–230
    PCR products, **59,** 232–233
    PCR reactions, **59,** 231–232
  general considerations, **59,** 247–248
  GFP-tagged zKLPs in cultured cells, **59,** 242–247
    DNA constructs, **59,** 244–246
    immunostaining of COS cells, **59,** 247
    materials, **59,** 243–244
    visualization, **59,** 246–247
  introduction, **59,** 227–229
Microtubule nucleation
  and γ-tubulin, **67,** 179–185
  at mitotic SPB, **67,** 171
  ectopic, effect of γ-tubulin, **67,** 185–189
  primary antibodies for, **67,** 328
  reconstitution by salt-inactivated centrosomes, **67,** 142–145
  scoring, **67,** 328–330
  solutions for, **67,** 331–333
  Spisula
    cosedimentation assay, **67,** 158–161
    disassembly and reassembly, **67,** 152–153
    immobilized centrosome recovery assay, **67,** 153–155
    IVEM tomography, **67,** 155–156
    linear sucrose gradient centrifugation, **67,** 157–158
  yeast SPBs, **67,** 169–170
Microtubule nucleation assays
  Dictyostelium centrosomes, **67,** 344–345
  for γTuSC and γTuRC, **67,** 19–20
  on touch preparations, **67,** 327–328
  preparation, **67,** 326
  technical consideration, **67,** 326–327
Microtubule reassembly assay, **67,** 182
Microtubules, **62,** 51–57, 95
  array, evolution, **67,** 298
  aster assembly, **67,** 228–229
    mitotic extract preparation, **67,** 226–228
    motor function, **67,** 229–233
  cytoplasts
    Cy3-labeled tubulin preparation, **67,** 45–46
    enucleation, **67,** 47–48
    image acquisition, **67,** 48–49
    image analysis, **67,** 49–51
    microinjection, **67,** 47–48
  during conjugation, **62,** 95–96
  in plant cells, **67,** 194–196
  longitudinal bands, **62,** 81
  pericentrin recruitment onto centrosomes, **67,** 60–61
  polymerization assays, **67,** 241–243
  protein kinase, **67,** 310–311
  sea urchin, **67,** 151–152
  sedimentation, pericentrin, **67,** 65–66
  visualized by immunofluoresence microscopy, **62,** 32, 49–50, 81, 291–310

Xenopus oocytes, **70,** 398, 402, 411
Xenopus oocytes and embryos, **70,** 383–386
Microtubules, high-resolution multimode digital analysis, **56,** 186, 212–214
Microtubules, kinesin molecule movement observation, low-background total internal reflection fluorescence microscopy, **55,** 123
Microtubules, rotational stiffness, **68,** 25–27
Midbrain–hindbrain boundary, zebrafish, **59,** 259–260
Midline mesodermal tissues, zebrafish, **59,** 165–172
Migration
antigen, **63,** 261
cell, *see* Cell migration
phosphatidylserine, in apoptotic protoplasts, **63,** 523
Miller chromatin spreading
advantages of technique, **53,** 472–474
development of technique, **53,** 472, 474
disadvantages of technique, **53,** 474
overview, **53,** 472
Xenopus oocytes
braking pipettes, **53,** 480–481
comparison to other systems, **53,** 474
optimization, **53,** 490–491
solution preparation, **53,** 477–478
technique, **53,** 485–486
Millicell inserts, **71,** 92
Millipore Corporation, **71,** 312
Mims-enhanced yellow fluorescent protein
HeLa cells, **65,** 375
titration curve, **65,** 377
mindbomb, mutation effects in development, **60,** 79–80
Miniature inverted-repeat transposable elements
Angel, **60,** 113, 115
integration mechanism, **60,** 113
size, **60,** 112
Minimal residual disease
cell culture, **64,** 397
cytogenetics, **64,** 397–401
detection
requirements for, **64,** 386
tissue sources for, **64,** 387
flow cytometry, **64,** 401–402
immunophenotyping, **64,** 388–397
PCR, **64,** 402–414
technical problems, **64,** 414–415
Mirrors, multiple trap synthesis, mechanical deflectors, **55,** 208
MITEs, *see* Miniature inverted-repeat transposable elements
Mithramycin
staining of plant chromosomes, **64,** 15
stock solution, **64,** 10
Mito Tracker dyes, **66, 84**
Mito Tracker green, **66, 38**
Mito Tracker orange, **66,** 381
Mito Tracker red, **66,** 381
Mitochondria, **62,** 32, 49–50, 81
ATP, **66,** 382
cellular autofluorescence measurements, **65,** 92–93
cytochrome *c* release, **66,** 368–369
fluorescent dyes, **66,** 379t, 380
function, **66,** 379–386
Holey study, **66,** 365–391
isolated
cytochrome *c* release, **66,** 376–377
in vitro assays, **65,** 76–87
inheritance cycle, **65,** 334–335
marker antigens
yeast, **65,** 265
mass, **66,** 382
membrane potential, **66,** 379–380
fluorescence measurements, **65,** 87–90
mouse liver, **66,** 375–376
outer membrane permeability, **66,** 384–386
permeability transition, **66,** 383–384
proteins released, **66,** 377–378
proton pumping measurements, **65,** 86–87
reactive oxygen species assays, **65,** 80–87
respiratory enzyme redox assay, **65,** 76–80
role during apoptosis, **63,** 470–472
ROS, **66,** 382–383
sedimentation assay, **65,** 167
swelling assay, **65,** 46
swelling, **63,** 471–472
toxins, **66,** 386
Mitochondria-actin binding
analysis, **65,** 170–172
mitochondrial purity, **65,** 172
scanning densitometry, **65,** 170–171
Mitochondria, fluorescent detection, **64,** 118
cardiolipin dye, **64,** 120, 124–125
cell preparation for, **64,** 122
critical aspects, **64,** 125–126
dyes for, **64,** 121–122
mitochondrial oxidative turnover, **64,** 123–124
mitochondrial protein level, **64,** 124
normalized mitochondrial membrane potential, **64,** 122–123, 126

Mitochondria, fluorescent detection (*cont.*)
reduced dyes, **64,** 120–121
subcellularly targeted GFP, **64,** 121
symmetrical and asymmetric carbocyanine dyes, **64,** 119–120
xanthylium dyes, **64,** 119
Mitochondria, green fluorescent protein targeting
chimera construction and expression, **58,** 78
expression in primary cultures, **58,** 80–83
fluorescence microscopy, **58,** 79–80, 83–84
multiple organelle labeling with blue-shifted protein, **58,** 81
transfection, **58,** 83
Mitochondria, pCa_i measurement, **70,** 464
Mitochondrial ATP synthesis
animal cells, **65,** 133–143
assay specificity, **65,** 136
ATP detection, **65,** 135–136
ATP measurement, **65,** 138–141
cell culture, **65,** 138
cell permeabilization, **65,** 135
experimental procedures, **65,** 136–141
media and reagents, **65,** 137
Mitochondrial calcium measurements
aequorin, **65,** 358
Mitochondrial cotransformants, **65,** 385
Mitochondrial cytochrome *c* release
cell-free systems, **66,** 375–377
cultured cells, **66,** 369–372
mitochondria, **66,** 368–369
single cells, **66,** 372–375
Mitochondrial-dependent apoptosis pathway, **66,** 367
Mitochondrial disorders, **65,** 312–313
Mitochondrial DNA (mtDNA)
binding protein, **65,** 420
competitive PCR, **65,** 418–419
COX 1 gene
histochemistry, **65,** 325
DAPI, **65,** 415–417
detection, **65,** 415–419
EtBr, **65,** 399–400
gene replacement
Saccharomyces cerevisiae mitochondria, **65,** 388–392
human molecule
description, **65,** 414
immunolocalization
respiratory chain, **65,** 317–318
localization
immunofluorescence, **65,** 327
mutations, **65,** 312–313
nuclear DNA-encoded subunits
fluorochromes, **65,** 326–327
PCR
COX 1 gene, **65,** 325
recombination
Saccharomyces cerevisiae mitochondria, **65,** 384
replication
fundamental aspects, **65,** 414–415
Saccharomyces cerevisiae mitochondria, **65,** 383–384
transcription-dependent, **65,** 414–415
replication defects
diagnosis, **65,** 414–419
sequences
homologous double crossovers, **65,** 388–389
Southern hybridization, **65,** 417–418
transfer
transmitochondrial hybrid cells, **65,** 407–409
Mitochondrial encephalopathy, **65,** 136
Mitochondrial encephalopathy, lactic acidosis, and stroke-like episodes (MELAS), **65,** 312, 316
Mitochondrial genome
human
description, **65,** 312
size, **65,** 420
Mitochondrial green fluorescent protein
fusion markers, **65,** 278–281
Mitochondrial green fluorescent protein markers
cells expressing
fixation, **65,** 281
Mitochondrial imaging
cryotechniques, **65,** 251
Mitochondrial localization
protein import assaying
cofractionation studies, **65,** 203–204
membrane integration determination, **65,** 200–201
mitochondrial swelling experiments, **65,** 202–203
preprotein protease degradation, **65,** 201
soluble protein extraction, **65,** 200
Mitochondrial matrix
reporter targeting, **65,** 355
Mitochondrial megachannel, **65,** 148
Mitochondrial membrane potential, **65,** 285–306
collapse, **63,** 472–473
fluorescent probes, **63,** 471
FRET-based assay, **65,** 296–306
in apoptotic protoplasts, **63,** 523

in early apoptotic phases, **63,** 468
measured with
dye, **63,** 12
JC-1 staining technique, **63,** 479–480
TPP electrode, **63,** 469
measurement, **65,** 287–288
novel fluorescence resonance energy transfer, **65,** 294–306
single potentiometric dyes, **65,** 288–294
Mitochondrial membrane potential, normalized, **64,** 122–123, 126
Mitochondrial membranes
integrity analysis, **65,** 46
Mitochondrial OXPHOS
digitonin titration, **65,** 127
in situ analysis, **65,** 126–130
data analysis, **65,** 128–130
P/O ratio assay, **65,** 127–128
Mitochondrial pathway, **66,** 454
Mitochondrial preparations
contaminants, **65,** 4–5
differential centrifugation, **65,** 3
gradient centrifugation, **65,** 4
homogenization, **65,** 3
isolation medium, **65,** 2–3
purity criteria, **65,** 5
storage, **65,** 4
Mitochondrial proteins
immunolocalization
ABC method, **65,** 327–328
paraffin-embedded samples, **65,** 319–321
intermembrane, **65,** 151
membrane bound, **65,** 47
soluble, **65,** 47
Mitochondrial respiratory complex activity
diagnostic investigations, **65,** 115–116
fibroblasts, **65,** 100–101
linked assays, **65,** 112–114
mitochondrial fraction preparation, **65,** 98–101
muscle tissue, **65,** 98–101
muscle mitochondrial fraction, **65,** 99–100
muscle post-600 g supernatant, **65,** 99
muscle tissue
reagents, **65,** 98
myoblasts, **65,** 100–101
Mitochondrial RNA (mtRNA)
core human, **65,** 420
human
core, **65,** 420
polymerase, **65,** 415
Mitochondrial-targeted pH-sensitive green fluorescent protein, **65,** 374–375
Mitochondrial targeting peptides (mTPs), **65,** 175–183
automatic classification, **65,** 177–178
automatic identification, **65,** 178–183
classes, **65,** 176
cleavage site production, **65,** 182–183
description, **65,** 176–177
distribution, **65,** 178
subcellular location discrimination, **65,** 179–182
Mitochondrial transcription
in vivo analysis, **65,** 419–423
human cells, **65,** 421–422
standard assays, **65,** 421–423
yeast, **65,** 422–423
initiation, **65,** 419–421
Mitochondrial transformants
bombardment, **65,** 387–388
identification, **65,** 388
Mitochondrial translation
products
in organello labeling, **65,** 431–432, 435
in vivo labeling, **65,** 431
SDS-PAGE, **65,** 435–436
protein assembly, **65,** 433
protein degradation, **65,** 433–434
protein folding, **65,** 432–433
protein sorting, **65,** 434–435
Mitochondrial transmembrane potential, **66,** 80–84
flow cytometry, **66,** 81–82
LSC, **66,** 82
measurement, **65,** 155–156
Mitogen-activated protein kinases (MAPKs), **66,** 187–188
Mitoplasts
cultured cells, **65,** 30–31
MitoProt II, **65,** 179–181
Mitosis, **62,** 74–76
Mitosis, extracts from cultured cells, **67,** 226–228
Mitosis, laser micromanipulation, **55,** 83–85
Mitotic cells
fluorescence analysis, **63,** 346
immunofluorescence microscopy, **63,** 347–349
markers, comparison with anti-H3-P monoclonal antibody, **63,** 350–353
preparation and staining, **63,** 345–346
Mitotic cells with unreplicated genomes (MUGs), *see* Centromere fragmentation
Mitotic detachment, cell cycle synchronization, **57,** 241–242, 244
Mitotic index, **64,** 362

Mitotic index, cell cycle phase duration determination, **57,** 233, 235–237
Mitotic kinesin-related proteins, *see* Antisense technology
Mitotic spindle function, inhibitors, **64,** 175
MitoTracker dye, detection of DY, **63,** 475–476
MitoTracker dyes, **65,** 341–342; **70,** 357
Mito Tracker dyes, **70,** 357
Mito Tracker Green, mitochondrial dye, **64,** 120
Mixed cultures, in *Xenopus* spinal neurons, **71,** 139
Mixing, sample, for flow cytometry, **63,** 171–175
MMP2, *see* Matrix metalloproteinase 2
Modeling, SPBs and organelles, **67,** 37–40
Models
  atomic, AFM topograph comparison
    bacteriorhodopsin surface, **68,** 276–278
    OmpF porin, **68,** 275
  basement membranes, **69,** 114
  finite element, AFM interfacing, **68,** 182
  forced protein unfolding
    Monte Carlo simulations, **68,** 327
    pathway questions, **68,** 324–325
    single-parameter model, **68,** 325–327
  synthetic provisional matrix, **69,** 155
  worm-like chain, protein folding, **68,** 17
Modified gene product, **65,** 258–265
  cassette preparation, **65,** 258–262
  isolation and characterization, **65,** 262–265
Modified strains
  functional analysis, **65,** 262–264
  PCR screening, **65,** 262–264
  protein expression, **65,** 262–264
  selection, **65,** 262
Molecular adhesion, force spectroscopy, **68,** 14–15
Molecular analysis, monitoring MRD by, **64,** 412–414
Molecular biology, PCNA, **63,** 420–421
Molecular cloning, centrosome-associated protein cDNA
  bacterial fusion protein antibodies, **67,** 131–132
  clone characterization, **67,** 129–130
  expression library
    immunoscreening, **67,** 128–129
  mammalian cell expression, **67,** 132–133
Molecular force probing
  basic technique, **68,** 223–224
  modifications, **68,** 224
  pulling, **68,** 224
  sample preparation, **68,** 225
  typical experiment, **68,** 225
Molecular genetic methods, **65,** 328–329
  DNA
    single-fiber PCR, **65,** 328–329
  ISH to mtDNA, **65,** 328–329
Molecular genetics, integrin-mediated signaling pathways
  antisense RNA, **69,** 381
  dominant inhibitory mutants, **69,** 379
  protein introduction, **69,** 381–382
  transfection, **69,** 379–381
Molecular handshakes, between PTPs and their cargo, **71,** 329–330
Molecular interactions, PFM studies, **68,** 196–199
Molecular mass
  native protein complexes, **65,** 239–240
Molecular motion, SFM of DNA, **68,** 374–376
Molecular motors, kinesins, PFM studies
  basic structure, **68,** 24–25
  mechanical behavior, **68,** 26
  mechanical properties, **68,** 25–27
Molecular probes
Molecular probes, **71,** 132
  acid sphingomyelinase, **66,** 152
  ceramide synthase, **66,** 157
  neutral sphingomyelinase, **66,** 154–155
  sphingomyelin synthase, **66,** 158–159
Molecular pulling/force spectroscopy
  $\alpha_5\beta_1$ immobilization, **69,** 179–181
  bioartificial membranes, **69,** 177–179
  force measurements, **69,** 181–182
  overview, **69,** 174
  unfolding tenascin, **69,** 175–177
Molecular recognition studies
  applications, **68,** 116
  dynamic recognition force
    microscopy, **68,** 134–137
  lateral force mapping, **68,** 133–134
Molecules of equivalent soluble fluorochrome, **63,** 313, 315–317, 320–322, 324–325
Molluscan phylum, large neuron size in, **71,** 158
Moloney murine leukemia virus, *see* Retrovirus-mediated insertional mutagenesis, Zebrafish
Monitors, *see* Video monitors
Mono Q chromatography
  plant mitochondria, **65,** 67
Mono Q Column
  ion-exchange chromatography
    Apaf-1 purification, **66,** 22
Monoamine biogenesis, **71,** 267
Monoblastic cells, component of bone marrow infiltration, **64,** 349, 351

Monoclonal antibodies
instructive ECM
antigens and immunization, **69,** 41–42
cloning, **69,** 47
example, **69,** 29–32
fusion and hybridoma preparation, **69,** 42–43
initial characterization, **69,** 47
primary screens, **69,** 43–44
purification, **69,** 47–48
secondary screens, **69,** 44–47
integrin-associated proteins
characterization, **69,** 235
screening hybridoma clones, **69,** 235–237
Monoclonal antibodies, against *Dictyostelium* centrosomes, **67,** 347–351
Monoclonal antibody
anti-H3-P, comparison with other mitotic cell markers, **63,** 350–353
binding desired epitope, **63,** 229, 231
GNS1, **63,** 285–286
identification of lymphocyte subsets, **63,** 438
Ki-67, cell cycle distribution, **63,** 399–400
Monoclonal antibody, *see* Hybridoma
Monocot species, **61,** 426
Monocytic cells, PECY5 specific binding to, **63,** 236–237
Monokinetids, **62,** 53
Monolayer assays
cell–cell adhesion, **63,** 603–605
cell–ligand adhesion, **63,** 606–607
Monolayer study, high-resolution video light microscopy, perfusion chamber, **56,** 253–271
assembly, **56,** 270–271
chamber design, **56,** 256–260, 265–266
construction, **56,** 260–265, 267–269
historical perspectives, **56,** 256
overview, **56,** 253–256
temperature regulation, **56,** 269–270
Monoscanning confocal microscope, **70,** 47–49
Monovalent intracellular ions
flow cytometry, **66,** 57–63
cell preparation, **66,** 58
data analysis, **66,** 59–63
fluorescent ionic dyes, **66,** 58
instruments, **66,** 58
intracellular ion changes, **66,** 59
Monster phenotype, **62,** 423
Monte Carlo simulations, forced protein unfolding, **68,** 327, 334
Morone labrax, **65,** 28
Morphogenetic movements, analysis, **59,** 179–202
confocal analysis, **59,** 195–200
cell movements, analysis, **59,** 198–199
cellular trajectories, **59,** 200
embryo labeling, **59,** 196
time-lapse data sets, **59,** 196–198
confocal imaging of embryos, **59,** 181
embryo imaging, **59,** 201
imaging procedures, **59,** 193–195
confocal image acquisition, **59,** 194–195
optics selection, **59,** 193–194
time-lapse recordings, storage and analysis, **59,** 195
introduction, **59,** 180–181
mounting embryos for imaging, **59,** 192–193
imaging chambers, **59,** 192
pH stabilization, **59,** 193
spatial orientation, **59,** 192–193
summary, **59,** 201–202
visual information, distribution, **59,** 200
vital staining, general principles, **59,** 181–192
addition vital stains, **59,** 191–192
materials, **59,** 189–190
procedure, **59,** 190–191
zebrafish embryos, **59,** 182–189
Morphological analysis
neural-enriched cultures for, **71,** 143
Purkinje neuronal cultures, **71,** 96–100
Morphological similarity, chick forebrain and rat hippocampal neurons, **71,** 54–56
Morphological types
identifying, **71,** 98
Purkinje neurons, **71,** 96–97
quantitating, **71,** 99
Xenopus varieties, **71,** 143
Morphology, control of, in avian Purkinje cultures, **71,** 90–104
Morphology, invadopodia and phagosomes, **63,** 614–615
Mosaic analysis with repressible cel marker (MARCM), **71,** 222–223
Mosaic analysis, neural development, techniques, **59,** 263–267
Motility
assays, in B35 cell lines, **71,** 299–300
comparison among neuronal cultures, **71,** 9
monitoring via Transwel migration assays in B35 cells, **71,** 299–300
Motion analysis, **71,** 320–321
Motion, *see* Vibration isolation

Motor function, MT, in aster
assembly, **67,** 229–233
Motorneuron mutation screening, *see* Early pressure screens, Zebrafish
Motor neurons
enrichment by density gradient centrifugation, **71,** 79–80
experimental uses of, **71,** 80–82
induction via notochord, in chick embryos, **71,** 68
Mounting
Drosophila embryos, **70,** 364–365
thick tissue sections, **70,** 331–332
Xenopus oocytes and eggs, **70,** 395–397
Mounting embryos, imaging, **59,** 192–193
Mouse
mammary epithelial cell cultures, **69,** 409
mutant, fibroblast derivation, **69,** 314
Mouse cortical neurons, **71,** 5
culturing, **71,** 111–126
Mouse embryo fibroblasts (MEFs), **66,** 198–202
analysis, **66,** 210–222
cell isolation, **66,** 200–201
controls, **66,** 208
culture, **66,** 201–202
freezing, **66,** 202
infection protocol, **66,** 205–210
maintenance, **66,** 201–202
multiple gene infection, **66,** 210
packaging cell culture, **66,** 205
packaging cell transfection, **66,** 205–206
variables, **66,** 208–210
viral supernatant collection, **66,** 206–207
Mouse hippocampal neurons, culturing, **71,** 111–126
Mouse liver mitochondria
isolation, **66,** 375–376
Mouse mutants, cerebellum as study model for, **71,** 91
Mouse skin fibroblasts, **66,** 223
Mouse/zebrafish somatic cell hybrid
applications, **60,** 304, 319–320
chromosome morphology, **60,** 312–313
construction
cell lines and culture, **60,** 304–306
chromosome tagging with *neo* resistance gene, **60,** 306–307
fusion of cells, **60,** 307
fluorescence *in situ* hybridization characterization
chromosome and slide preparation, **60,** 310–311
chromosome number, **60,** 313
denaturation, **60,** 311
fluorescence detection and amplification, **60,** 312
hybridization, **60,** 311
probes
Alu Type Ib, **60,** 308–309
labeling, **60,** 309
painting probe, **60,** 308
SRI, **60,** 308
washes, **60,** 311–312
freezing, **60,** 307–308
morphology and growth rate, **60,** 308
polymerase chain reaction characterization
amplification reaction, **60,** 313–314
available markers, **60,** 313
DNA/mermaid polymerase chain reaction, **60,** 317–318
hybrid panel composition, **60,** 314
Southern blot analysis, **60,** 314, 316–317
stability, **60,** 318–319
Movement
mechanism of axonal, **71,** 305
observing axonal, **71,** 314–321
Movies, *see* Adobe Photoshop, Fluorescence microscopy, Laser scanning confocal microscopy
Mowiol mounting medium, preparation, **53,** 365
MRD, *see* Minimal residual disease
mRNA and protein trafficking and targeting, **71,** 112
MRNA polymerase
core yeast, **65,** 420
MtDNA, *see* Mitochondrial DNA
MtHsp 70 defects
yeast mutants, **65,** 207
MTOCs, *Xenopus* oocytes, **70,** 412
MTPs, **65,** 175–183
MtRNA
core human, **65,** 420
human
core, **65,** 420
polymerase, **65,** 415
MTS, **65,** 180
MTs, *see* Microtubules
MTT assay, applied to Boyden chamber assay, **63,** 557
MTT, apoptosis assay, **57,** 261–262
Mucin, **62,** 350
Mucocysts, *see* Dense-core secretory granules
Mucosa, proliferation rates, **64,** 467

Multicolor detection, PRINS, **64,** 60–61
Multicolor laser scanning confocal microscopy,
see also Confocal laser scanning microscopes
air-cooled examples, **70,** 228–231
argon ion lasers, **70,** 187–188
detector aperatures, **70,** 190
detector–double-labeled specimen
cross-talk, **70,** 213–217
detectors, **70,** 190–191
double-labeled specimen imaging, **70,** 225–228
excitation/emission wavelength differences
chromatic aberration, **70,** 235–237
optical sectioning depth, **70,** 237
resolution effects, **70,** 233–235
spherical aberration, **70,** 235
fluorophores and labeling reagents
blue fluorophores, **70,** 177–178
BODIPY, **70,** 171–173
cyanine fluorophores, **70,** 176
fluorescein, **70,** 171
Lucifer yellow, **70,** 173
overview, **70,** 167–171
phthalocyanine derivatives, **70,** 176–177
phycobiliproteins, **70,** 174–176
rhodamine fluorophores, **70,** 173
fluorophore visualization
specificity, **70,** 209–212, 221–225
helium–cadmium lasers, **70,** 189
helium–neon lasers, **70,** 189
image constrast, **70,** 238
image display, **70,** 239–240
image registration, **70,** 238
immunofluorescence staining techniques
background staining, **70,** 184–185
fluorophore visualization specificity, **70,** 181
overview, **70,** 178–180
primary antibody specificity, **70,** 183–184
secondary antibody specificity, **70,** 181–183
krypton–argon ion lasers, **70,** 188
krypton ion lasers, **70,** 188
laser light source overview, **70,** 185–187
multiple fluorophore single-wavelength
excitation
cross-talk compensation, **70,** 195–201
overview, **70,** 192
two-color LSCM with 514-nm argon ion
laser, **70,** 193–195
optical filters, **70,** 189–190
overview, **70,** 166–167, 191–192
three-color LSCM, **70,** 206–207, 217–221
tunable dye lasers, **70,** 189
two-color LSCM
argon ion lasers, **70,** 201–204
helium–neon lasers, **70,** 204
krypton–argon ion laser, **70,** 204–206
overview, **70,** 201, 212–213, 225
Multicomponent extracellular matrix
invasion assays, **69,** 198, 201
matrix degradation, **69,** 197–198
matrix preparation, **69,** 196–197
matrix protein components, **69,** 201–204
Multifraction studies, spheroids for, **64,** 226–227
Multiimmersion objective lenses, **70,** 266
Multimode digital microscopy,
see also Digital image processing, **56,** 185–214
applications, **56,** 209–214
anaphase, *in vitro*, **56,** 210–211
digitally-enhanced differential interference
contrast images, **56,** 209–210
fluorescence photoactivation, **56,** 211–212
green fluorescent protein–dynein-labeled
microtubules, **56,** 186, 212–214
immunofluorescent specimens, **56,** 214
migrating tissue cells, **56,** 212
cooled charge-coupled device
camera, **56,** 199–207
background reference image
subtraction, **56,** 205–206
camera features, **56,** 199–200
cooling advantages, **56,** 203–205
dark reference image subtraction, **56,** 205–206
linearity, **56,** 203–205
noise, **56,** 203–205
quantum efficiency, **56,** 200–202
readout, **56,** 202–203
shading, **56,** 206
signal-to-noise ratio, **56,** 206–207
design criteria, **56,** 187–190
digital imaging system
computer control, **56,** 207–208
image analysis, **56,** 208–209
image processing, **56,** 208–209
software, **56,** 209
microscope components, **56,** 191–199
epi-illumination optics, **56,** 194–197
focus control, **56,** 198–199
imaging optics, **56,** 191–194
trans-illumination optics, **56,** 197–198
vibration isolation, **56,** 199
overview, **56,** 186–187
Multiparameter data analysis
algorithms, **64,** 488–489
asthma, **64,** 490, 494–502
cell gating approach, **64,** 308

Multiparameter data analysis (*cont.*)
HIV infection, **64,** 490
immunophenotyping study, **64,** 346–347
melanoma, **64,** 489–490, 493–494
numbers of analysis parameters, **64,** 309, 311
Multiparametric analysis
DNA content in, **63,** 292–293
H_3 phosphorylation in U937 cells, **63,** 348–349
lymphocyte activation, **63,** 434
using FITC conjugated MIB-1 antibody, **63,** 414
Multipass system,
immunophenotyping, **64,** 336–340
Multiphoton autofluorescence imaging, **63,** 592
Multiphoton excitation imaging
calcium dynamics visualization, **70,** 141–142
cancerous tissue *in vivo*, **70,** 141
imaging penetration depth, **70,** 133–134
laser choices, **70,** 140
long-term, embryo development, **70,** 142–143
microscope optimization, **70,** 139–140
photoactivation, **70,** 144
principles, **70,** 131–133
probes, **70,** 138–139
pulse-width considerations, **70,** 137–138
specimen viability, **70,** 135–137
three-photon excitation, **70,** 138
Multiphoton microscopy
advantages and disadvantages, **70,** 288
confocal microscopy alternative, **70,** 69–73
overview, **70,** 288
suitable objectives, **70,** 288–289
Multiphoton microscopy, elimination of aperture in, **63,** 586–587
Multiple cell exclusion, LSCM list mode feature, **63,** 79–80
Multiple conductance channel, **65,** 148
Multiple drug resistance
as phenotypic characteristic, **64,** 193
drugs, **64,** 197
identification methods, **64,** 194–196
Multiple-frequency fluorescence lifetime imaging microscopy, **69,** 273–275
Multiple gene infection
MEFs, **66,** 210
Multiple-label imaging
chromatic aberration corrections, **70,** 278–279
confocal microscopy, **70,** 56
Multiple mutations
rho$^+$ recipients, **65,** 390
Multiple myeloma, detection of MRD, **64,** 396
Multiple traps, synthesis, **55,** 205–215
acousto-optic deflectors, **55,** 208–209
computerized position control, **55,** 212–213
mechanical deflectors, **55,** 208
noise, **55,** 210–212
overview, **55,** 205–206, 213–215
position, **55,** 210–212
trapping configurations, **55,** 206–208
object holding, **55,** 206
object manipulation, **55,** 206–207
static force exertion, **55,** 207–208
Multivalent cations, DNA on mica
overview, **68,** 361–362
soluble cation effects, **68,** 362–363
Murine ecotropic retroviral receptor, **66,** 224–225
Murine hepatitis virus, **63,** 103
Muscle fiber
dissection, **65,** 323–324
Muscle mitochondria, **65,** 19–20
Mutagenesis, **62,** 189, 264, 282
Mutagenesis, in flies, **71,** 202
Mutant analysis, *see* Genetic analyses
Mutant isolation, **62,** 156–158, 279–287
Mutant phenotype rescue neural development, techniques, **59,** 267–268
Mutant screens, **71,** 196
Drosophila, **71,** 199
Mutant strains
cell cycle (*chp*), **62,** 280
cell division (*cdaA*), **62,** 73–74, 81
ciliary reversal, **62,** 364–365
ciliogenesis (*dcc*), **62,** 280
conjugation (*cnj*), **62,** 96–97
cortical organization, **62,** 80–81
cycloheximide (*chx1*), **62,** 520
cytokinesis (*dcc*), **62,** 280
dynein (*oad*), **62,** 280
exocytosis, **62,** 189
isolation, **62,** 156–158
lysozymal enzymes, **62,** 190
methylpurine (*mpr1*), **62,** 519
oral apparatus, **62,** 190
paromomycin (*pmr*), **62,** 543
patterning (*hpo*), **62,** 78, 84–88
patterning (*jan*), **62,** 84, 98
phagocytosis, **62,** 43, 190
rDNA (*rmm*), **62,** 139, 143
secretion (*sec-1*), **62,** 44
sexual maturity (*em*), **62,** 133
Mutation, nomenclature, **62,** 562
Mutations
affecting vascularization and hematopoiesis, **59,** 327–332
chemically induced, **71,** 199, 202

clonal techniques for analyzing, **71,** 222–223
generating in genes of interest, **71,** 201–202
integrin adhesion molecule gene, **69,** 314–317
integrin-mediated signaling pathways, **69,** 379
phenotypes of early lethal, **71,** 216
zebrafish fin, growth control, **59,** 303–307
Myc, immortalization of cells, **57,** 74–76
Mycobacteria, diagnostic assays, **64,** 562–563
Mycoplasma, culture contamination
containment and elimination, **57,** 63
effects and consequences, **57,** 57–60
prevention and control, **57,** 58, 60, 63–64
sources, **57,** 56–57
testing, **57,** 41–43, 55, 60–63, 153–154
types of contaminants, **57,** 56
Myc-tagged proteins, immunoprecipitation, **67,** 86–87
Myelin-associated glycoprotein, myelination marker, **57,** 185
Myeloma, fusion techniques, **69,** 42
Myoblasts
mitochondrial respiratory complex activity, **65,** 100–101
Myoclonus epilepsy with ragged-red fibers (MERRF), **65,** 312, 316
ISH to mtDNA, **65,** 323
olivary nucleus
immunostaining, **65,** 320
Myosin, **62,** 63
Myosin light chain–green fluorescent protein fusion protein
construction
primer design, **58,** 242–243
terminus selection for protein fusion, **58,** 241–242
fluorescence microscopy
charge-coupled device camera, **58,** 256
culture chambers, **58,** 255–256
time-lapse acquisition and image processing, **58,** 256
microinjection, **58,** 247
photobleaching, **58,** 258
subcellular distribution, **58,** 252
transfection of cells
calcium phosphate, **58,** 246–247
efficiency, **58,** 246
embryonic avian cardiomyocyte preparation, **58,** 243–244
embryonic avian myoblast preparation, **58,** 243, 245–246
lipofection, **58,** 246
troubleshooting, **58,** 257–258
Myosin molecule, ATP turnover observation, low-background total internal reflection fluorescence microscopy, **55,** 123–125
Myxothiazol, **65,** 105–106

N

N_2-based media, **71,** 34
for peripheral neuron cultures, **71,** 23
N-cadherin, in *Xenopus cultures*, **71,** 146
N_9 neuronal growth supplement, **71,** 61
N_3 supplement, **71,** 84
n-Propyl gallate, **62,** 303
NA, *see* Numerical aperature
Na^+/D-glucose cotransporter, dynamic force microscopy
epitope mapping, **68,** 131–133
membrane-sidedness and functionality, **68,** 131
overview, **68,** 130–131
Nab2p, fluorescence image microscopy, **53,** 59
Nab3p, Western blotting of yeast nucleolus-enriched fraction, **53,** 43
NADH
cell culture, **66,** 309
plasma membrane purification, **66,** 309–310
NADH-cytochrome c reductase
complex I assays, **65,** 78
NADH-ferricyanide reductase assay
complex I assays, **65,** 77–78
NADH-Q reductase
complex I assays, **65,** 76
NADH, mitochondrial, **64,** 122
NADPH/NADH consumption assay, **66,** 310
NADH-ubiquinone oxidoreductase, **65,** 101–104
rotenone sensitivity, **65,** 103
Names
of cell lines, **71,** 3
of neurons, **71,** 3
Nanosep microconcentrators, **71,** 138
Narishege IM-200 microinjector, **71,** 138
National Cancer Institute Cooperative Human Tissue Network, **66,** 412
Native protein
BN-PAGE
electroelution, **65,** 236–237
semidry electroblotting, **65,** 237
complexes
molecular mass determination, **65,** 239–240
NBD-phospholipid transport measurements, **66,** 356–357

Ncd–green fluorescent protein, fusion protein studies of *Drosophila* spindle dynamics
advantages over direct cytoskeletal protein fusion, **58,** 154–155, 162
confocal microscopy imaging
embryo preparation, **58,** 156–158
filters, **58,** 158–159
oocyte preparation, **58,** 156–157
time-lapse imaging, **58,** 158
construct design, **58,** 155–156
mutant analysis, **58,** 160
photoactivation applications, **58,** 160–161
subcellular localization, **58,** 159–160
transgene expression, **58,** 154
NCI-Frederick Cancer Research and Development Center, **71,** 372
Nearest-neighbor distribution function, statistical analysis of particle distribution, **53,** 89, 91, 93–95
Necrosis
light microscopy, **66,** 397
morphological features, **66,** 395–396
vs. apoptosis
flow cytometry, **66,** 71
Necrosis, vs apoptosis, **63,** 535–536
Needles, for *Drosophila* centrosome analysis, **67,** 116–118
Neffs medium, see Culture media
Neighborhood-based processing, **70,** 107–108
Neomycin, **66,** 207, 424, 426, 427–428, 431
Neonatal rat
central nervous system, **71,** 287
superior cervical sympathetic ganglia, dissection of, **71,** 25–27
Neoplasms, Ki-67 protein as prognostic marker, **63,** 400
Nernst equation, permeant cationic dye redistribution
plasma membranes, **70,** 434–436
single mitochondria membrane potential, **70,** 434–436
Nerve growth factor
for peripheral neuron cultures, **71,** 22
in PC12 cell line development, **71,** 268
Nervous system development, *Drosophila*
cell labeling, overview of techniques, **58,** 166–167
cell types, **58,** 166
green fluorescent protein expression
β-galactosidase fusion protein, **58,** 170
GAL4 system in targeting, **58,** 167–168, 174
imaging
confocal microscopy, **58,** 178–180
epifluorescence, **58,** 176–178
fixed samples, **58,** 175–176
live samples, **58,** 175–176
movement imaging, **58,** 179
photobleaching minimization, **58,** 179
time-lapse imaging, **58,** 179–180
maternal and early embryonic expression of fusion proteins, **58,** 174
tau fusion proteins, **58,** 170–173
UAS vector and green fluorescent protein fusions, **58,** 168–170
Nervous system injury, *in vivo* delivery of reagents in therapy, **71,** 326
Neural cell adhesian molecule (CAM) function, **71,** 291–292
Neural crest mutation screening, *see* Early pressure screens, Zebrafish
Neural development, techniques, **59,** 253–269
anterior–posterior pattern, organizers in neural plate, **59,** 253–259
organizing center specification, **59,** 253–256
signals, **59,** 256–259
introduction, **59,** 253–254
neural patterning mutants, **59,** 262–268
mosaic analysis, **59,** 263–267
mutant phenotype rescue, **59,** 267–268
neuroanatomy, **59,** 262–263
whole-mount RNA *in situ* hybridization, **59,** 263
zebrafish mutations, **59,** 259–262
forebrain, **59,** 260–261
hindbrain, **59,** 261–262
midbrain–hindbrain boundary, **59,** 259–260
Neural induction, in chick embryos, **71,** 68
Neural patterning mutants, neural development, techniques, **59,** 262–268
Neural tube, injection of retrovirus into, **71,** 381–382
Neural-enriched cultures, *Xenopus* spinal neurons, **71,** 143–145
Neurite branching
studies of, **71,** 9
using motor and spinal cord neuron cultures, **71,** 80–82
toxicologic effect of lead on, **71,** 290
Neurite diameter, as criterion for choice of neuronal culture, **71,** 2
Neurite initiation and extension, in hippocampal and cortical neurons, **71,** 112
Neurite length, as criterion for choice of neuronal culture, **71,** 2

Neurites, length and number of, **71,** 9–10
Neuroanatomy, of chick ciliary ganglion, **71,** 38
Neurobasal medium, in chick ciliary neuron cultures, **71,** 43
Neuroblastoma, **71,** 3
Neurodegenerative diseases, cortical and hippocampal involvement in, **71,** 112
Neurofilament 200 monoclonal antibodies, **71,** 95
Neuromuscular junction studies, **71,** 41
  in *Drosophila*, **71,** 230–232
Neuron choice, factors influencing, **71,** 2
Neuronal precursor cells, migration, **69,** 351
Neuron properties, affecting choice of culture, **71,** 2
Neuron size
  advantages of *Heliosoma* species, **71,** 157
  in Tibia-1 pathway, **71,** 172
  problematic in *Drosophila* recording, **71,** 228–230
Neuron-specific enolase (NSE), **71,** 95
Neuronal cell death, **71,** 287–288
Neuronal characterization, Purkinje neuronal cultures, **71,** 95–96
Neuronal culture systems
  choosing, **71,** 8–14
  properties of, **71,** 2
  relative infectivity of, **71,** 408–411
Neuronal fate, in *Drosophila* studies, **71,** 196
Neuronal guidepost cells, in Tibia-1 pathway studies, **71,** 174–176
Neuronal labeling
  immunolabeling in *Drosophila*, **71,** 217–218
  in *Xenopus* cultures, **71,** 131–139
  of Tibia-1 pathway neurons, **71,** 185–186, 190
Neuronal migration, cerebellum as study model for, **71,** 91
Neuronal morphology, neural-enriched cultures for study of, **71,** 143
Neuronal outgrowth, Tibia-1 pathway studies in, **71,** 171–191
Neuronal properties, study through hippocampal and cortical neurons, **71,** 111
Neurons
  ease and cost of obtaining and growing, **71,** 13–15
  purity of, **71,** 11–12
Neurons, live, AFM studies, **68,** 38
Neurotransmission, in *Drosophila* studies, **71,** 196
Neurotransmitter metabolism, **71,** 91
Neurotransmitter receptor assembly, **65,** 241–242
Neurotrophin action, **71,** 267
Neutral sphingomyelinase, **66,** 152–161
  materials, **66,** 153
  molecular probes, **66,** 154–155
  principle, **66,** 153
  procedure, **66,** 154
  signaling regulation, **66,** 153
Neutron scattering, complementation with scanning transmission electron microscopy, **53,** 143
NF-κB, LSCM analysis, **64,** 437–438
NH_3 permeability, cortical collecting tubule, **70,** 422–423
Nichols Institute Diagnostics, **71,** 281
nic1, gamma-ray mutagenesis, **60,** 65
Nick translation reaction, *Tetrahymena thermophila* rDNA FISH, **70,** 346
NIH 3T3 cells, AFM studies, **68,** 38
NIH-Image
  availability, **58,** 190, 268, 351
  features and applications, **58,** 96, 161, 191, 268, 270, 350, 360
NIH Image V1.55, **65,** 347
Nikon
  electronic scan zoom and digital resolution, **70,** 252
  multiphoton imaging objectives, **70,** 288–289
  objective lenses, **70,** 283
  slow-scan CLSM, **70,** 33–36
Nikon TE300 inverted microscope, **71,** 318
Nitrocellulose, use in *Xenopus* cultures, **71,** 147
Nitrogen cavitation
  mitochondria, **66,** 370–371
Nitrous oxide, cell cycle synchronization, **57,** 240
NLS, *see* Nuclear localization signal
NMR, *see* Nuclear magnetic resonance
NNPSL, **65,** 180,182
no arches
  gene cloning, **60,** 92, 94–95
  retrovirus-mediated insertional mutagenesis, **60,** 92
no tail, gamma-ray mutagenesis, **60,** 65
Nocodazole, cell cycle synchronization, **57,** 239–240
Nocodazole, *see* Drug role in analyzing mitosis
Noise
  charge-coupled device cameras, high-resolution multimode digital microscopy, **56,** 203–205
  dark noise
    reference image subtraction, **56,** 205–206
    sensitivity enhancement, **56,** 52, 54
  instrumental noise, sources, **55,** 147–154

Noise (*cont.*)
amplifier noise, **55,** 59–60, 151–152
backreflections, **55,** 60, 152–153
beam-pointing, **55,** 59, 62, 153
electronics, **55,** 58–62, 147–152
Johnson noise, **55,** 149–151
power fluctuations, **55,** 58, 152
shot noise, **55,** 58–59, 149
vibration, **55,** 58, 154
low-light electronic cameras
background noise, **56,** 27–28
readout noise, **56,** 26
shot noise, **56,** 27
micromechanical limitations
dynamic probe interactions, **55,** 146–147
force clamp experiments, **55,** 143–146
overview, **55,** 129–131, 154
position clamp experiments, **55,** 140–143
multiple trap synthesis, **55,** 210–212
reduction, digital image processing, **56,** 95–101
arithmetic removal, **56,** 97–99
digital mask, **56,** 99
filters, **56,** 100–101
spatial methods, **56,** 96–101
temporal averaging, **56,** 95–96
signal-to-noise ratio, *see* Signal-to-noise ratio
Noise, background, and higher wavelengths, **63,** 93
Noise distribution, for low background condition, **63,** 314–315
Noise, in fluorescence microscopy, **64,** 76–77
Non-Hodgkin's lymphoma, **64,** 386
immunophenotyping, **64,** 396
PCR analysis, **64,** 413
Non-intercalating DNA-binding agent facilitation, **61,** 76
Noncentrosomal aster formation, **67,** 243
Nonfluorescent standard, defining, **63,** 317
Nonimmortal cells, **66,** 198–199
analysis, **66,** 210–222
apoptosis assays, **66,** 211–214
gene confirmation, **66,** 210
growth assays, **66,** 219–222
transformation assays, **66,** 216–219
viability assays, **66,** 214–216
Nonlinear optics
membrane potential
methods, **70,** 446–448
second harmonic generation, **70,** 441–444
second harmonic generation enhancement, **70,** 444–446
two-photon excited fluorescence, **70,** 441–444
new developments, **70,** 144–145
Nonlinearity, differential, problem in ADCs, **63,** 157
Nonneuronal contamination, reducing, **71,** 412–413
in peripheral neuron cultures, **71,** 29–33
Nonproteinase enzymes
proteinase inhibitors, **66,** 263–264
Nonrespiring cytoductants
marker rescue, **65,** 390–392
Non-small cell lung cancer
BrdUrd labeling, **64,** 464
proliferation markers and prognosis, **64,** 372
proliferative activity, **64,** 365
Nonyl acridine orange, detection of mitochondrial mass, **63,** 469, 473
Nonyl acridine orange, mitochondrial dye, **64,** 120
Nop1p
fluorescence image microscopy, **53,** 59
Western blotting of yeast nucleolus-enriched fraction, **53,** 43
Nop2p, Western blotting of yeast nucleolus-enriched fraction, **53,** 43
NOR, *see* Nucleolar Organizer Regions
NORAN instruments, video-rate CLSM, **70,** 39–42
Normarski prism, high-resolution multimode digital microscopy, **56,** 193–194
nosedive, mutation effect in development, **60,** 81
Nosepiece, light microscopy, **56,** 139
Notochord, in chick embryos, **71,** 68
gene cloning, **60,** 92, 94–95
retrovirus-mediated insertional mutagenesis, **60,** 92
Novel fluorescence resonance energy transfer
assay validation, **65,** 301–306
cell selection, **65,** 298–299
dyes and protocol, **65,** 299–301
mitochondrial membrane potential, **65,** 294–306
NPC, *see* Nuclear pore complex
NPL3–green fluorescent protein
nuclear import assay *in vivo*, **58,** 114–115
shuttling assay between nucleus and cytosol, **58,** 115–117
NR-13, **66,** 36
Nsr1p, Western blotting of yeast nucleolus-enriched fraction, **53,** 43
Nuclear calcium flux, **66,** 233–234
Nuclear damage, affecting biolistics performance, **71,** 364–366

Nuclear DNA
immunolocalization
respiratory chain, **65,** 317–318
Nuclear DNA-encoded subunits
mitochondrial DNA
fluorochromes, **65,** 326–327
Nuclear envelope, *see also* Nuclear pore complex
assembly
cell-free systems, *see* Cell-free nuclear assembly
overview, **53,** 368–369, 391–393
chromatin alignment, **53,** 80
membranes, **53,** 4
nuclear lamina, **53,** 5, 24
perinuclear space, **53,** 4
preparation for immunolabeling
oocytes, amphibian, **53,** 292–293
rat liver, **53,** 291–292
tissue culture cells, **53,** 292
subfractionation from nuclei, *see* Karyoskeletal protein-enriched fraction
Nuclear fluorescence
LSC, **66,** 74
Nuclear fragments, misclassification as apoptotic cells, **63,** 533–534
Nuclear injection, transfection of neurons by, **71,** 310–313
Nuclear localization signal (NLS)
nucleoplasmin, **53,** 522
protein binding, **53,** 518
protein conjugation, **53,** 523–524, 549–550
sequence, **53,** 523, 546
Nuclear magnetic resonance, structure studies of green fluorescent protein, **58,** 7–8
Nuclear matrix protein, content, flow cytometric analysis, **64,** 272–273
Nuclear matrix stability assay, after heat shock, **64,** 274–276
Nuclear pore complex (NPC)
assembly
cell-free systems, *see* Cell-free nuclear assembly
overview, **53,** 368–369, 391–393
components, **53,** 5–6, 287–288
confocal scanning laser microscopy
antibody labeling, **53,** 87–88, 96
cell culture, **53,** 86
imaging parameters, **53,** 88
model system
correction for photobleaching, **53,** 84–85
electron microscopy verification, **53,** 85
optimization of imaging conditions, **53,** 83–84
point spread function approximation, **53,** 84
single particle localization, **53,** 85–86
smoothing, **53,** 85
specimen preparation, **53,** 82–83
permeabilization
digitonin, **53,** 87
mechanical method, **53,** 87
statistical analysis of particle distributions
cluster analysis, **53,** 95–96
nearest-neighbor distribution function, **53,** 89, 91, 93–95
numerical computation of estimators, **53,** 92
pair correlation function, **53,** 91–95
density in various systems, **53,** 96
diameter, **53,** 96–97, 287, 546
distribution during cell division, **53,** 80–81
genome architecture maintenance, **53,** 80
nucleoporin localization with immunogold electron microscopy
cell permeabilization
digitonin, **53,** 289–290
freeze-thawing, **53,** 290–291
mechanical perforation, **53,** 290
immunolabeling
cryosections, preparation and processing, **53,** 294
oocytes, *in vivo* labeling in amphibians, **53,** 293–294
permeabilzed cells and isolated nuclei, **53,** 293
plastic sections, preparation and processing, **53,** 295
mAb414 studies, **53,** 297
mapping strategies, **53,** 288–289, 296
nucleus and nuclear envelope isolation
oocytes, amphibian, **53,** 292–293
rat liver, **53,** 291–292
tissue culture cells, **53,** 292
precision of localization, **53,** 297–299
principle, **53,** 288
transmission electron microscopy, **53,** 296
phosphorylation of proteins, **53,** 369
structure, **53,** 5, 24, 81, 97
transport role, **53,** 79–80, 97
Nuclear probes, *Xenopus* oocytes, **70,** 410–413
Nuclear protein, content
flow cytometric analysis, **64,** 271–272
temperature effect, **64,** 281

Nuclear stains, *Xenopus* oocytes, **70,** 413
Nuclear transformants, **65,** 385
Nuclear transport
components of import and export, **58,** 108
fluorescence microscopy of green fluorescent protein fusion proteins
colocalization of fusion proteins with organelles
dye specificity in organelle staining, **58,** 110–111
fixed cells, **58,** 111–114
living cells, **58,** 111
design of *in vivo* localization experiments, **58,** 117–118
excitation source, **58,** 109
expression in yeast, **58,** 110, 116–117
filters, **58,** 109
NPL3–green fluorescent protein
import assay *in vivo*, **58,** 114–115
shuttling assay between nucleus and cytosol, **58,** 115–117
objective lenses, **58,** 109
screening for fusion protein mislocalization, **58,** 118
time-lapse imaging
Metamorph software, **58,** 119–120
photobleaching and cell viability, **58,** 118–119
steps in protein import, **58,** 108
Nucleation
assay for γTuSC and γTuRC, **67,** 19–20
MT, *see* Microtubule nucleation
Nuclei
in mitochondrial preparations, **65,** 4
nuclear condensation, **66,** 173
Nuclei assays, **66,** 171–175
chromatin condensation assays, **66,** 172–174
DNA fragmentation assays, **66,** 174–175
nuclei preparation, **66,** 171–172
Nuclei, *Dictyostelium discoideum*, **67,** 340–342
Nuclei preparation
nuclei assays, **66,** 171–172
Nucleic acid dyes, in flow cytometric detection of bacteria, **64,** 514–516
Nucleic acids, injection into *Xenopus* spinal neurons, **71,** 130, 137–138
Nucleolar Organizer Regions, zebrafish, **60,** 335
Nucleolar organizing region, **63,** 12
Nucleoli, measurement with LSCM FISH, **64,** 438–441
Nucleolus
acidic toluidine blue staining, **53,** 305–306
actinomycin D
inhibition of nucleolar function, **53,** 316
segregation of nucleolus, **53,** 314–315
α-amanitin, inhibition of nucleolar function, **53,** 317
activation model, **53,** 312–313
applications of purified nucleoli, **53,** 34, 46
azure C staining, **53,** 305
characterization of yeast fraction
microscopy, **53,** 43–44
proteins, **53,** 43–44
dense fibrillar component
components and functions, **53,** 304–305, 311–312
immunolabeling, **53,** 311–312
5,6-dichloro-β-D-ribofuranosylbenzimidazole
inhibition of nucleolar function, **53,** 316
segregation of nucleolus, **53,** 315–316
fibrillar center
components and functions, **53,** 304–305, 309–311
immunolabeling, **53,** 309–311
granular component
components and functions, **53,** 304–305, 312
immunolabeling, **53,** 312
isolation from *Saccharomyces cerevisiae*
cell growth
conditions, **53,** 37
medium, **53,** 35
nuclei-enriched fraction preparation, **53,** 38–40
optimization
buffer ionic strength, **53,** 43
DNase I treatment, **53,** 42
heparin/nuclear protein ratio, **53,** 42–43
sedimentation conditions, **53,** 43
protein assay, **53,** 40
reagents, **53,** 36–37
solution and buffer preparation, **53,** 35–36
spheroplast preparation, **53,** 37–38
sucrose density gradient centrifugation, **53,** 41, 43
yeast strains, **53,** 35
isolation from HeLa cells, **53,** 317–319
monoclonal antibody preparation, **53,** 46
nucleolus organizer region silver staining
blotted proteins, **53,** 308–309
cytological preparations, **53,** 306–307
ribosome subunit assembly role, **53,** 33–34, 304–305
Nucleolus, protoplast, swelling, **63,** 507–508

Nucleoplasm, staining intensity, **63,** 410
Nucleoplasmin
chromatin remodeling role, **53,** 498–499
fluorescence labeling, **53,** 522–523
nuclear localization signal, **53,** 522
Nucleoporin
characterization, **53,** 6
localization with immunogold electron microscopy
cell permeabilization
digitonin, **53,** 289–290
freeze-thawing, **53,** 290–291
mechanical perforation, **53,** 290
immunolabeling
cryosections, preparation and processing, **53,** 294
oocytes, *in vivo* labeling in amphibians, **53,** 293–294
permeabilized cells and isolated nuclei, **53,** 293
plastic sections, preparation and processing, **53,** 295
mAb414 studies, **53,** 297
mapping strategies, **53,** 288–289, 296
nucleus and nuclear envelope isolation
oocytes, amphibian, **53,** 292–293
rat liver, **53,** 291–292
tissue culture cells, **53,** 292
precision of localization, **53,** 297–299
principle, **53,** 288
transmission electron microscopy, **53,** 296
Nucleosome
atomic force microscopy, **53,** 182
transmission electron microscopy
in situ
buffer system selection, **53,** 169
cryoultramicrotomy, **53,** 175
data collection, **53,** 175–176
Lowicryl resin handling and sectioning, **53,** 170–173
overview, **53,** 168
low-temperature embedding and ultraviolet polymerization, **53,** 172
progressive lowering of temperature embedding, **53,** 169–171
rapid freezing and freeze substitution, **53,** 169–172
staining of sections, **53,** 173–175
in vitro
adhesion to substrate, **53,** 178–179
cryoelectron microscopy, **53,** 180–183
fixation, **53,** 177–178
heavy metal shadowing, **53,** 178
negative staining, **53,** 180
nuclei isolation, **53,** 176
overview, **53,** 168
positive staining, **53,** 178–180
soluble chromatin preparation, **53,** 176–177
Nucleosome, *see* Chromatin
Nucleotide-dependent intermediates, **65,** 208–209
Nucleotide sequence, two-hybrid vectors, **67,** 76–79
Nucleus
assembly in cell-free systems, *see* Cell-free nuclear assembly
imaging, *see* Confocal scanning laser microscopy; Fluorescence microscopy; Transmission electron microscopy
protein import, *see* Protein import, Nucleus
purification protocols
buffer selection, **53,** 176
chicken liver, **53,** 13–14
Chironomus tentans salivary gland, **53,** 128–131
Drosophila embryo
method I, **53,** 25–26
method II, **53,** 26–27
frog liver, **53,** 13–15
guinea pig liver, **53,** 12–13
rat liver
centrifugation, **53,** 10–11
homogenate preparation, **53,** 7–10, 291–292
liver removal, **53,** 6–7
protease inhibitors, **53,** 8–9
quantification by absorbance, **53,** 11
storage, **53,** 12
sucrose density gradient centrifugation, **53,** 11
Saccharomyces cerevisiae
Ficoll gradient centrifugation, **53,** 38–40
growth, **53,** 35, 37
materials, **53,** 35–37
spheroplast preparation, **53,** 37–38
tissue culture cells grown in suspension, **53,** 131–133, 292
Xenopus oocyte, **53,** 126–128, 292–293
RNA transport, *see* RNA transport, nucleus
Nucleus, *see also* Macronucleus, Micronucleus, isolation, **62,** 382–385
Nucleus, bare, isolation, **63,** 409

Nucleus, green fluorescent protein targeting
chimera construction and
expression, **58,** 78, 210–212
expression in primary cultures, **58,** 80–83
fluorescence microscopy, **58,** 79–80, 83–84
multiple organelle labeling with blue-shifted
protein, **58,** 81
transfection, **58,** 83
transport of proteins, *see*
Nuclear transport
Nucleus, pCa_i measurement, **70,** 461
Nucleus size, chick ciliary neurons, **71,** 39
Nucleus, visualization, **62,** 291–310,
319–321
Nude mice transplantation
transformation assays, **66,** 219
Nuf2–green fluorescent protein, fusion protein
studies in yeast, **58,** 100–101
Null cell lines
fibronectin-null fibroblasts, **69,** 321
integrin-null ES cells, **69,** 320
α3 integrin–null mouse, **69,** 311–313
$\alpha3\beta1$ integrin–null cells, **69,** 319–320
α4 integrin–null fibroblasts, **69,** 320
α5 integrin–null ES cells, **69,** 320–321
α5 integrin–null fibroblasts, **69,** 320
β1 integrin–null ES cells, **69,** 317–318
β1 integrin null-somatic cell lines, **69,**
318–319
syndecan-4–null fibroblasts, **69,** 321
thrombospondin 2–null
fibroblasts, **69,** 321–322
Nullisomic, **62,** 159–162
strains, **62,** 190–191
strains (table), **62,** 265
NuMA protein, in SPB organization, **67,**
224–225
Number of neurons, disadvantage of chick ciliary
ganglia, **71,** 41
Numerical aperture
depth discrimination, **70,** 156–157
lens selection, **70,** 155–156
objective lens, **70,** 258–259
Xenopus oocyte microscopy studies,
70, 398–401
NuSerum, **71,** 377
Nutritional requirements, **62,** 35, 195
Nycodenz density gradient centrifugation
yeast mitochondria isolation, **65,**
43–45
Nyquist frequency, power spectrum
interpretation, **55,** 133

O

O ring, AFM in aqueous fluids, **68,** 215–216
Objective
fluorescence microscopy, **56,** 191
light microscopy, **56,** 139
multimode digital microscopy, **56,** 191
video camera–microscope mating optical
considerations, **56,** 151–152
resolution, **56,** 147–151
video-enhanced differential interference
contrast light microscopy, **56,** 173–174
Objective lens selection
bone and dental cell biology, **70,** 273
brain tissue, neuroscience, and retinal
imaging, **70,** 273–274
cytoskeleton, **70,** 272
genetics, **70,** 273
immunogold and silver grains, **70,** 272–273
microbiology, **70,** 275–278
ocular imaging, **70,** 274
UV imaging, **70,** 269–271
visible Ca^{2+} imaging, **70,** 271–272
Objective lenses
achromat, **70,** 256
application markings, **70,** 260
chromatic aberration, **70,** 262–263
cleaning, **70,** 263–264
color code, **70,** 258
correction collar, **70,** 259
cover glass, **70,** 259
cover-glass correction, **70,** 259
dry lens, **70,** 263
epi-objective, **70,** 266–268
intact tissue problems, **70,** 303
Leica, **70,** 286
multiimmersion objectives, **70,** 266
multiphoton imaging, **70,** 288–289
Nikon, **70,** 283
numerical aperature, **70,** 258–259
oil immersion lens, **70,** 263
Olympus, **70,** 285
overview, **70,** 15–16, 248–249
plan achromat, **70,** 256–257
plan aprochromat, **70,** 257–258
plan fluor, **70,** 257
practical applications, **70,** 266
reflected light type, **70,** 266–268
retraction stopper, **70,** 260
spherical aberration, **70,** 261–262, 264–266
super fluor, **70,** 258
transmission data, **70,** 289–290
water dipping, **70,** 258

water immersion type, **70,** 264–265
working distance, **70,** 259
Xenopus oocyte cytoskeleton, **70,** 398–399
Zeiss, **70,** 284
Ocular imaging, confocal objective lens selection, **70,** 274
Ocular myopathy (OM), **65,** 312, 316
OHC, *see* Outer hair cells
Oil immersion objective lens, **70,** 263
Olfactory bulb, migration from subventricular zone
brain slice preparation, **69,** 349
DCC, **69,** 351–353
neuronal precursor cells, **69,** 351
overview, **69,** 349
Olfactory learning, in *Drosophila*, **71,** 237
Oli-green reagent, **61,** 330
Oligodendroglioma, *see* Glioma
Oligohymenophorea, **62,** 30
Oligomerization
Apaf-1 function, **66,** 23–25
Oligomerization, COCO system proteins, **69,** 78
Oligomerization, membrane proteins, AFM studies, **68,** 270–272
Oligomycin, **65,** 137
Oligonucleotide primers
for PCR, **64,** 106–107, 405
in PRINS, **64,** 55, 59–60
Olivary nucleus
immunostaining
MERRF, **65,** 320
Olympus
objective lenses, **70,** 285
slow-scan CLSM, **70,** 37–39
Olfactory learning, in *Drosophila*, **71,** 237
Olympus Fluoview 500 laser-scanning confocal microscope, **71,** 150–151
OM, **65,** 312, 316
OMM, **65,** 357–358
OmpF porin, AFM studies
conformational changes, **68,** 282–285
topograph–atomic model comparison, **68,** 275
On-glass nucleation assay, for γTuSC and γTuRC, **67,** 20
Oncogenes, immortalization of cells, **57,** 74–76, 79–81
Oncogenic mutations
apoptosis, **66,** 198
Oncology, cell–matrix adhesion studies, **69,** 19
one-eyed pinhead
gamma-ray
mutagenesis, **60,** 66–67
mutation detection, **60,** 250
positional cloning, **60,** 176, 261
Oocytes, *Spisula solidissima*, lysate, **67,** 149
Oocytes, *Xenopus*, confocal microscopy
antibody incubations, **70,** 391–392
bleaching, **70,** 391
borohydride reduction, **70,** 391
brightness and contrast extremes, **70,** 402
chromosome staining, **70,** 393–394
clearing, **70,** 394–395
cytoskeletal and nuclear probes, **70,** 410–413
digital image processing, **70,** 403–404
handling procedures, **70,** 390–391
image brightness optimization, **70,** 400–402
image scale optimization, **70,** 399
keratin filaments, **70,** 386–388
microfilaments, **70,** 388–390
microtubule fixation, **70,** 383–386
mounting procedures, **70,** 395–397
objective choice, **70,** 398–399
optical sectioning, **70,** 399–400
overview, **70,** 380–382
photobleaching, **70,** 397
presentation, **70,** 404–408
recipes and reagents, **70,** 409–410
secondary antibody choice, **70,** 392–393
Oocytes, laser micromanipulation, **55,** 88–92
OPA derivatives
ceramide, **66,** 148–149
Open reading frames (ORFs)
yeast genome
web site, **65,** 446
Operational amplifier, **64,** 297–298
Optical filters, LSCM, **70,** 189–190
Optical filtration
for four-color compensation, **64,** 303–307
for two- or three-color compensation, **64,** 301–303
PECY5 conjugated antibodies: problems, **64,** 300–301
problem of overlapping fluorochrome spectra, **64,** 295–298
Optical force microscopy
atomic force microscope-like applications
intermolecular force measurements, **55,** 101–104
optical force microscopy, **55,** 100–101
experimental design, **55,** 105–113
calibration, **55,** 111–112
detection, **55,** 108–111
probe selection, **55,** 106–108

Optical force microscopy (*cont.*)
scanning, **55,** 108
signal processing, **55,** 113
future research directions, **55,** 113–115
overview, **55,** 99–100
Optical force, *see* Gradient trap force, Optical force microscopy
Optical imaging, **65,** 311–329
histochemistry, **65,** 313–317
methods, **65,** 326
immunohistochemistry, **65,** 317–321
methods, **65,** 326–328
ISH to mtDNA, **65,** 321–322
molecular genetic methods, **65,** 328–329
single-fiber PCR, **65,** 322–325
Optical interference, **63,** 116
Optical memory disk recorders, video-enhanced differential interference contrast light microscopy, **56,** 178–179
Optical microscopy, AFM combination
infected monkey kidney cells, **68,** 8–9
living cell imaging, **68,** 7–8
patch–clamp combination, **68,** 19
pipetting concepts, **68,** 10–11
Optical properties, **71,** 6–7
Optical resolution, **70,** 249–251
Optical sections
excitation/emission wavelength effects, **70,** 237
Xenopus oocyte cytoskeleton, **70,** 399–400, 404–408
Optical signals, converting to current, **63,** 118–123
Optical traps, *see* Laser tweezers
Optics
DIC microscopy, **68,** 202
laser trap, PFM, **68,** 200–203
laser tweezer system, **55,** 35–36, 50–53
single-beam gradient laser trap, dielectric sphere force calculations, **55,** 1–25
arbitrary trap location force, **55,** 11–16, 23–25
index of refraction effects, **55,** 19–20
light forces, **55,** 4–8
mode profile effects, **55,** 16–20
overview, **55,** 1–4, 20–21
ray force, **55,** 21–22
Y axis trap focus force, **55,** 10–11, 22–23
Z axis trap focus force, **55,** 9–10
Optimization
conditions for protein labeling, **63,** 187–190
experimental conditions for cell adhesion, **63,** 609–611
Optoporation, laser micromanipulation, **55,** 85–87
Oral apparatus, **62,** 33, 43, 53, 83–86, 314, 335
Oral filaments, **62,** 57–60
Oral primordium, **62,** 68, 70, 77–78, 84
Orcein staining, **61,** 339
Oregon green BAPTA dextran, **71,** 132
ORFs
yeast genome
web site, **65,** 446
Organ of Corti
AFM setup, **68,** 151–153
AFM/patch–clamp setup, **68,** 150–151
characteristics, **68,** 143–144
preparation, **68,** 146–147
Organelle traffic, in dendrites, **71,** 8
Organelle transport, **71,** 8
Organelles, *in vivo* manipulation, **55,** 195–202
optical tweezer setup, **55,** 196–197
overview, **55,** 195–196
reticulomyxa, **55,** 197–199
artificial membrane tubes, **55,** 198–199
cell preparation, **55,** 197
organelle movement, **55,** 197–198
Spirogyra cytoplasmic streaming, **55,** 199–202
Organelles, in yeast cells, **67,** 37–40
Organic solvents, SFM of DNA, humidity issue, **68,** 366–367
Organogenesis, zebrafish system, **59,** 6–7
Ori5 transcripts
analysis, **65,** 424
Orthopox virus, infected monkey kidney cells, **68,** 8–9
Oscillating mode, AFM cell imaging, **68,** 58–61
Osmium amine B, nucleic acid staining, **53,** 106–107, 173–175
Osmium tetroxide, **62,** 317, 325–326
Osmium tetroxide, cell postfixation for transmission electron microscopy, **57,** 300–301
Osmotic lysis, separation of erythrocytes from leukocytes, **63,** 211
Osteoblasts, AFM studies, **68,** 38, 184
Osteoclasts, AFM studies
$\alpha_v\beta_3$ integrin adhesion force, **68,** 182–184
receptor mapping, **68,** 184
Outer hair cells
AFM studies, lateral link effects, **68,** 155
organ of Corti, **68,** 144, 146
transduction channel gating, **68,** 161–165
Outer membrane permeability
mitochondria, **66,** 384–386

Outer membrane vesicle purification
  yeast mitochondria, **65,** 48–49
Outer membrane, bacteria, **64,** 556
Outer mitochondrial membrane (OMM), **65,** 357–358
Outer ring, Aufderheide rotocompressor device, **70,** 349
Outgrowth rates, comparison of, **71,** 9
Ovarian carcinomas, BrdUrd incorporation, **64,** 463
Ovulation, frog, **67,** 276
Oxidants
  caspase activity, **66,** 313–315
Oxidative enzyme stains, **65,** 314
Oxidative phosphorylation (OXPHOS), **65,** 120
  beef heart mitochondria, **65,** 6
  BN-PAGE, **65,** 238
Oxidative stress, in apoptotic protoplasts, **63,** 522–523
Oxidative stress, on hippocampal neurons, **71,** 13
Oxidative turnover, mitochondrial, **64,** 123–124
Oxonol VI, **65,** 85
OXPHOS, **65,** 120
  beef heart mitochondria, **65,** 6
  BN-PAGE, **65,** 238
Oxygen free radicals, apoptosis triggered by, **63,** 468
Oxygen level, specific to neuronal species, **71,** 13
Oxygen monitors, **71,** 13
Oxygen scavenger system, **65,** 343
Oxyrase, **65,** 343; **71,** 317

P

P^+ cells
  transmitochondrial cybrid generation, **65,** 406–407
P_0 cells
  examples, **65,** 401
  generation, **65,** 399–407
  transmitochondrial cybrid generation, **65,** 401–406
p21, complex with PCNA, **63,** 421
P_1-derived artificial chromosome
  comparison with other artificial chromosome systems, **60,** 236–237
  overview of cloning system, **60,** 238–239, 272–273
  positional cloning
    candidate gene isolation
      direct cDNA selection, **60,** 248
      direct hybridization of clones to cDNA libraries, **60,** 249, 279, 281–282
      direct sequencing, **60,** 249–250, 282–284, 278–280
      exon trapping, **60,** 248
      gene density in genome, **60,** 247
      validation of genes, **60,** 250, 280–281
    chromosomal walking
      end-isolation of inserts, **60,** 244–246
      large-insert clone characterization, **60,** 246–247
      overview, **60,** 244, 273
    genetic mapping, **60,** 243–244
    overview of steps, **60,** 243
  zebrafish genomic libraries
    availability, **60,** 239–241
    hybridization-based screening, **60,** 241, 273, 276–277
    polymerase chain reaction-based screening, **60,** 241
P50/dynamitin
  production
    disruption of dynactin complex in *Xenopus* egg extracts
      demonstration, **61,** 140
      procedure, **61,** 141
      sedimentation patterns, **61,** 140–141
    disruption of spindle poles in *Xenopus* egg extracts, **61,** 142
    procedure
      ammonium sulfate precipitation, **61,** 139
      peak elution fraction, **61,** 140
      Superose purification, **61,** 140
    solutions, **61,** 138
  role in dynactin complex, **61,** 137–138, 140–142
P-element-generated mutants, **71,** 200
p53 gene, LSCM analysis, **64,** 437–438
P-glycoprotein, methotrexate as substrate, **64,** 194
p24, HIV: measurement, **64,** 583
p-Hydroxybenzylideneimidazolidinone, *see* Chromophore, Green fluorescent protein
P53, immortalization of cells by mutant genes, **57,** 76–77, 105–106
P/O ratio assay, **65,** 127–128
  digitonin, **65,** 129
PAB 416, affinity of, **63,** 284–287
PAC, *see* P_1-derived artificial chromosome
pAcGP67A.coco design, **69,** 71–73
Packaging cell culture
  MEFs, **66,** 205

Packaging cell transfection
MEFs, **66,** 205–206
Packaging systems, **66,** 203
Paclitaxel, **66,** 211
pADEasy system, adenoviral plasmid
generation in *Escherichia coli*, **69,** 410–413
HEK293 E1 cells, **69,** 415–416
high-titer viral stocks, **69,** 418–420
screening, **69,** 413–415
virus titration, **69,** 417–418
pAdTrack shuttle vector, **71,** 391–392
Painting, chromosome, combined with PRINS, **64,** 61–64
Pair correlation function, statistical analysis of particle distribution, **53,** 91–95
Pal Filtron, **71,** 138
PAN3, Na^+/D-glucose cotransporter, **68,** 131–133
Pancreatic cancer, inhibition via *in vivo* therapies, **71,** 337
Pancreatiaacinar cells, fusion pores, **68,** 40, 43
Papillomavirus, immortalization of cells, **57,** 71–72
Paracentrotus lividus, **65,** 28
Paraffin-embedded sections
mitochondrial proteins
immunolocalization, **65,** 327–328
Paraformaldehyde
fixation, **63,** 404–405
of cell suspensions, **63,** 406
preparation in PBS, **63,** 402, 404
Ultrapure formaldehyde, **63,** 243, 249
Paraformaldehyde, preparation, **53,** 365
Parafusin, **62,** 47
Paramecium tetraurelia, **65,** 29
Parasomal sacs, **62,** 42–43, 48
Paromomycin, *see* Selectable markers
Paroral membrane, **62,** 83
Parthenogenesis
centrosomes, **67,** 217–220
Xenopus, vs. fertilization, **67,** 212–214
Particle attachment strength, as parameter of biolistics, **71,** 355
Particle bombardment, *see* Transformation
Particle damage, affecting biolistics performance, **71,** 355
Particle density, affecting biolistics performance, **71,** 355, 364
Particle design improvements, in biolistics, **71,** 365
Particle immobilization, in biolistics, **71,** 356
Particle propulsion, in biolistics, **71,** 357
Particle size and composition, as parameter of biolistics, **71,** 355
Particle speed and spread, as parameter of biolistics, **71,** 355, 357
Particles
calibration in MESF units, **63,** 320–322
cellular, loss during nuclear isolation, **64,** 272
centering in illuminating beam, **63,** 26–33
comparing fluorescence from, **63,** 325–326
fluorochrome-stained, standards, **63,** 320
for standardizing fluorescence
intensity, **63,** 319
illumination, by flow cytometer, **63,** 20–26
in bacteria test samples, **64,** 515–516
light signals from, collection, **63,** 33–37
measurement, standardization, **63,** 301
monodispersal fluorescent, **64,** 549
sizing and concentration, **63,** 328–330
weakly fluorescent, detection, **63,** 314–317
Passaging
animal cell culture, **57,** 15, 99, 177–180
embryonic stem cells, **57,** 283
invertebrate cell culture
cell harvest, **57,** 196–197
passaging without splitting, **57,** 197
Passivated force sensor, living cell force spectroscopy, **68,** 95
Pat Brown laboratory microarray guide
web site, **65,** 440
Patch–clamp techniques, AFM
combination technique, **68,** 19
instrumentation, **68,** 11–12
membrane patch, **68,** 12
patch–clamp setup, **68,** 153
protein unfolding, **68,** 318–319
setup, **68,** 150–151
voltage-sensitive ion channels, **68,** 12–13
Pathfinding, **71,** 172
in *Drosophila* studies, **71,** 196
Pattern formation, cerebellum as study model for, **71,** 91
Pattern stamping, micropatterned culture surfaces, **69,** 393, 397–398
Patterning, **62,** 79–88
Pawn, **62,** 365
Paxillin, **66,** 263
PBS, *see* Phosphate-buffered saline
PC, **66,** 341
PC-12 cell line, **66,** 419, 420–423, 425, 429–430
transfection, **66,** 429–430
PC-12 cell lines, **71,** 3, 8
cell biological studies based on, **71,** 267
cell labeling and harvesting, **71,** 271

cell permeabilization, **71,** 272
developmental properties of, **71,** 5
high-efficiency transfection of, **71,** 268
immunocytochemical detection of cytosolic proteins in, **71,** 284–285
immunostaining of, **71,** 283
isolation of stable transfectants in, **71,** 281
propagation and culture of, **71,** 270
secretion assays
using continuous electrochemical detection of released monoamines, **71,** 276–278
using human growth hormone, **71,** 278–279
using radioactive catecholamines, **71,** 271
PC-12 cells, AFM studies, **68,** 38
pCa_i, *see* Intracellular calcium
PCD, *see* Programmed cell death
PCNA, *see* Proliferating cell nuclear antigen
pCOCO, recombinant proteins
characterization, **69,** 77
expression and purification, **69,** 76–77
protein production, **69,** 74–75
recombinant baculovirus, **69,** 75–76
PCR
Bcl-2 family cloning, **66,** 34
Bcl-w, **66,** 34
PCR protocol, **61,** 121
PCR, *see* Polymerase chain reaction
pCS2 expression vector, **71,** 132
PDMS stamps
creation, **69,** 392–393
micropatterned culture surface, **69,** 395
PE, **66,** 341
Peak detection
apoptotic, **63,** 516
device timing and performance, **63,** 153
in flow cytometry, **63,** 127
Peaks
CFSE, cell sorting on, **63,** 384, 386
individual positions, **63,** 387–388
PECY5
CD64 binding, **64,** 578
compensation requirements, **64,** 305–307
plus CD8, **64,** 303
properly compensated instrument for, **64,** 300–301
Pedestal, Ringlien differential screw thread compressor, **70,** 356
PEG, *see* Polyethylene glycol
Penetration depth, required for ballistics, **71,** 361
Pepsin digestion, as variable step, **63,** 363, 370
Peptidase nomenclature
web site, **66,** 249
Peptide sequences, promoting cellular uptake, **71,** 326
Peptide sequencing
TGase-mediated glutamine deamidation, **66,** 130–131
Peptides
assisted delivery into live neurons, **71,** 325–337
blocking, in Tibia-1 pathway cultures, **71,** 191
injection into *Xenopus* cultures, **71,** 131
noncovalently associated, promoting cellular uptake, **71,** 327
Peptides, affinity cross-linking, **69,** 249–250
Peptidoglycan layer
gram-positive bacteria, **64,** 556–557
inhibition by β-lactams, **64,** 558–559
Peptidylvinyl sulfones, **66,** 255
Percentage of labeled mitotic figures, **63,** 355–356
Percoll density gradient
potato tuber mitochondria, **65,** 60
Percoll gradients
density gradient purification, **65,** 57–59
Perfusion chamber
high-resolution video light microscopy, vertebrate cell monolayer study, **56,** 253–271
assembly, **56,** 270–271
chamber design, **56,** 256–260, 265–266
construction, **56,** 260–265, 267–269
historical perspectives, **56,** 256
overview, **56,** 253–256
temperature regulation, **56,** 269–270
intracellular ion ratio imaging, **56,** 249–250
Perfusion chambers, fluorescence microscopy, **58,** 185–186, 215–216, 231, 255–256, 348–349
Pericentrin, **62,** 253
centrosomal localization, **67,** 54–58
enrichment, **67,** 63–65
level elevation, **67,** 66–67
MT-dependent sedimentation, **67,** 65–66
recruitment to centrosomes, **67,** 58–63
staining, for centrosome defects, **67,** 67–68
Pericentrin–green fluorescent protein, fusion protein analysis of centrosome dynamics
construction of vector, **58,** 225
controls following imaging
cell viability assay, **58,** 237
functional assay of centrosomes, **58,** 236–237
microtubule integrity assays, **58,** 237
movements of cells or cytoplasm, **58,** 238
structural comparisons of centrosomes in fixed cells, **58,** 236

Pericentrin–green fluorescent protein, fusion protein analysis of centrosome dynamics (*cont.*)
subcellular localization by immunofluorescence, **58,** 235–236
expression in COS-7 cells
confirmation, **58,** 226
electroporation, **58,** 226
Lipofectamine transfection, **58,** 225–226
high-speed fluorescence microscopy
charge-coupled device camera, **58,** 230–231
deconvolution of images, **58,** 231–234
dynamic properties of centrosomes, **58,** 234–235
heated cell chamber, **58,** 231
image capture, **58,** 234
lasers, **58,** 229
overview of instrumentation, **58,** 228–229
piezoelectric focus drive, **58,** 229
subcellular localization, **58,** 227–228
Peripheral ganglia, dissection of, **71,** 13
Peripheral neurons
and microinjection, **71,** 17
and quantity of yield, **71,** 11
buccal ganglion, **71,** 3
chick sensory, **71,** 17
culture maintenance, **71,** 28–29
culture types, **71,** 31–34
growing and working with, **71,** 17–34
methods for working with, **71,** 19–30
rat sympathetic, **71,** 17
sensory, **71,** 3
stereotyped in Tibia-1 pathway, **71,** 172
sympathetic, **71,** 3
transfection via electroporation, **71,** 339–350
Periplaneta americana, **65,** 24
Peritubular cell, establishment of cell lines
cell preparation
fractionation over discontinuous Percoll gradient, **57,** 96–97
isolation, **57,** 96
clone selection, **57,** 98–99
histology of immortalized cells, **57,** 100–101
immunodetection of simian virus 40 large T antigen, **57,** 99–100
medium for immortalized cell culture, **57,** 99
overview, **57,** 94–95
passaging, **57,** 99
seminiferous cord formation *in vitro* with other testicular cells, **57,** 101–103, 105
transfection
calcium phosphate transfection with simian virus 40 large T antigen gene, **57,** 97–98
principles, **57,** 95–96
Permeability transition (PT), **66,** 80, 368
mitochondria, **66,** 383–384
Permeability transition (PT) pores, **65,** 148
Permeability transition pores, mitochondrial, **63,** 470
Permeabilization reagents
and cell fixation, **63,** 254–259
for intracellular staining, **63,** 242–243
properties, **63,** 276–277
Permeabilization, yeast SPBs
cell culture, **67,** 167–168
cell wall digestion, **67,** 168–169
solutions, **67,** 167
with Triton X-100, **67,** 169
Permeabilized human skin fibroblasts
ATP synthesis rates, **65,** 140
luminescence curve, **65,** 139
Peroxisomes, **62,** 50–51
in mitochondrial preparations, **65,** 4
Perturbation strategies, zebrafish, **59,** 87–108
pescadillo
gene cloning, **60,** 92, 94–95
retrovirus-mediated insertional mutagenesis, **60,** 92
PET vectors
recombinant caspase expression, **66,** 7–8
Petri dishes, untreated, for peripheral neurons, **71,** 19
PFGE, *see* Pulsed-field gel electrophoresis
PFGE, *see* Pulsed field gel electrophoresis
PFM, *see* Photonic force microscopy
PGD, *see* Preimplantation genetic diagnosis
pH
intracellular, **65,** 372
reporter targeting, **65,** 372–376
pH-sensitive fluorescent proteins, **65,** 373–374
pH, control of media, **57,** 22
pH, *see* Intracellular pH
pH, intracellular ratio imaging, **56,** 237–251
experimental chamber, **56,** 249–250
fluorescence signal calibration, **56,** 245–247
imaging workstation components, **56,** 247–249
indicator properties, **56,** 239–244
incorporation into intact living cells, **56,** 242–243
pitfalls, **56,** 243–244
signal-to-noise ratio, **56,** 243

spectral characteristics, **56,** 239–242
structural characteristics, **56,** 239–242
overview, **56,** 237–238, 251
perfusion system, **56,** 249–250
rationale, **56,** 238
pH, optimal, for protein labeling, **63,** 187–189
Phagocytosis
apoptotic cells, **66,** 358–362
macrophage preparation, **66,** 359
assay, **66,** 361
Drosophila, **66,** 330–332
inhibitors, **66,** 361
lipid-symmetric erythrocytes, **66,** 360
PS-expressing D011.10 cells, **66,** 360
PS vesicles preparation, **66,** 361
Phagocytosis, **62,** 43, 283
Phagocytosis, FACS assay, **63,** 619–623, 625–626
Phagolysosomes, *see* Food vacuoles
Phagosomes, **62,** 42
Phalloidin, **62,** 64
Phalloidin staining
Drosophila embryo confocal imaging, **70,** 373–374
Xenopus oocytes and embryo microfilaments, **70,** 388–389
Pharmaceuticals, AFM applications, **68,** 188–190
Pharmacological agents, *see* Drugs
delivery to neuronal cultures, **71,** 333
use in *Xenopus* live cell cultures, **71,** 148
Phase contrast imaging, fluorescence microscopy, **56,** 190, 193
Phase-sensitive detector, two channels, **63,** 142
Phase suppression, **63,** 140
Phatidylethanolamine (PE), **66,** 341
Phe-64, thermal stability of green fluorescent protein site-directed mutants, **58,** 27
Phenotype, asynchronous, detection in B-lineage ALL, **64,** 393
Phenotype, cell-specific, in 3-D ECM system, **63,** 575
Phenotypes
Saccharomyces cerevisiae
mitochondria, **65,** 382–383
Phenotypic assortment, **62,** 152–153, 494, 520, 527
Phenylmethylsulfonyl fluoride (PMSF), **65,** 39–40
Phenylmethylsulfonyl fluoride, proteoglycans, **69,** 56–57
Phenythiourea (PTU)
ε-(γ-glutamyl)-lysine cross link density, **66,** 117–118
Pheochromocytoma, **71,** 3
Pheochromocytoma-derived cell lines, *see* PC12 cell lines
pHi, *see* Intracellular pH
Phoenix packing cells, **66,** 203–210
Phosphate analysis
SM mass levels, **66,** 142–143
Phosphate-buffered saline
paraformaldehyde preparation in, **63,** 402, 404
with sodium azide, **63,** 249
Phosphatidylcholine (PC), **66,** 341
Phosphatidylinositol glycans, **62,** 37
Phosphatidylserine (PS), **66,** 340
cell surface exposed
in vivo detection, **66,** 349–354
rapid determination
adherent cells, **66,** 348
cells in suspension, **66,** 347
Phosphatidylserine, migration, early indicator of apoptosis, **63,** 523
Phosphocellulose chromatography, **61,** 159
Phospholipid-enriched mitoplasts
rat liver mitochondria, **65,** 18–19
Phospholipid measurement
metabolic labeling, **66,** 141
Phospholipid, mitochondrial membrane, **64,** 124–125
Phospholipids, **62,** 48
Phosphonolipids, **62,** 36, 42
Phosphorylating electron transfer particles, **65,** 7–9
Phosphorylation, histone H_3, during cell cycle, **63,** 343–344
Phospho-specific antibodies, cell–matrix adhesion assays, **69,** 374–375
Photinus pyralis firefly tails
luciferase, **65,** 366
Photoactivation, green fluorescent protein
applications in protein dynamics studies, **58,** 134–135, 156, 160–161
characteristics, **58,** 134–135, 156
Photoactivation, multiphoton excitation imaging, **70,** 144
Photobleaching, **65**; **71,** 305–321
fluorescence, **65,** 345
Photobleaching, fluorescence microscopy, **56,** 190, 285–287
Photobleaching, fluorescent images, **64,** 75
Photobleaching, fluorochrome, in confocal microscopy, **63,** 97, 99
Photobleaching, green fluorescent protein, *see also* Fluorescence loss during

photobleaching; Fluorescence recovery after photobleaching
controls in fluorescence microscopy, **58,** 33
stability, **58,** 32–33, 36
Photobleaching, *Xenopus* oocytes and eggs, **70,** 397
Photodetector, **70,** 20–22
Photodetector, for side scatter detection, **63,** 36–37
Photodiode
avalanche, **63,** 121–122
lens focusing light onto, **63,** 33–36
Photodiodes, characteristics, **63,** 119
Photodynamic therapy, **66,** 368
Photolithographic master
creation, **69,** 392–393
micropatterned culture surface, **69,** 394–395
Photomask
creation, **69,** 392–393
micropatterned culture surface, **69,** 394
Photomultiplier tubes
Bio-Rad Laboratories slow-scan CLSM, **70,** 26–27
Carl Zeiss CLSM, **70,** 27
characteristics, **63,** 119–121
confocal microscopy, **70,** 20–22
fluorescence 1 and 2, **64,** 297–298
for spectral shift analysis, **64,** 186
gain, **63,** 126
in CRM, **63,** 588–589
in LSC, **64,** 337–338
LSCM, **70,** 190–191
Nikon CLSM, **70,** 36
Photon counting cameras, **65,** 367
Photon counting cameras, low-light microscopy
sensitivity enhancement, **56,** 52–53
signal-to-noise ratio, **56,** 54–56
Photonic force microscopy
contact recording mode, **68,** 208–209
design principles, **68,** 200–203
experimental considerations, **68,** 199–200
fast 3D single-particle tracking, **68,** 210
feedback control, **68,** 202–203
force sensor, calibration, **68,** 207
instrumentation, **68,** 20–24
kinesins
basic structure, **68,** 24–25
mechanical behavior, **68,** 26
mechanical properties, **68,** 25–27
local viscosity, **68,** 28–29
molecular interactions, **68,** 196–199
plasma membrane, **68,** 194–196
position detection, **68,** 203–204
probes
material and size, **68,** 204–205
surface modifications, **68,** 205–206
recording mode
contact mode, **68,** 208–209
fast 3D single-particle tracking, **68,** 210
tapping mode, **68,** 209–210
resolution, **68,** 207–208
sample preparation, **68,** 210–211
tapping recording mode, **68,** 209–210
Photons
reaching cell, **63,** 110, 112
single, counting, **63,** 122
Photoreactive cross-linkers, **65,** 226
Photoshop, *see* Adobe Photoshop
Phototransduction studies, in *Drosophila,* 238
Phototoxicity
excitation light, **65,** 345
Phragmoplast, **61,** 414, 415, 418
Phragmoplastin dynamics, **61,** 416
Phthalocyanine derivatives, **70,** 176–177
Phycobiliproteins, **70,** 174–176
Phycobiliproteins, labeling of proteins, **63,** 194–201
Phycoerythrin
activation/proliferation markers labeled with, **63,** 443
CD11b-conjugated, **63,** 80–81
complex with antibody, **63,** 136
plus FITC: histograms, **64,** 490–491
plus Texas Red: fluorochromes, **64,** 339
R and B forms, as fluorescent labels, **63,** 195
scattergrams, **64,** 324–326
tandem conjugate with Cy5, **63,** 109, 234–237
Phycoerythrinated goat anti-mouse Ig, **63,** 240
Phylogeny, **62,** 19–21, 28–32
PI, **66,** 346, 482
apoptosis assays, **66,** 211–213
cell size
apoptosis, **66,** 54–57
flow cytometry
DNA fragmentation assay, **66,** 87–88
Picture processing, *see* Digital image processing
Pinhole diameter, achieving, **63,** 91
Pipes-based buffer hazards, **61,** 462
Pipette techniques
AFM concepts, **68,** 10–11
patch–clamp, *see* Patch–clamp techniques
Pipette tips, preparing for blastomere injections, **71,** 136–139
Pisaster ochraceus, see Spindle birefringence
Pit-1, dimerization studies by fluorescence resonance energy transfer, **58,** 307–309

Pits, plasma membranes, AFM studies, **68,** 39–40
Pixel resolution, confocal microscopy 3D reconstruction, **70,** 63
Pixel values, in LSCM, **63,** 58–61
Pixels
binning
high-resolution multimode digital microscopy, **56,** 203
low-light microscopy, **56,** 54, 61
definition, **56,** 67–68
digital video imaging, **56,** 66–70, 82, 92–94
small object visualization, **56,** 69–70
spatial noise reduction, **56,** 96–101
Placenta, native basement membrane proteins, **69,** 115–116
Plan achromat, **70,** 256–257
Plan aprochromat, **70,** 257–258
Plan fluor, **70,** 257
Plant cells
analysis with flow cytometer, **63,** 509–511
autofluorescence, **63,** 511
changes in scattering properties, **63,** 512, 514
chromatin condensation, **63,** 514–516
fluorescence energy transfer, **63,** 512
MT arrays, **67,** 194–196
nDNA fragmentation, **63,** 516–520
physiological changes during apoptosis, **63,** 520–523
probe loading problem, **63,** 511–512
γ-tubulin
antibodies against, **67,** 200–206
cDNA and gene cloning, **67,** 196–200
functional studies, **67,** 207
immunofluorescence microscopy, **67,** 206–207
Plant development, intercellular interactions, **58,** 139–140
Plant mitochondria
blue native-PAGE, **65,** 67–68
centrifugation, **65,** 55–56
density gradient purification, **65,** 57–61
cell culture preparation, **65,** 59–61
enzyme activities, **65,** 64
growth and preparation, **65,** 54–55
in organello translation products, **65,** 70–71
membrane integrity determination, **65,** 62
membrane separation, **65,** 64–69
mitochondrial function assays, **65,** 63–64
mono Q chromatography, **65,** 67
proteome analysis, **65,** 68–70
purity determination, **65,** 61–62
SDS-PAGE separation, **65,** 66
storage, **65,** 62–63
yield calculations, **65,** 61
Plant senescence
chlorophyll fluorescence, **66,** 449–450
color scanning, **66,** 449
colorimetry, **66,** 447
hyperspectral imaging, **66,** 449
PCD
nondestructive measurements, **66,** 447–450
spectrometry, **66,** 449
visual scoring, **66,** 447
Plants
flow cytogenetics
chromosome sorting, **64,** 26–28
problems with, **64,** 4–5
PCD assays, **66,** 437–450
seeds
germination media, **64,** 13
with reconstructed karyotypes, **64,** 6
Plants, apoptosis in, **63,** 506–509
Plasma membrane
ECM degradation mediated by, **63,** 614
integrity, changes during apoptosis, **63,** 520, 536
Plasma membrane-associated oxidases, **66,** 308
Plasma membrane-bound NADPH oxidase, **66,** 308–309
Plasma membrane, *see* Cell membrane mechanics
Plasma membrane, green fluorescent protein targeting
chimera construction and expression, **58,** 79
expression in primary cultures, **58,** 80–83
fluorescence microscopy, **58,** 79–80, 83–84
multiple organelle labeling with blue-shifted protein, **58,** 81
transfection, **58,** 83
Plasma membrane potential, laser scanning microscopy
methods, **70,** 437–441
plasma membrane, **70,** 434–436
single mitochondria, **70,** 436–437
Plasma membrane purification
NADH, **66,** 309–310
Plasma membranes
apical structures, AFM studies, **68,** 39–40
PFM imaging, **68,** 194–196
Plasma membranes, pCa_i measurement, **70,** 464
Plasmid-encoded green fluorescent protein fusions
structures, **65,** 278–279
Plasmid rescue, **66,** 489
Plasmids
apoptosis proteins

Plasmids (*cont.*)
yeast, **66,** 456
yeast, **66,** 458
Plasmids, expression cloning, **69,** 219
Plasmids, IFI value, **64,** 90
Plasmid DNA, **71,** 133
coated particles in biolistics, **71,** 354–355
Plasmid RNA, **71,** 133
Plasmodium falciparum
mitochondrial genome
size, **65,** 420
Plasticware, animal cell culture, **57,** 13, 54–55
Plating efficiency
chick forebrain neuronal cultures, **71,** 62
hippocampal and cortical neurons, **71,** 117–119
in neural-enriched cultures, **71,** 145
long-term glia feeder cultures, **71,** 123
Plated ELISA method
TGase assay, **66,** 115
Platelet mitochondria
human, **65,** 25–28
Platelets
transmitochondrial cybrid
generation, **65,** 405–406
Plaque assay, **71,** 403
Pleiotropic murine systems, **66,** 203
Ploidy
in surgical oncology, **64,** 447–448
measurements, as prognostic markers, **64,** 448
Plugs
containing YAC, **64,** 39
preparation by PFGE, **64,** 105
Plunge freezing, **67,** 32
PMC, *see* Pollen mother cells division studies
PMSF, **65,** 39–40
PMSF, *see* Phenylmethylsulfonyl fluoride
PMT, *see* Photomultiplier tubes
PNB, *see* Prenucleolar body
pNH_3, *see* NH_3 permeability
Point mutations, **71,** 202
Point-spread function, **70,** 253
Point-spread function, quantitative video microscopy, **56,** 126–127
Polarity, establishment of, **71,** 4
Polarization microscopy, for centrosome reproduction studies, **67,** 271–275
Polarized light microscopy
advantages, **61,** 177–178
basic setup and observations
components
analyzer, **61,** 178
axis orientation, **61,** 179
compensator effects, **61,** 179–180
compensator types, **61,** 180
extinction orientation, **61,** 179
polarizer, **61,** 178
high extinction importance
Airy disk, **61,** 187
contrast reversal, **61,** 187
extinction coefficient, **61,** 186
high extinction achievement, **61,** 187
light leakage, **61,** 186, 187
Maltese cross patterns, **61,** 186, 187
image contrast
background light levels, **61,** 184
electronic light detectors, **61,** 184
intensity difference, **61,** 183
liquid crystal variable retarders
advantages, **61,** 188
features, **61,** 188
mechanism, **61,** 188
nematic liquid crystals, **61,** 188
Pol-scope
features, **61,** 188–190
orientation measurement, **61,** 190
retardance magnitude measurement, **61,** 190
retardance
bias retardance, **61,** 183
Brace–Kohler compensator retardance, **61,** 182
calculation, **61,** 182
definition, **61,** 181
image intensities and compensation angle, **61,** 182
specimen retardance calculation, **61,** 183
signal to noise ratio
CCD detector noise, **61,** 184
compensator angles, **61,** 185
equation, **61,** 185
maximizing signal to noise ratio, **61,** 185–186
noise sources, **61,** 184
photon counts, **61,** 185
polarization aberrations, **61,** 185
shot noise reduction, **61,** 184
birefringence of mitotic figures, **61,** 175–178
Polarizers, video-enhanced differential interference contrast light microscopy, **56,** 171–173
Pollen mother cells (PMC) division studies
culture methods
Lilium meiosis staging, **61,** 426
slide chambers, **61,** 426

sucrose solution requirements, **61,** 427
findings
fluorescence microscopy studies
confocal microscopy, **61,** 427
deconvolution, **61,** 427
endoplasmic reticulum visualization, **61,** 427
microfilament visualization, **61,** 427
phase contrast and DIC studies, **61,** 427
polarized light microscopy studies, **61,** 427
monocot species, **61,** 426
Pol-scope, *see* Polarized light microscopy
Poly-D-lysine, **71,** 48, 92
in hippocampal and cortical neuron cultures, **71,** 118
promoting neuron adhesion with, **71,** 19
Poly-D-ornithine, **71,** 47
Poly-DL-ornithine, **71,** 47
Poly-L-lysine, **71,** 92
preparing coverslips for Tibia-1 pathway studies, **71,** 188
Poly(ADP-ribose) polymerase (PARP)
apoptosis, **66,** 97–100, 291–300
in vitro cleavage, **66,** 296–299
activity Western Blot, **66,** 294–296
cleavage characterization, **66,** 299–300
endogenous cleavage, **66,** 291–296
Western Blot procedure, **66,** 292–294
cleavage, **66,** 311–312
detection, **66,** 409
cleavage assay, **66,** 175–176
data analysis, **66,** 99–100
flow cytometry, **66,** 98
LSC, **66,** 98–99
materials, **66,** 97–98
schematic representation, **66,** 293
vs. DNA fragmentation, **66,** 99
vs. TUNEL assay, **66,** 99–100
Polyacrylamide bead, **69,** 337–338
Polyacrylamide gel
mechanical properties, **69,** 331–332
preparation, **69,** 327–329
protein coating, **69,** 330–331
Polyacrylamide substratum
bead preparation, **69,** 337–338
deformation analysis, **69,** 334–335
deformation data, **69,** 334
gel preparation, **69,** 327–329
glass surface preparation, **69,** 327
image collection, **69,** 333–334
matrix protein conjugation, **69,** 329–330
mechanical stimulations, **69,** 336–337
microscope setup, **69,** 332–333
preparation, **69,** 332
stiffness gradient, **69,** 337
traction forces, **69,** 334–335
Polyadenylation sites, **62,** 554
Polycation-coated substrates, **71,** 11–12
Polyclonal antibodies, in SPB immunoprecipitation, **67,** 87
Polyethylene glycol, cell hybridization, **57,** 112, 115–116, 124
Polylinkers, in *Xenopus* spinal cultures, **71,** 132
Polylysine
as coverslip coating for *Heliosoma* cultures, **71,** 163
as culture surface for chick forebrain neurons, **71,** 52, 59
in *Xenopus* cultures, **71,** 145
Polymerase chain reaction
analysis of MRD, **64,** 402–414
technical considerations, **64,** 406–412
clone-specific markers, **64,** 404–406
degenerate oligonucleotide-primed, **64,** 106–107
estimation of chromosome purity, **64,** 18
for gene cloning into two-hybrid vectors, **67,** 75–76
in HIV detection, **64,** 584
inter-Alu, **64,** 106
plant γ-tubulin cDNA, **67,** 199
premix, **64,** 11–12
real-time quantitative, **64,** 411–412
tumor-specific markers, **64,** 403–404
Polymerase chain reaction (PCR), **71,** 11, 199
competitive
mtDNA, **65,** 418–419
modified strains, **65,** 262–264
mtDNA
COX 1 gene, **65,** 325
single-fiber, **65,** 322–325
deleted-mtDNA, **65,** 325
DNA, **65,** 328–329
yeast, **65,** 261
Polymerase chain reaction (PCR), amplification of base-unpairing region binding protein ligands, **53,** 346, 350
Polymerase chain reaction-based gene modification
yeast, **65,** 259
Polymerase chain reaction, $\alpha 3$ integrin null mouse, **69,** 312

Polymerase chain reaction, polymorphism amplification in cell line authentication, **57,** 212–213
Polymerase chain reaction, *see also* Amplified fragment length polymorphisms, Random amplified polymorphic DNAs, Simple sequence length polymorphisms, Single strand conformation polymorphisms
full-length cDNA clone isolation, **60,** 226–226
genomic DNA templates, **60,** 271
hamster/zebrafish radiation hybrid characterization
agarose gel electrophoresis of products, **60,** 294, 301
amplification reactions, **60,** 293–294
DNA stock plate preparation, **60,** 293, 301
materials, **60,** 292–293
scoring of results, **60,** 294, 300
haploid screen, **60,** 56, 68
insert-flanking genomic DNA, **60,** 91
insert length determination in clones, **60,** 216
labeling of probes, **60,** 309, 330
mouse/zebrafish somatic cell hybrid characterization
amplification reaction, **60,** 313–314
available markers, **60,** 313
DANA/mermaid polymerase chain reaction, **60,** 317–318
hybrid panel composition, **60,** 314
screening of genomic libraries, **60,** 241
mutagenesis of promoter/enhancer constructs, **60,** 145–146
mutation detection in candidate genes by reverse transcription, **60,** 250, 282–283
promoter isolation, **60,** 228, 230
repetitive element polymorphism detection in genomes, **60,** 117–118, 124, 247
sequencing of DNA, **60,** 217
transgenic zebrafish
founder fish identification, **60,** 141–142
offspring identification, **60,** 89
Polymerization assay, basement membrane self-assembly, **69,** 129–130
Polymerization assays, MT, **67,** 241–243
Polypeptide growth factors, HS regulation, **69,** 83–84
Polypeptide quantitation, green fluorescent protein, **58,** 31–32
Polypeptides, loop replacement
membrane protein AFM imaging, **68,** 265–267
polypeptide end removal, **68,** 267–270
polypeptide loop removal, **68,** 267
Porcine adrenal cortex mitochondria, **65,** 24–25
Position clamp experiments, noise limitations, **55,** 140–143
Positional cloning
candidate gene isolation
direct cDNA selection, **60,** 248
direct hybridization of clones to cDNA libraries, **60,** 249, 279, 281–282
direct sequencing, **60,** 249–250, 282–284
exon trapping, **60,** 248
gene density in genome, **60,** 247
validation of genes, **60,** 250, 280–281
chromosomal walking
end-isolation of inserts, **60,** 244–246
large-insert clone characterization, **60,** 246–247
overview, **60,** 244
complementation by injection of genomic clones, **60,** 275–276, 280–281
critical region definition, **60,** 274–279
difficulty, **60,** 87
genetic mapping, **60,** 243–244
genomic clone isolation, **60,** 272–274
initiation of positional cloning project
genomic DNA isolation from embryos, **60,** 270–271
high-resolution mapping, **60,** 268–270
isolation of tightly linked genetic markers, **60,** 263–264, 266–267
mutant allele generation, **60,** 262–263
overview of steps, **60,** 243, 261–262
Posttranslational modifications
Bcl-2 family, **66,** 42–45
Potassium iodide, centrosome inactivation, **67,** 143–145
Potassium ions, **65,** 287
Potato mitochondria
inner membrane separation, **65,** 67
Potato tuber mitochondria
Percoll density gradient, **65,** 60
Potential doubling time
adenocarcinomas, **64,** 462–463
and cell loss factor, **64,** 472
pretreatment, **64,** 469–470
squamous carcinomas, **64,** 456, 459
Potentiometric dyes
disadvantages, **65,** 289–290
environmental variables, **65,** 295–296
fast membrane staining indicators, **70,** 430–431

mitochondrial membrane potential, **65,** 288–294
optimal dye-loading protocols, **65,** 293
overview, **70,** 429–430
selection, **65,** 291–293
slow redistribution indicators, **70,** 432–434
Poultry connection (www.poultryconnection.com), **71,** 70
Power spectrum
analysis
Brownian motion, **55,** 136–138
calculation, **55,** 134–135
interpretation, **55,** 132–134
trap calibration, **55,** 138–139
instrumental noise fluctuations, **55,** 58, 152
Preaxogenesis neurons, as guidepost cells, **71,** 176
Preimplantation genetic diagnosis, **64,** 98–99
selection of probes for, **64,** 103
YAC probes, **64,** 112
Prenucleolar body (PNB), components, **53,** 312
Preparation requirements, **71,** 6–7
Preparation time, reduced in biolistic transfection, **71,** 354
Preprotein
location, **65,** 199
Preprotein synthesis
ammonium sulfate precipitation, **65,** 195–196
protein import assaying, **65,** 190–196
template DNA, **65,** 191–193
PCR, **65,** 191–193
in vitro transcription, **65,** 193–194
in vitro translation, **65,** 194–196
Presorting, improving resolution by, **63,** 380–381
Presynaptic terminal research, **71,** 39
Primary culture
animal cell culture, **57,** 14–15
glioma, **57,** 150–151
Schwann cells, **57,** 177, 180
Primary epithelial cells, **66,** 223–224
Primary fibroblast, cell culture, **59,** 39–42
Primary neurons, infection with replication-deficient adenovirus, **71,** 406–413
Primed *in situ* labeling
applications, **64,** 56
chemical and instruments for, **64,** 56–58
critical aspects, **64,** 66–67
protocols, **64,** 58–66
reaction mix, **64,** 11
PRINS, *see* Primed *in situ* labeling
Printers, video-enhanced differential interference contrast light microscopy, **56,** 178–179
Prisms, *see* Beam-splitting prisms
Proapoptotic Bcl-2 family, **66,** 30
schematic diagrams, **66,** 32
subgroups, **66,** 32–33
Proapoptotic genes, **66,** 204–205
Proapoptotic protein assays
yeasts, **66,** 464
Probe loading, problem in analysis of plant cells, **63,** 511–512
Probe microscopy
bioartificial membranes, **69,** 177–179
cantilevers and tips, **69,** 167
collagen degradation, **69,** 171
collagen III assemblies, **69,** 174
ECM macromolecules, **69,** 183–185
FN–heparin–gold complexes, **69,** 185–188
force measurements, **69,** 181–182
$\alpha_5\beta_1$ immobilization, **69,** 179–181
molecular pulling, **69,** 168
overview, **69,** 174, 182–183
sample imaging, **69,** 167–168
unfolding tenascin, **69,** 175–177
Probe spots
contours, in LSCM, **63,** 61–62
counting feature in LSCM, **63,** 78
Probe synthesis, protein and gene expression, analysis, **59,** 64–67
Probes
AFM studies
cell adhesion receptors, **68,** 307–308
isolated components, **68,** 120–121
membranes and cells, **68,** 121
overview, **68,** 36
streptavidin–biotin unbinding, **68,** 306–307
cloned, mapping onto DNA molecules, **64,** 50–51
fluorochrome-conjugated probes, **70,** 76–77, 315–321, 330–331
generation
by *in vitro* DNA amplification, **64,** 43–44
from cloned DNA fragments, **64,** 41–43
labeling via random priming, **64,** 44–45, 107–108
local
AFM instrumentation, **68,** 2–3
definition, **68,** 1–2
STM instrumentation, **68,** 2
membrane potential, **65,** 88
molecular force
basic technique, **68,** 223–224
modifications, **68,** 224
pulling, **68,** 224
sample preparation, **68,** 225
typical experiment, **68,** 225

Probes (*cont.*)
nuclear probes, **70,** 410–413
pCa_i, **70,** 456
PFM studies
material and size, **68,** 204–205
surface modifications, **68,** 205–206
preparation for chromosome translocation study, **64,** 99–112
SFM, DNA, **68,** 367–368
unlabeled, as primer in PRINS, **64,** 55–56
Probes, *see also* Scanning probe microscopy
noise limitations, **55,** 146–147
Procaspase 3
Apaf-1 apoptosome assay, **66,** 20
Procaspase 9
Apaf-1, **66,** 464
Apaf-1 apoptosome assay, **66,** 19
Apaf-1 function, **66,** 23
Procaspases
processing, **66,** 11
Production
of dendrites, **71,** 10
of synapses, **71,** 10
Prodomains
caspases, **66,** 4
Programmed cell death, **71,** 69
Profilin, **62,** 57
Prognosis
acute leukemia, **64,** 356
proliferation markers and, **64,** 367–372
Prognostic markers
S-phase fraction measurements as, **64,** 448–449
ploidy measurements as, **64,** 448
Prognostic studies, in surgical oncology, **64,** 447–448
Programmed cell death, *see* Apoptosis
Programmed cell death (PCD), **66,** 394, 438
Dictyostelium
induction, **66,** 476–480
molecular mechanisms, **66,** 472–473
morphotype, **66,** 472
plant senescence
nondestructive measurements, **66,** 447–450
plants
assays, **66,** 437–450
DNA fragmentation, **66,** 438–447
TUNEL, **66,** 443–447
Proliferating cell nuclear antigen
analysis of
cell proliferation, **63,** 427–428
DNA repair, **63,** 426
complex with
p21, **63,** 421
replication factor C, **63,** 420
as empirical marker, **63,** 419–420
flow cytometric analysis, **63,** 423–425
image cytometric analysis, **63,** 425–426
immunochemical detection, **63,** 421–423
immunofluorescence, **63,** 280
in simultaneous staining, **63,** 264–266
Proliferation
activity of tumor cell subpopulations, **64,** 360–361
cytometric studies, **64,** 448–449
data, and biological aggressiveness, **64,** 473
human tissues and tumors, overview of data on, **64,** 470–474
in surgical oncology, **64,** 447–448
in transition tissues, **64,** 467
lymphocyte, in HIV infection, **64,** 583–584
research with halogenated pyrimidines, **64,** 449–452
TGF-β effect, **64,** 145–147
Proliferation-associated antigens, **64,** 364
LSCM analysis, **64,** 430–434
Proliferation markers
first generation translational studies, **64,** 377–378
human solid tumors, **64,** 361
Ki-67, **64,** 364
and prognosis, **64,** 367–372
and response to systemic treatments, **64,** 372–377
Proliferative fraction, cell cycle phase duration determination, **57,** 232–233
PROLYSIS
web site, **66,** 249
Pronuclear transfer, **62,** 96
Pronuclei assembly, *see* Cell-free nuclear assembly
Pro-ox model 110 (Biospherix), **71,** 12
Prooxidant measurement
reagents and staining procedure, **64,** 280–281
special considerations, **64,** 281
Prooxidant production, heating protocol-dependent, **64,** 283, 285
Propidium iodide
characterization of nuclear changes in protoplasts, **63,** 507–508
counterstaining with, **63,** 66
fluorescence intensity decrease with cell age, **63,** 515
stock solution, **63,** 250
Propidium iodide (PI), **66,** 346, 482

apoptosis assays, **66,** 211–213
cell size, **66,** 54–57
Propidium iodide, as contouring parameter in LSCM, **64,** 318
Prostate adenocarcinoma
cystolic calcium
spectrofluorimetric measurement, **66,** 238–239
intracellular calcium
singe-cell measurement, **66,** 241–242
Protargol, *see* Microscopy, Light
Protease activity, neural-enriched cultures for study of, **71,** 143
Protease assay
caspase assays, **66,** 178–179
Protease inhibitor
selection and application, **66,** 259–265
Protease inhibitors, **62,** 352, 382, 403
Proteasome, scanning transmission electron microscopy, **53,** 154
Proteasomes, **66,** 255–259
Protein
complex with DNA, precipitation, **64,** 181–182
mitochondrial, **64,** 124
nucleolar, stained by silver methods, **64,** 439
Protein A
beads, antibody coupling to, **67,** 84
fusion proteins, and SPB precipitation, **67,** 87–88
Protein A-sepharose beads, **65,** 221–222, 224
Protein adsorption, force spectroscopy, **68,** 14–15
Protein and gene expression, analysis, **59,** 63–84
double-fluorescent *in situ* hybridization, **59,** 77–78
fixation, **59,** 67–68
zebrafish embryos for *in situ* fertilization, **59,** 68
in situ hybridization to RNA, **59,** 63–64
hybridization to whole-mount embryos, **59,** 68–71
zebrafish embryos, post-hybridization washes, **59,** 70–71
zebrafish embryos, pretreatments to, **59,** 68–70
immunolocalization of probes, **59,** 71–73
zebrafish whole-mounts, staining with diaminobenzidine, **59,** 72–73
zebrafish whole-mounts, staining with NBT/BCIP, **59,** 71–72
probe synthesis, **59,** 64–67
antisense RNA probe synthesis, **59,** 65–66
plasmid and probe synthesis, **59,** 66–67
RNase-free conditions, **59,** 64–65
solutions and reagents, **59,** 83–84
transcription and translation gene products, localization, **59,** 78–82
two-color *in situ* hybridization, **59,** 73–77
incubation, **59,** 73–75
sequential incubation, **59,** 75–77
whole-mount embryos, embedding and sectioning, **59,** 82–83
Protein assay, signal transduction, **69,** 369–370
Protein-bound NADPH, **65,** 92–93
Protein coating, polyacrylamide gel, **69,** 330–331
Protein complex isolation
applications, **65,** 238–242
blue-native electrophoresis, **65,** 232–237
blue-native gels, **65,** 231–242
cell lines, **65,** 240–241
chemicals, **65,** 232
chloroplasts, **65,** 238–244
SDS-PAGE, **65,** 237–238
total tissue homogenates, **65,** 240–241
Protein concentrations
cell-free systems, **66,** 171
Protein degradation, multicomponent ECM matrix proteins, **69,** 202
Protein detection
Bcl-2 family
antibodies, **66,** 37
Protein dynamics
immunocytochemical detection in PC12 cell lines, **71,** 283–285
monitoring in *Drosophila,* 226–228
Protein engineering, AFM studies, **68,** 314–316
Protein expression, brefeldin A effect, **63,** 227
Protein expression, in *Drosophila,* **71,** 203, 216–220
Protein expression, overexpression, **62,** 486, 495, 555
Protein expression systems
pAcGP67A.coco design, **69,** 71–73
rationale, **69,** 71
requirements, **69,** 70–71
Protein folding, force spectroscopy, **68,** 15–19
Protein folding, *see* Folding, Green fluorescent protein
Protein function
blocking, in Tibia-1 pathway studies, **71,** 191
studying with B35 cell transfection and motility assays, **71,** 295–297
Protein G, beads, antibody coupling to, **67,** 84
Protein import assaying, **65,** 189–213
blue-native electrophoresis

Protein import assaying (*cont.*)
protein assembly, **65,** 208–209
import intermediates formation, **65,** 207–208
fusion protein intermediates, **65,** 208
low-temperature intermediates, **65,** 208
nucleotide-dependent intermediates, **65,** 208–209
mitochondrial localization, **65,** 199–204
cofractionation studies, **65,** 203–204
membrane integration determination, **65,** 200–201
mitochondrial swelling experiments, **65,** 202–203
preprotein protease degradation, **65,** 201
soluble protein extraction, **65,** 200
pathways, **65,** 205–208
bypass import, **65,** 205
receptor inhibition, **65,** 205–206
preprotein import energetic requirements, **65,** 204–205
preprotein synthesis, **65,** 190–196
protocols, **65,** 196–199
import time course experiment, **65,** 196–199
mitochondria isolation, **65,** 196
purified recombinant proteins, **65,** 209–213
detection, **65,** 212–213
import reaction, **65,** 212
preprotein preparation, **65,** 210–212
yeast mutants, **65,** 206–207
Hsp70 defects, **65,** 207
TIM machinery defects, **65,** 206–207
TOM machinery defects, **65,** 206
Protein import, nucleus
assays, general methodology
digitonin
cell permeabilization, **53,** 521–522
stock solution preparation, **53,** 520–521
energy-regenerating system, **53,** 533–534
flow cytometry assay, **53,** 539
fluorescence labeling of proteins, **53,** 522–523
fluorescence microscopy
background in rapid screening, **53,** 535–536
imaging techniques, **53,** 538–539
indirect immunofluorescence, **53,** 537–538
incubation conditions for import
extracts, **53,** 534–535
recombinant factors, **53,** 535
nuclei recovery, **53,** 536–537
small nuclear ribonucleoprotein particle preparation, **53,** 524–527
transport buffer preparation, **53,** 522
GTPases, **53,** 519–520, 532–533
importins
purification of recombinant proteins, **53,** 530–532
roles, **53,** 518–519
types, **53,** 518
kinetic assays in *Saccharomyces cerevisiae*
export assay, **53,** 553
fluorescence microscopy, **53,** 552–553
import assay, **53,** 553
materials, **53,** 552
nuclear localization signal–green fluorescent protein substrate
construction, **53,** 549–550
fluorescence properties of mutants, **53,** 550
passive import and export, **53,** 547, 551
scoring cells, **53,** 553–554
temperature effects, **53,** 547
transport competence assessment, **53,** 549
yeast mutants, **53,** 547, 551–552
mammalian cells, cytoplasmic extract preparation, **53,** 529–530
nuclear localization signal
nucleoplasmin, **53,** 522
protein binding, **53,** 518
protein conjugation, **53,** 523–524, 549–550
sequence, **53,** 523, 546
reticulocyte lysates in cell-free systems, **53,** 530
small nuclear ribonucleoprotein particle pathway, **53,** 519–520, 546
trimethylguanosine cap in signaling, **53,** 519–520
Xenopus systems
egg extract preparation, **53,** 528–529
oocyte extract preparation, **53,** 527–528
overview, **53,** 518
Protein interactive cloning
Bcl-2 family, **66,** 35
Protein kinase C antagonists, **66,** 264
Protein kinase inhibitors, **66,** 368
Protein kinases
indirect immunofluorescence microscopy, **67,** 306–308
MT *vs.* centrosome localization, **67,** 310–311
subcellular fractionation, **67,** 311
subcellular localization, **67,** 305–306
tagged, microscopy, **67,** 308–310
Protein labeling
with organic fluorescent dyes
choice of reactive group, **63,** 193–194
conjugation chemistry, **63,** 186–187
optimizing conditions, **63,** 187–190
specific dyes used, **63,** 190–193
with phycobiliproteins

heterobifunctional reagents, **63,** 195–201
light absorbance, **63,** 194–195
Protein loading
affecting choice of neuronal culture, **71,** 2
and specific species, **71,** 9
as single-cell injections, **71,** 134–135
in *Xenopus* spinal cultures, **71,** 130, 131
Protein measurement
metabolic labeling, **66,** 141
Protein–protein interactions, **65,** 217–229
antibody cross-linking, **65,** 222
Bcl-2 family, **66,** 34–36
coimmunoprecipitation, **65,** 222–225
Cox II immunoprecipitation, **65,** 219–222
cross-linking, **65,** 225–228
unstability, **65,** 218
Protein–protein interactions, force spectroscopy, **68,** 292
Protein sequencing
TGase reactive glutamine residues, **66,** 125–126
Protein storage
annexin V purification, **66,** 345
Protein substrate assay
caspase, **66,** 16–17
cell-free proteins
in vitro, **66,** 17
Protein synthesis inhibitors, **66,** 368
Protein tagging, **62,** 549–557
Protein trafficking, **71,** 267
Protein transduction domains, **71,** 327
exhibiting minimal toxicity, **71,** 328
Protein transduction peptides, exhibiting minimal toxicity, **71,** 320
Protein transglutamylation
apoptosis, **66,** 111–132
antidinitrophenyl antibodies, **66,** 131–132
buffer, **66,** 131
ε-(γ-glutamyl)-lysine cross link density, **66,** 115–119
reagents, **66,** 131–132
Protein unfolding, AFM studies
biological implications, **68,** 328
calibration, **68,** 332
data acquisition, **68,** 331–332
data fitting procedures, **68,** 323
double-sensor-stabilized AFM, **68,** 328–331
force–clamp experiments, **68,** 318–319
force curves, **68,** 317–318
force spectroscopy mode, **68,** 317
forced unfolding, **68,** 321, 331–332
general scheme, **68,** 312–313
Monte Carlo simulations, **68,** 327, 334
multiple pickup, **68,** 322–323
pathway questions, **68,** 324–325
refolding, **68,** 319–321
single-parameter model, **68,** 325–327
sorting data, **68,** 321
Protein unfolding, tenascin, **69,** 175–177
Proteinase, **66,** 247–249
classification, **66,** 248–249, 248
web site, **66,** 249
Proteinase inhibitors, **66,** 249–259
aspartic, **66,** 251
commercial sources, **66,** 252
guidelines, **66,** 265–266
nonproteinase enzymes, **66,** 263–264
specificity, **66,** 263
web site, **66,** 249
Proteins
assembly, **65,** 433
assisted delivery into live neurons, **71,** 325–337
bacterial fusion, **67,** 131–132
basement membrane, *see* Basement membrane proteins
cell-assembled ECM, **69,** 152–153
cellular, *see* Cellular proteins
centrosome-associated, *see* Centrosome-associated proteins
centrosome, *see* Centrosome proteins
chimeric, regulatory HS binding probes, **69,** 97–98
COCO system, **69,** 78–80
containing PTDs, puchasing and producing, **71,** 330
degradation, **65,** 433–434
detection, in SPB precipitation, **67,** 88
DNA interactions, SFM studies, **68,** 368–369, 376–377
ECM, **69,** 8–9, 17
expression, **65,** 262–264
folding, **65,** 432–433
FRET by FLIM, **69,** 262
HA-tagged proteins, **67,** 86–87
immunolocalization
ABC method, **65,** 327–328
paraffin-embedded samples, **65,** 319–321
inclusion in fibrin–FN matrix, **69,** 148–150
individual, AFM studies, **68,** 291
integrin-associated, *see* Integrin-associated proteins
integrin-mediated signaling pathways, **69,** 381–382

Proteins (*cont.*)
intermembrane, **65,** 151
intracellular, immunocytochemical detection, **63,** 346–347
labeling, **69,** 262
load, **65,** 234–235
loss, from formaldehyde fixation, **63,** 279–280
matrix, *see* Matrix proteins
membrane bound, **65,** 47
membrane, *see* Membrane proteins
metabolically labeled coimmunoprecipitation, **66,** 41
modular structure, force spectroscopy, **68,** 15
Myc-tagged proteins, **67,** 86–87
NuMA protein, **67,** 224–225
soluble, **65,** 47
SPB, *in vitro* binding, **67,** 89–92
in spindle assembly, **67,** 247–251
studies with recombinant ECM, **69,** 70
yeast mitochondria, **65,** 430
Proteins, laser tweezer plasma membrane movement studies
bound *vs.* unbound receptor molecules, **55,** 189–192
fence effects, **55,** 188–192
laser calibration, **55,** 179–181
membrane skeleton protein transport, **55,** 188–189
overview, **55,** 174–176
receptor molecule dragging obstacles, **55,** 181–185
representative experiments, **55,** 181–183
trajectory analysis, **55,** 183–185
single-particle tracking
elastic intercompartmental barriers, **55,** 185–188
future research directions, **55,** 192
gold labeling particle preparation, **55,** 176–178
protein motion, **55,** 176
video microscopy, **55,** 178–179
tether effects, **55,** 188–192
Proteoglycans, **71,** 12
anion-exchange chromatography, **69,** 57–59
basic interactions, **69,** 53–54
categories, **69,** 54
ELISAs, **69,** 63–65
extraction, **69,** 55–57
gel filtration chromatography, **69,** 60–61
hydrophobic interaction chromatography, **69,** 59–60
immunoaffinity chromatography, **69,** 61–63
radiolabeling, **69,** 57
Proteolytic digestion, real-time AFM imaging, **69,** 172–174
Proteome analysis
plant mitochondria, **65,** 68–70
Proteomics, in centrosome kinase isolation, **67,** 312–313
Protomers, bacterial pore unzipping, **68,** 287
Proton–motive Q cycle, **65,** 105–106
Proton pumping measurements, **65,** 86–87
Protoplast
flow cytometry, **58,** 326, 331–333, 337
preparation, **58,** 322–323
recombinant green fluorescent protein construct preparation, **58,** 323
transfection, **58,** 323–324
Protoplasts
isolation from whole tissue, **63,** 509–510
nuclear changes in, **63,** 507–508
Protozoan kinetoplasts mitochondria, **65,** 28–29
Provisional matrix
advantages, **69,** 147
contraction, **69,** 150
disadvantages, **69,** 147
fibrinogen preparation, **69,** 148
immunofluorescence, **69,** 150–151
other protein inclusion, **69,** 148–150
synthesis *in vitro*, **69,** 148
troubleshooting, **69,** 151
PS, **66,** 340
cell surface exposed
in vivo detection, **66,** 349–354
rapid determination
adherent cells, **66,** 348
cells in suspension, **66,** 347
PS-expressing D011.10 cells
phagocytosis, **66,** 360
PS vesicles preparation
phagocytosis, **66,** 361
Pseudotime, flow cytometric, **63,** 179
PSF, *see* Point-spread function
Protrusive activity, of growth cone, **71,** 3–4
PSORT I, **65,** 180–181
PSORT II, **65,** 180–181
PT, **66,** 80, 368
PT pore opening
atractyloside, **65,** 155
PT pores, **65,** 148
PTU
ε-(γ-glutamyl)-lysine cross link density, **66,** 117–118

Pulse characterization
by digital signal processing, **63,** 159–161
with analog hybrid circuits, **63,** 150–154
Pulse-chase labeling, cell cycle phase duration determination, **57,** 235–237
shuttle vector, **71,** 391, 392
Pulse duration, as electroporation parameter, **71,** 345
Pulse shaping, integration by, **63,** 154
Pulse width
measurement circuits, timing and performance, **63,** 153–154
measurement in flow cytometry, **63,** 127
Pulsed field gel electrophoresis
for preparation of DNA from YAC, **64,** 105
for QDFM, **64,** 39–40
Pulsed-field gel electrophoresis (PFGE), meiotic recombination intermediate analysis in *Saccharomyces cerevisiae*
contour-clamped homogeneous electric field electrophoresis, **53,** 267–269
DNA preparation, **53,** 268
double-strand break detection, **53,** 266–268
principle, **53,** 267
Pump-probe microscopy, temporal imaging, **56,** 298–301
applications, **56,** 300–301
instrumentation, **56,** 299–300
Purification
recombinant caspases
folding method, **66,** 13–14
Purification, histidine-tagged green fluorescent protein
expression in *Escherichia coli*, **58,** 46
nickel affinity chromatography, **58,** 46–47
plasmid construct, **58,** 46
Purified plasma membrane
superoxide anion production assay, **66,** 310–311
Purines, **62,** 35
Purity of neurons, **71,** 10–11
importance for biochemical studies, **71,** 130
Purity, protected by aborting of drop charging, **63,** 47–48
Purity, sorted fraction of chromosomes, **64,** 17–18, 20–21, 25–26
Purkinje neuron monolayer, **71,** 90
Purkinje neuronal cultures, **71,** 3, 89–105
astrocyctes in, **71,** 93
astrocyte effects on, **71,** 101–102
cerebellar cell culture of, **71,** 92
coculture variations, **71,** 94
glutamate transport in, **71,** 105
granule cells in, **71,** 93–94
identifying, **71,** 96
immunohistochemistry of, **71,** 94
materials for culturing, **71,** 92
procedures for culturing, **71,** 92–93
Puromycin, **66,** 207t, 431
[^{14}C]putrescine method
TGase assay, **66,** 114–115
Pyramidal neurons, **71,** 112
in chick forebrain cultures, **71,** 53–54
in hippocampal cultures, **71,** 119
Pyrimidines, **62,** 35
Pyrophosphate treatment, *Dictyostelium discoideum,* **67,** *340–342*
Pyruvate, **65,** 137

Q

QDFM, *see* Quantitative DNA fiber mapping
QPD, *see* Quadrant photodiode
Quadrant photodiode, PFM studies, **68,** 203–204
Quality control
data management for HIV evaluation, **64,** 575–576
flow cytometric analysis of drug–DNA interactions, **64,** 178–182
Quality, antibody, K_a as descriptor, **63,** 223–224
Quality control, identifying time-dependent artifacts, **63,** 178–179
Quantification, coexistence with simplification, **63,** 161–167
Quantitation, aggregation assays, **63,** 601–602
Quantitation limit, **63,** 310–311, 315–316
Quantitative assay, for β-galactosidase, **67,** 81–82
Quantitative DNA fiber mapping
applications, **64,** 34
measuring clone overlap, **64,** 51
chemicals and buffers for, **64,** 35–37
critical aspects, **64,** 46–47
digital image analysis, **64,** 46
FISH, **64,** 45–46
generation of probes, **64,** 41–44
instruments for, **64,** 37–38
preparation of DNA fibers, **64,** 40–41
probe labeling via random priming, **64,** 44–45
protocols, **64,** 38–40
Quantitative fluorescence cytometry, **63,** 311–312
Quantitative indirect immunofluorescence assay, **63,** 332–333
Quantitative microscopy, *see* Video microscopy, Quantitative imaging
Quantitative Western blotting

Quantitative Western blotting (*cont.*)
TNF ligands, **66,** 507–508
Quantity, of neurons by species, **71,** 11
Quantum efficiency
charge-coupled device cameras
fill factor, **56,** 24–26
high-resolution multimode digital microscopy, **56,** 200–202
spectral response, **56,** 23
units, **56,** 23–24
sensitivity enhancement, **56,** 50, 54, 62
Quantum simply cellular beads, **63,** 333
Quenching, in protein labeling, **63,** 190
Quick development assay
tilted flask, **66,** 479
Quicktime movies, **58,** 96, 191, 362
Quiescent cells, centrosome activity, **67,** 215–217
Quin-2, **66,** 235
Quinacrine mustard staining, chromosome analysis, **57,** 209, 211

R

R-PE-CY5.18, *see* R-phycoerythrin/cyanine 5.18
R6G, **65,** 406
Rad, somitic migration, **69,** 348
Radiation hybrid, *see* Hamster/zebrafish radiation hybrid
Radiation, ionizing
contact effect, **64,** 217–218
inhibition of DNA loop rewinding, **64,** 261–262
producing strand breaks, **64,** 245
Radioactive catecholamines, in PC12 cell studies, **71,** 271
Radioactive rubidium ions, **65,** 287
Radioactivity, cell attachment assays, **69,** 287
Radiolabeled precursor protein
cross-linking
import machinery component, **65,** 227–228
Radiolabeled substrate cleavage
caspase assay, **66,** 17–18
Radiolabeled tetraphenylphosphonium (TTP^+), **65,** 287–288
Radiolabeling, proteoglycans, **69,** 57
Ragged-red fiber (RRF), **65,** 313–314
Ran
nuclear protein import role, **53,** 519–520
purification of recombinant protein, **53,** 532–533
Random amplified polymorphic DNAs
advantages and disadvantages as polymorphic markers, **60,** 170–171
codominant alleles, **60,** 168–169
genetic map construction, **60,** 169–170
heredity of markers, **60,** 169
markers closely linked to mutations, identification, **60,** 172–174, 176
mutation identification, **60,** 165–166
positional cloning project initiation, **60,** 263–264, 266–267
primers and polymerase chain reaction, **60,** 166, 168, 171–172, 176
sequence tagged site conversion, **60,** 176, 271
strain-specific differences, **60,** 166, 168
Random priming, probe labeling via, **64,** 44–45, 107–108
RAPDs, **62,** 161–163, 267–276, 562
RAPDs, *see* Random amplified polymorphic DNAs
Rapid fixation and staining, in live-cell imaging studies, **71,** 151
Rapid freezing
high-pressure freezing, **67,** 32
plunge freezing, **67,** 32
process, **67,** 29–31
techniques, **67,** 31
Ras, immortalization of cells, **57,** 74–76
Rat brain cytosol, **71,** 272
Rat cerebellum, Purkinje neuron cultures from, **71,** 90
Rat cortical neurons
developmental properties of, **71,** 5
relative infective efficiency of, **71,** 48
Rat dorsal root ganglia, quantity of yield from, **71,** 11
Rat hippocampal neurons, **71,** 4
contrasting with chick forebrain neurons, **71,** 54–57
culturing, **71,** 111–126
developmental properties of, **71,** 5
growth and development characteristics of, **71,** 53
infective efficiency of, **71,** 48
quantity of yield from, **71,** 11
small size of, **71,** 8
Rat liver mitochondria, **65,** 13–19
frozen tissues, **65,** 17–18
gradient-purified rat liver mitochondria, **65,** 15–16
phospholipid-enriched mitoplasts, **65,** 18–19
SMP, **65,** 16

standard preparation, **65,** 13–14
TEM, **65,** 247–248
Rat sympathetic neurons, **71,** 17
developmental properties of, **71,** 5
long-term cultures of, **71,** 33–34
organelle transport in, **71,** 8
quantity of yield from, **71,** 11
short-term cultures of, **71,** 32–34
Rate of development, **71,** 10
Rat thymocytes
sorting, **66,** 66f, 67
Ratio imaging, **56,** 217–234
detector systems, **56,** 231–233
digital image processing, **56,** 233–234
emission control, **56,** 230–231
illumination control, **56,** 230–231
instrument choice, **56,** 219–222
intracellular ion measurement, **56,** 237–251
experimental chamber, **56,** 249–250
fluorescence signal calibration, **56,** 245–247
imaging workstation components, **56,** 247–249
indicator properties, **56,** 239–244
incorporation into intact living cells, **56,** 242–243
pitfalls, **56,** 243–244
signal-to-noise ratio, **56,** 243
spectral characteristics, **56,** 239–242
structural characteristics, **56,** 239–242
overview, **56,** 237–238, 251
perfusion system, **56,** 249–250
rationale, **56,** 238
methodology
emission ratio method, **56,** 223–225
excitation ratio method, **56,** 222–223
multiple fluorophores, **56,** 225–226
optical considerations, **56,** 226–230
chromatic aberration, **56,** 227–230
spherical aberration, **56,** 226–227
overview, **56,** 217–219, 234
Ratiometric method, **65,** 440–441
Ray optics regime, single-beam gradient laser trap, dielectric sphere force calculations, **55,** 1–25
arbitrary trap location force, **55,** 11–16, 23–25
index of refraction effects, **55,** 19–20
light forces, **55,** 4–8
mode profile effects, **55,** 16–20
overview, **55,** 1–4, 20–21
ray force, **55,** 21–22
Y axis trap focus force, **55,** 10–11, 22–23
Z axis trap focus force, **55,** 9–10
rDNA, **62,** 139–140, 142–144, 164–171, 470, 477
antisense ribosomes, **62,** 533–545
vectors, **62,** 166–168
rDNA, *Tetrahymena thermophila*, FISH
nick translation reaction, **70,** 346–347
probe synthesis, **70,** 346
sample preparation, **70,** 344–346
RCAS virus preparation, **71,** 376
Reactive group, choice: for protein labeling, **63,** 193–194
Reactive oxygen species
living cells, **65,** 91–92
mitochondrial staining, **65,** 92
Reactive oxygen species (ROS), **66,** 307–318
methods, **66,** 309–318
apoptosis sensitivity, **66,** 309–313
mitochondria, **66,** 382–383
sources, **66,** 308
Reactive oxygen species assays, **65,** 80–87
DCFDA assay, **65,** 81–83
membrane potential measurements, **65,** 83–85
Reagents
effectiveness and cross-reactivity of, by species, **71,** 9–10
screening, in PC12 cell lines, **71,** 273
Reagents
choices, for immunophenotyping, **64,** 577
for plant flow cytogenetics, **64,** 6–12
gating, CD45 as, **64,** 573
labeling of DNA strand breaks, **64,** 427–428
multicolor fluorescence immunophenotyping, **64,** 578, 580
Receptor-mediated ligand precipitation
tumor necrosis factor-related ligands, **66,** 514–516
Receptor-mediated signalling, **71,** 12
Receptor–cytoskeletal responses, basement membrane components, **69,** 135–136
Receptor molecules
bound *vs.* unbound receptor molecules, plasma membrane movement, laser tweezer study, **55,** 189–192
laser tweezer dragging, obstacles, **55,** 181–185
representative experiments, **55,** 181–183
trajectory analysis, **55,** 183–185
Recombinant adenovirus, methods for making, **71,** 389–391
Recombinant baculovirus
expression
insect Sf9 cells, **66,** 221
Recombinant caspases
bacterial overexpression, **66,** 7–9

Recombinant caspases (*cont.*)
expression
caspase induction, **66,** 9
caspase purification, **66,** 9–12
PET vectors, **66,** 7–8
purification
folding method, **66,** 13–14
Recombinant cytoductants
isolation, **65,** 389–390
mating, **65,** 389
Recombinant expression
targeted GFP mutants, **65,** 375
Recombinant extracellular matrix
pCOCO, **69,** 74–77
role in protein studies, **69,** 70
Recombinant involucrin protein
cross-linking, **66,** 128
Recombinant retrovirus particles, **66,** 202–203
Recombinant Rho-family GTPases, **71,** 335
Recombinant soluble forms
tumor necrosis factor-related ligands, **66,** 503
Recombinant targeted luciferase
immunolocalization, **65,** 369
Recombination
macronuclear, **62,** 514–515
meiotic crossing over, **62,** 163
Recombination nodule (RN)
electron microscopy
sample preparation from avian oocytes, **53,** 248–249
staining
mammalian nodules, **53,** 248
plant nodules, **53,** 249
synaptonemal complex association, **53,** 236, 247
Reconstitution
MT nucleation, **67,** 142–145
γTuSC in Sf9 cells, **67,** 12–18
Recorders, video-enhanced differential interference contrast light microscopy, **56,** 178–179
Recording mode photonic force microscopy
contact mode, **68,** 208–209
fast 3D single-particle tracking, **68,** 210
tapping mode, **68,** 209–210
Recovering cells, **71,** 13
Recruitment assay, for centrosome proteins, **67,** 317–318
Recurrence, local, proliferating tumors, **64,** 369
Recyling of neurons, **71,** 5–8
Red cells, fetal nucleated, isolation, **64,** 144–145
Red fluorophore, **70,** 213–215
Red helium neon laser, **71,** 151
Red maximal pixel
apoptotic cells
LSC, **66,** 78f, 79
Redox state
caspase, **66,** 315–318
Redox status, integrins, **69,** 256–258
Reference point, in time-lapse studies of cell migration, **63,** 553
Reflected light objective lens, **70,** 266–268
Refractive index effects, confocal microscopy, **70,** 150–155
REGA-1, **71,** 182–183
Regrowth assay, **66,** 480–481
Regulatory heparan sulfate
binding probes
chimeric proteins, **69,** 97–98
FGF production, **69,** 95–97
fixed cell monolayers, **69,** 99
heparin binding on 96-well plates, **69,** 98–99
specificity, **69,** 100–101
tissue sections, **69,** 101–102
identification considerations, **69,** 95
Regulatory volume decrease (RVD), **66,** 51
Regulatory volume increase (RVI), **66,** 51
Released monoamines, secretion assays of, in PC12 cell lines, **71,** 276–277
Reimaging, green fluorescence, **63,** 293
Released monoamines, secretion assays of, in PC12 cell lines, **71,** 276–277
Relocalization, for light microscopy and FISH, **64,** 327, 329, 331–333, 335–336
Relocation functionality, LSCM, **63,** 71,74
REMI, **66,** 487–490
transfection protocol, **66,** 488–489
Renilla reniformis, green fluorescent protein comparison to *Aequorea victoria*, **58,** 1–2, 8
Repeated-PRINS, stronger PRINS signals obtained by, **64,** 64
Repetitive elements
applications
chromosomal marking, **60,** 115
genetic mapping, **60,** 114–115
overview, **60,** 113–114
polymorphism detection in genes, **60,** 115, 117
polymorphism detection in genomes, **60,** 117–118
transposons, **60,** 118–122, 124
C_0t analysis, **60,** 100
microsatellite repeat sequences, **60,** 104–105

mouse/zebrafish somatic cell hybrid, polymerase chain reaction characterization
amplification reaction, **60,** 313–314
available markers, **60,** 313
DANA/mermaid polymerase chain reaction, **60,** 317–318
hybrid panel composition, **60,** 314
overview of types in zebrafish, **60,** 101–102
satellite elements
Alu element homology, **60,** 103
type I elements, **60,** 103–104, 308–309
type II elements, **60,** 104
telomeres, **60,** 104
transposons
DNA transposons
frequency in genome, **60,** 105
miniature inverted-repeat transposable elements, **60,** 112–113
Tc1-like transposable elements, **60,** 109–112
mutation effects on phenotypes, **60,** 118
retrotransposons
classification, **60,** 106
DANA, **60,** 107–108
frequency in genome, **60,** 105
LINEs, **60,** 108
SINEs, **60,** 106–107
unclassified and composite elements, **60,** 108
tagging, **60,** 118
transformation vectors, **60,** 121–122, 124
transgenic fish generation, **60,** 119
unclassified tandem repeat elements, **60,** 105
Replication-competent viruses, testing viral stocks for, **71,** 406
Replication-deficient adenovirus, **71,** 387
Replica plating, **62,** 198–199, 230–231
Replicon, *see* DNA replication site, mapping *in situ*
Reporter cells
instructive ECM, **69,** 47
response, characterization, **69,** 48–49
response, instructive ECM
adhesion assays, **69,** 37–40
coculturing, **69,** 37
deposited ECM, **69,** 35–36
ECM testing, **69,** 34–35
overview, **69,** 33–34
soluble ECM, **69,** 36–37
Reporter targeting, **65,** 353–377
ATP, **65,** 365–371
calcium, **65,** 357–365
IMS, **65,** 355–356
mitochondrial matrix, **65,** 355
pH, **65,** 372–376
strategy, **65,** 354–357
Reproducibility, as advantage of transfection by electroporation, **71,** 341
Reproducibility in isolating tissue, explant culture assay, **59,** 132
Reproduction, centrosomes
assay monitoring, **67,** 300–301
aster duplication *in vitro* studies, **67,** 281–283
duplication assay logic, **67,** 289–290
duplication assay preparation, **67,** 290–292
duplication assay steps, **67,** 292–294
manipulation *in vitro,* **67,** *283*
polarization microscopy studies, **67,** 271–275
spindle assay, **67,** 294–295
video microscopy system, **67,** 275–276
Xenopus egg S-phase extract preparation, **67,** 270
Reproductive system, laser micromanipulation, **55,** 87–94
oocytes, **55,** 88–92
preembryos, **55,** 92–94
sperm, **55,** 87–88
Repulsive guidance cues, **71,** 177
Research applications
for B35 and B50 cell lines, **71,** 289–294
for Purkinje neurons, **71,** 100–105
making proteins into drugs, **71,** 336–337
Resistance
drug
genetic or kinetic basis, **64,** 227
in large spheroids, **64,** 223
kinetic, in multifraction exposures, **64,** 225–227
Resolution
ADCs, **63,** 158–159
coefficient of variation as measure, **63,** 309–310
fluorescence, **63,** 313–317
improvement by presorting, **63,** 380–381
in fluorescence microscopy, **64,** 77
increased due to changes in fluorescence signals, **64,** 189
of time axis, **63,** 177
Resolution, *see* Spatial resolution, Temporal imaging
Resolution, confocal microscopy
excitation/emission wavelength effects, **70,** 233–235
lenses, **70,** 324–326

Resolution, confocal microscopy (*cont.*)
  microscope, **70,** 321–324
  overview, **70,** 10–11
Resolution, PFM studies, **68,** 207–208
Resonance energy transfer, *see* Fluorescence resonance energy transfer
Respiratory chain
  brain mitochondria, **65,** 22
  broken SMP, **65,** 11
  immunolocalization
    DNA, **65,** 317–318
  nuclear DNA-encoded subunits
    mitochondrial DNA, **65,** 326–327
Respiratory enzyme redox assay, **65,** 76–80
  complex I assays, **65,** 76–78
  complex II assays, **65,** 78–79
  complex III assays, **65,** 79–80
  complex IV assays, **65,** 80
Rest for recovery, as electroporation parameter, **71,** 348
Restriction fragment length polymorphism analysis, cell line authentication, **57,** 212
Retention
  anthracycline, **64,** 195
  fluorochromes, **64,** 201–206
Reticulomyxa, organelle manipulation, *in vivo*, **55,** 197–199
  cell preparation, **55,** 197
  organelle movement, **55,** 197–198
Retina, development, **59,** 273–295, 278–285
  introduction, **59,** 273–275
  summary, **59,** 294–295
  wild-type and mutant visual systems
    histological analysis, **59,** 280–281
    molecular marker expression, **59,** 281–283
    mosaic analysis, **59,** 283–285
    other approaches, **59,** 285
  zebrafish, **59,** 275–278
    early morphogenetic events, **59,** 275–276
    genetic analysis, **59,** 285–294
    neurogenesis, **59,** 276–277
    retinotectal projections, **59,** 277–278
Retinal cell biology, **70,** 326–327
Retinal imaging, **70,** 273–274
Retraction stopper, objective lens, **70,** 260
Retrospective studies, ploidy and proliferation in surgical oncology, **64,** 447–448
Retroviral-mediated gene transfer, **66,** 202–205
Retroviral transduction
  target cells, **66,** 209–210
  viral concentration, **66,** 208–209
Retroviral vectors
  diagram, **66,** 204
Retroviral vectors
  building, **71,** 371–378
  compared with electroporation, **71,** 385
  concentrating, titering, and storing, **71,** 378
  confirming orientation and sequence of insert, **71,** 374
  disadvantages of, **71,** 385
  expressing transgenes through, **71,** 369–387
  infecting chick embryos via injection, **71,** 379–382
  transfection with, **71,** 310–314
Retrovirus-mediated insertional mutagenesis, zebrafish
  efficiency, **60,** 87–89, 95
  large-scale screening, **60,** 95, 97
  limitations, **60,** 121
  Moloney murine leukemia virus vector, **60,** 88, 120
  pilot screen
    cloning affected genes, **60,** 92, 94–95
    finding insertional mutants, **60,** 89, 91–92
  principle, **60,** 88, 120, 260
  tropism of virus, **60,** 88
Reverse genetics, **71,** 196, 200–201
  in PC12 cell lines, **71,** 270
Reverse transcriptase PCR (RT-PCR)
  caspase assay, **66,** 6–7
RFLP analysis, *see* Restriction fragment length polymorphism analysis
Rh123
  mitochondrial transmembrane potential
    apoptosis, **66,** 81–82
Rho
  mitochondrial transformants
    selection, **65,** 393
  recipients
    multiple mutations, **65,** 390
  Rho$^+$ cells
  transformation, **65,** 392–394
Rhodamine, **66,** 381
Rhodamine 123, probe for mitochondria, **64,** 119
Rhodamine 6G (R6G), **65,** 406
Rhodamine B hexyl ester, **65,** 341
Rhodamine fluorophores, **70,** 173
Rhodamine, protein labeling, **63,** 193
Rhodamine 123, study of lymphocyte activation/proliferation, **63,** 123
Rhodamine, **65,** 88, 90, 123, 341
  mitochondrial membrane potential, **65,** 291
Ribosomal DNA, see rDNA
Ringlien differential screw thread compressor

construction details, **70,** 354–356
operating instructions, **70,** 356–357
overview, **70,** 353–354
RL cells, adhesion force measurements, **68,** 100–104
RMS
discrepancy
antilog, **65,** 446
RMS, *see* Rostral migratory stream
RN, *see* Recombination nodule
RNA
as *Xenopus* culture injections, **71,** 131
metabolism studies in *Drosophila,* 203
preparing injection samples of, **71,** 132–133
RNA
BrdUrd-substituted, **64,** 130–131
isolation, **62,** 427–429, 537
labeling with BrdUrd, **64,** 134–135
ribosomal, content of microorganisms, **64,** 524–525
synthesis, flow cytometric analysis, **64,** 129–137
RNA polymerase
yeast mitochondrial
core, **65,** 420
RNA transport, nucleus
processing of RNA, **53,** 560
RNA-binding proteins, **53,** 582
small nuclear ribonucleoprotein particle pathway, **53,** 519–520, 546
Xenopus oocyte system
advantages, **53,** 560, 584
enucleation, manual techniques, **53,** 572–573
immunoprecipitation analysis of RNA-associated proteins, **53,** 574–575
kinetic analysis
competition analysis, **53,** 581
initial rates, **53,** 579–580
rate-limiting step, **53,** 577
RNA cap effects, **53,** 577
transport time course, **53,** 579
microinjection
cytoplasmic injection technique, **53,** 572
incubator, **53,** 569
microinjector, **53,** 569
micromanipulator, **53,** 569
needle puller and beveler, **53,** 569
nuclear injection technique, **53,** 571
sample preparation, **53,** 570–571
stereomicroscope, **53,** 568
oocyte production
frog colony maintenance, **53,** 560, 562
isolation for microinjection, **53,** 564–565
ovariectomy, **53,** 562–564
RNA, analysis of extractions, **53,** 573–574
RNA substrate preparation
metabolic radiolabeling, **53,** 567
modified RNA synthesis, **53,** 566
purification, **53,** 566
template DNA, **53,** 565
transcription, **53,** 565–566
stability of injected RNA, **53,** 579
RNA, antisense, integrin-mediated signaling pathways, **69,** 381
Rna1p, nuclear protein import role, **53,** 519
RNase A, digestion of nuclei, **53,** 15–16, 27
RNase protection assay, **64,** 408–409
Robon-Beard cells, **71,** 140
Roche Laboratories, **71,** 137
Rodent cortical neurons, **71,** 4
Rodent sensory neurons, **71,** 3
Roller bottle
cell harvesting, **57,** 221
medium harvesting, **57,** 221
plating, **57,** 220–221
types, **57,** 220
Root mean square (RMS)
discrepancy
antilog, **65,** 446
Root tip cells, accumulation in metaphase, **64,** 13–14
ROS, **66,** 307–318
methods, **66,** 309–318
apoptosis sensitivity, **66,** 309–313
mitochondria, **66,** 382–383
sources, **66,** 308
Rose chamber method, **61,** 160, 223–224
Rostral migratory stream, cell migration, **69,** 351–353
Rotenone, **65,** 129, 137
Rotenone sensitivity
NADH-ubiquinone oxidoreductase, **65,** 103
Rous sarcoma virus vectors, **71,** 370
R-phycoerythrin/cyanine 5.18, **70,** 218–219
RRF, **65,** 313–314
RSGFPU green fluorescent protein, spectral properties, **58,** 14
RT-PCR
caspase assay, **66,** 6–7
RVD, **66,** 51
RVI, **66,** 51

S

S-100 extracts
caspase 9 depletion, **66,** 25
caspase assay, **66,** 14
S phase
duration, **64,** 471–472
incorporation of DNA precursors, **64,** 363
quantification of nDNA content, **64,** 363–364
S-phase fraction, measurements, as prognostic markers, **64,** 448–449
S-phase, *Xenopus* egg
arrested, **67,** 276–279, 281
for centrosome reproduction studies, **67,** 270
S175G green fluorescent protein, thermotolerance, **58,** 142
S205A green fluorescent protein, proton transfer, **58,** 13
S49 Neo cells
FSC
vs. PhiPhiLux contour plot, **66,** 60
vs. SSC dot plot, **66,** 54f–56
S65T green fluorescent protein
emission intensity compared to wild type, **58,** 23–25, 42, 50, 91, 109, 124, 156, 299
subcellular targeting, **58,** 76–79
total internal reflection fluorescence microscopy of single molecules
comparison to Cy3 emission, **58,** 55, 58–62
intensity, **58,** 60–61, 63–64
noise, **58,** 61
proton transfer, **58,** 63
reversible photobleaching, **58,** 61–64
SAC6–green fluorescent protein, fusion protein studies in yeast, **58,** 102–103
Saccharomyces cerevisiae, **66,** 455, 456
meiosis, essential gene identification
expression cloning, **53,** 263–264
mutant screening
dityrosine deposition, **53,** 260
ether sensitivity, **53,** 259, 261
heterozygote selection, **53,** 261
microscopic examination of sporulation, **53,** 259–260
mutagenesis, **53,** 259
principle, **53,** 258–259
recombination between ectopic sites, **53,** 261
recombination-deficient mutants, **53,** 261–263
transposon tagging, **53,** 264
nuclear transport kinetic assays
export assay, **53,** 553
fluorescence microscopy, **53,** 552–553
import assay, **53,** 553
materials, **53,** 552
nuclear localization signal–green fluorescent protein substrate
construction, **53,** 549–550
fluorescence properties of mutants, **53,** 550
passive import and export, **53,** 547, 551
scoring cells, **53,** 553–554
temperature effects, **53,** 547
transport competence assessment, **53,** 549
yeast mutants, **53,** 547, 551–552
nucleolus isolation
growth
conditions, **53,** 37
medium, **53,** 35
nuclei-enriched fraction preparation, **53,** 38–40
optimization
buffer ionic strength, **53,** 43
DNase I treatment, **53,** 42
heparin/nuclear protein ratio, **53,** 42–43
sedimentation conditions, **53,** 43
protein assay, **53,** 40
reagents, **53,** 36–37
solution and buffer preparation, **53,** 35–36
spheroplast preparation, **53,** 37–38
sucrose density gradient centrifugation, **53,** 41, 43
yeast strains, **53,** 35
nucleus isolation
Ficoll gradient centrifugation, **53,** 38–40
growth, **53,** 35, 37
materials, **53,** 35–37
spheroplast preparation, **53,** 37–38
pulsed-field gel electrophoresis analysis of recombination intermediates
contour-clamped homogeneous electric field electrophoresis, **53,** 267–269
DNA preparation, **53,** 268
double-strand break detection, **53,** 266–268
principle, **53,** 267
Southern blot analysis of recombination intermediates
cell growth and sporulation, **53,** 265–266
DNA preparation, **53,** 266
double-strand break detection, **53,** 264–265
synaptonemal complex
electron microscopy, **53,** 273–274
immunostaining of proteins, **53,** 274–276
preparation of meiotic cells, **53,** 269–270
spreading of nuclei with detergent, **53,** 270–273

whole-mount hybridization, fixation
cell wall removal, **53,** 217, 220
difficulty, **53,** 218–219
formaldehyde fixation for rDNA probe hybridization, **53,** 219–221
immunodetection of nuclear antigens, fixation protocol, **53,** 221–223
Saccharomyces cerevisiae, see also Yeast mitochondria
growth media, **65,** 335–336
immunofluorescent staining, **65,** 265–266
indirect immunofluorescence, **65,** 268, 272–274
mitochondrial genome
size, **65,** 420
mt transcription system, **65,** 420
paraformaldehyde, **65,** 266–269
synchronized cells, **65,** 336–337
two-hybrid analysis, **65,** 219
Saccharomyces cerevisiae mitochondria
gene replacement
mtDNA, **65,** 388–392
genetic transformation, **65,** 381–394
mitochondrial transformants, **65,** 384–388
mtDNA, **65,** 383–384, 388–392
phenotypes, **65,** 382–383
rho$^+$ cells, **65,** 392–394
mitochondrial transformants, **65,** 386–388
mtDNA recombination, **65,** 384
mtDNA replication, **65,** 382–384
phenotypes, **65,** 382–383
Safranine, **65,** 84
Salmonid ovarian fluid, zebrafish, **59,** 48–49
Salt exraction
yeast mitochondria, **65,** 47
Salt, SFM of supercoiled DNA, **68,** 385–388
Sample preparation
bacteria standard, **64,** 549–550
for flow cytometry
bacteria, **64,** 513–514
food and drink, **64,** 530–531
for telomere length measurement, **64,** 76, 79–80
whole blood, **64,** 569–570
Sample protocol, Ficoll-Hypaque density gradient separation, **63,** 213
Samples, *see also* Specimens
mixing and delivery: time role, **63,** 171–175
preparation, **63,** 361–364
critical aspects, **63,** 369–370
selective loss of apoptotic cells during, **63,** 537
time-point, single measurements, **63,** 367–369
SAPKs, *see* C-Jun N-terminal kinases
SAR, *see* Scaffold attachment region
Satellite DNA, probe synthesis for fluorescence *in situ* hybridization, **53,** 194–196
Satellite elements
Alu element homology, **60,** 103
type I elements, **60,** 103–104
type II elements, **60,** 104
SATP, *see* *N*-Succinnimidyl-3-(*S*-acetylthio) propinate, AFM tips, **68,** 119
SC, *see* Synaptonemal complex
Scaffold attachment region (SAR)
definition, **53,** 324
structure, **53,** 324
Scaffolds, 3-D, in ECM systems, **63,** 564
Scaleup, cell culture
applications of large-scale culture, **57,** 219–220
fermentors, **57,** 226–227
hollow fibers, **57,** 223–224
large plates, **57,** 220
microcarriers
cell growth, **57,** 222–223
types, **57,** 221–222
roller bottles
cell harvesting, **57,** 221
medium harvesting, **57,** 221
plating, **57,** 220–221
types, **57,** 220
suspension adaptation
approaches, **57,** 224
culture, **57,** 225–226
medium, **57,** 224
scaleup, **57,** 226
spinners, **57,** 224–225
Scan Array 3000, **65,** 441
Scan head, confocal microscopy, **70,** 20
Scanning densitometry
mitochondria-actin binding, **65,** 170–171
Scanning electron microscope, cryostage, **63,** 589
Scanning electron microscopy
cell processing hints, **57,** 310–311
monolayer cultures, **57,** 308–309
sputter coating, **57,** 310–311
suspension cultures, **57,** 310
Scanning electron microscopy (SEM)
resolution, **53,** 125–126
sample preparation for nuclear three-dimensional surface structure analysis
assembled nuclei *in vitro*, **53,** 134–137
Chironomus tentans salivary gland, **53,** 128–131
tissue culture cells
monolayers, **53,** 133

Scanning electron microscopy (SEM) (*cont.*)
suspensions, **53,** 131–133
Xenopus oocyte, **53,** 126–128
Scanning force microscopy
DNA
AC modes, **68,** 367
covalent, anchoring, **68,** 365–366
electron microscope legacy, **68,** 359
fragment ablation, **68,** 379
mobility modulation and solution dynamics, **68,** 371–374
molecular motion, **68,** 374–376
organic solvents, humidity issue, **68,** 366–367
protein and ligand interactions, **68,** 368–369
protein interactions, **68,** 376–377
secondary structure, **68,** 369–370
sharper probes, **68,** 367–368
single-molecule dissection, **68,** 377–379
single-molecule stretching, **68,** 388
solid-state sizing, **68,** 370
ss-DNA structure, **68,** 388
supercoiled, salt and surface effects, **68,** 385–388
water and buffers, **68,** 370–371
DNA curvature and flexibility
basic approaches, **68,** 380–384
ergodicity, **68,** 384–385
scanning forces, **68,** 54–56
Scanning nearfield optical microscopy, **68,** 4
Scanning probe microscopy
mechanical limitations, *see* Micromechanical limitations
optical force microscopy
atomic force microscope-like applications
intermolecular force measurements, **55,** 101–104
optical force microscopy, **55,** 100–101
experimental design, **55,** 105–113
calibration, **55,** 111–112
detection, **55,** 108–111
probe selection, **55,** 106–108
scanning, **55,** 108
signal processing, **55,** 113
future research directions, **55,** 113–115
overview, **55,** 99–100
Scanning transmission differential interference contrast, **70,** 282
Scanning transmission electron microscopy (STEM)
Brookhaven National Laboratory facility, **53,** 140, 149–150, 158–160
computer models of structure
types, **53,** 148
validation, **53,** 139–140, 148–149
historical development, **53,** 156–160
mass measurement
accuracy, **53,** 144–146
chromatin remodeling protein 1, **53,** 145–146, 156
complementary techniques, **53,** 142–143
effective range, **53,** 143
principle, **53,** 143–144
radial mass profile analysis, **53,** 147–148
VirE2, **53,** 146, 156
negative staining, **53,** 155
operation parameters, **53,** 149–150
resolution, **53,** 140, 142, 150
sample preparation
buffers, **53,** 151
concentration, **53,** 151
destruction of samples, **53,** 140, 151–152
purification, **53,** 140, 150–151
specimen selection for technique, **53,** 150
wet film method, **53,** 149
site-specific gold cluster labeling
advantages over colloidal gold, **53,** 152–153
labeling approaches, **53,** 153–154
silver enhancement, **53,** 153
types, **53,** 152
tobacco mosaic virus as control, **53,** 149–150
Scanning transmission phase contrast, **70,** 281–282
Scanning tunneling microscopy
bacteriophage T5, **68,** 4–7
instrumentation and function, **68,** 2
liquid crystals, **68,** 6
Scatter plot, colocalization in fluorescence images, **53,** 66–68
Scattergrams
cell location, **64,** 322
FITC *vs.* phycoerythrin, **64,** 324–326
immunofluorescence, **64,** 319
Schistocerca americana, **71,** 3
Schizosaccharomycetes pombe, **70,** 97
Schwann cell
conduction, **57,** 168–169
culture from human peripheral nerve trunks
dissociation of cells, **57,** 180
nerve harvesting, **57,** 179
passaging, **57,** 180
primary culture, **57,** 180
culture from rat dorsal root ganglia
adult
dissociation of cells, **57,** 177

dorsal root ganglia dissection, **57,** 175
embryo
dissociation of cells, **57,** 175–177
dorsal root ganglia dissection, **57,** 173–175
passaging, **57,** 178–179
primary culture, **57,** 177
development, **57,** 169
growth factors, **57,** 169
materials for culture
bovine serum albumin gradient preparation, **57,** 172–173
serum-free medium preparation and growth factor supplementation, **57,** 170–171
stock solution preparation, **57,** 171–172
substrate coating of culture vessels, **57,** 173
surgical instruments, **57,** 170
morphology of cultured cells, **57,** 181
myelination
remyelination in culture, **57,** 182–183, 185
status, **57,** 168
proliferation markers, immunocytochemistry, **57,** 182
Scintillation counter, bioluminescence detection, **58,** 286–287
Scramblase activity
assays, **66,** 356
Screening
chemical compounds in *Xenopus* egg, **67,** 250–251
Dictyostelium centrosomes, **67,** 351–352
suppressor, for SPBs, **67,** 97–102
synthetic lethal, for SPBs, **67,** 103–107
Screening, Mabs for instructive ECM identification, **69,** 43–47
SDH, **65,** 314, 326
SDS–gel electrophoresis, *Dictyostelium* centrosomes, **67,** 345–346
SDS-PAGE
mitochondrial translation products, **65,** 435–436
separation
plant mitochondria, **65,** 66
yeast actin purification, **65,** 163
Sea urchin egg mitochondria, **65,** 28
Sea urchin, MT isolation, **67,** 151–152
Sea urchin zygote mitosis studies
advantages
cell size, **61,** 441
optical characteristics, **61,** 441
quantity, **61,** 440
rapid cell cycle, **61,** 440
synchronous cell cycles, **61,** 440–441
annotated references
cell-free extracts, **61,** 470
classic texts, **61,** 464–465
colcemid use, **61,** 465
cytology and cytochemistry, **61,** 469–470
electrical and magnetic fields, **61,** 470
electron microscopy, **61,** 468–469
enucleation of zygotes, **61,** 465
isolation and *in vitro* assembly of cytoskeletal elements, **61,** 465–467
isolation of nuclei, **61,** 465
micromanipulation, microinjection and microtechniques, **61,** 470–471
molecular biology of development, **61,** 467–468
visualization of chromosome dynamics in living zygotes, **61,** 465
buffers mimicking intracellular conditions
formulations, **61,** 462
Pipes-based buffer hazards, **61,** 462
vs. conventional buffers, **61,** 462
enhanced visualization of cytoskeletal details
hexylene glycol augmentation, **61,** 463
procedure, **61,** 463
yolk granule clearing rationale, **61,** 463
fragmentation of zygotes
advantages, **61,** 463–464
fragment differentiation, **61,** 464
procedure, **61,** 464
maintenance of organisms
organisms
shipping, **61,** 441–442
temperature shock, **61,** 442
tanks
dolomite, **61,** 442
feeding, **61,** 442–443
nitrobacter, **61,** 442
organism overloading, **61,** 442
seawater salts, **61,** 442
water circulation, **61,** 442
water selection, **61,** 442
mounting cells
adhering zygotes to slides or coverslips
poly-L-lysine, **61,** 454
procedure, **61,** 455
protamine sulfate, **61,** 454–455
cleaning slides and coverslips, **61,** 451–452
long-term observation preparation
coverslip and slide sizes, **61,** 452
fluorocarbon preparation, **61,** 452
fluorocarbon sources, **61,** 454
inverted microscope use, **61,** 454

Sea urchin zygote mitosis studies (*cont.*)
procedure, **61,** 452–453
long-term *vs.* short-term observation, **61,** 448
microinjection preparations
chamber design, **61,** 458–459
zygote anchorage and flattening, **61,** 459
zygote introduction, **61,** 459
micromanipulation preparations
aluminum/stainless steel support slide, **61,** 455
applications, **61,** 455
chamber construction, **61,** 456–457
inverted microscope considerations, **61,** 455
time considerations, **61,** 457–458
zygote anchorage and flattening, **61,** 458
micropipettes
high filament heat and taper size, **61,** 460
loading chamber design, **61,** 460–462
upward tip curve, **61,** 460
microscopes
cleaning equipment, **61,** 450–451
grade of equipment, **61,** 450
short-term observation preparation
procedure, **61,** 451
wedge chamber, **61,** 451
tools
black Lucite work surface, **61,** 449
vacuum source, **61,** 449–450
obtaining gametes
eggs
storage times, **61,** 443
temperature considerations, **61,** 443
seawater, **61,** 443–444
spawning, **61,** 443
sperm, **61,** 443
zygotes
centrifuges
hand centrifuge commercial source, **61,** 447
motor *vs.* hand centrifuges, **61,** 446–447
time considerations, **61,** 446–447
fertilization
glassware conditions, **61,** 444
sperm and egg concentrations, **61,** 444
fertilization envelopes
calcium-free seawater use, **61,** 445
removal procedure, **61,** 444–445
pipettes, **61,** 448
storage
stirring, **61,** 446
water depth, **61,** 445
zygote concentration, **61,** 445–446
Second harmonic generation, **70,** 441–446
Secof-type cultures, **71,** 228
Second harmonic imaging microscopy, **70,** 445–448
Secretion assays, in PC12 cell lines, **71,** 272–274
Secretion granules
in mitochondrial preparations, **65,** 4
Secretory granule lumen, **70,** 464
Secretory granule surface, **70,** 464
Secretory granules, *see* Dense-core secretory granules
Secretory pathways, **62,** 40–49, 347–360
Secretory vessel dynamics, **71,** 267
PC12 studies of, **71,** 271–283
PC12 usefulness for, **71,** 268
Sectioning, *see* Transmission electron microscopy
Sectioning, for electron microscopy, **67,** 34–35
Sedimentation
MT-dependent, pericentrin, **67,** 65
velocity gradient, for pericentrin enrichment, **67,** 64
Sedimentation assays
actin, **65,** 167
mitochondria, **65,** 167
protocol, **65,** 168–169
reagents, **65,** 167–168
salt treatment, **65,** 169–170
yeast, **65,** 164
Seeds
germination media, **64,** 13
with reconstructed karyotypes, **64,** 6
Segment boundary epithelium, in Tibia-1 pathway studies, **71,** 174–176
Segmentation
chromosomes, **64,** 82, 84–85, 92–93
telomere, and IFI measurements, **64,** 80–82
Segmentation period, embryonic zebrafish, **59,** 24–25
Selectable markers
6-methylpurine, **62,** 491
anisomycin, **62,** 482
blasticidin S, **62,** 493
cycloheximide, **62,** 170, 470, 482, 510
neomycin/paromomycin, **62,** 170–171, 482, 493–494, 510
Selection, **62,** 220, 510
Self-assembly
basement membranes
affinity chromatography, **69,** 124–125
affinity coelectrophoresis, **69,** 125–127
affinity retardation chromatography, **69,** 125
cell–surface contributions, **69,** 135–136

electron microscopy, **69,** 133–135
equilibrium gel filtration, **69,** 128
genetic and molecular analyses, **69,** 137–138
polymerization assays, **69,** 129–130
solid-phase binding, **69,** 128–129
solution binding analysis, **69,** 125
ECM
molecules, **69,** 136–137
type I collagen paradigm, **69,** 111–112
SEM, *see* Scanning electron microscopy
SEM, *see* Microscopy, Electron
Semaphorins, **71,** 177–181
Semiconductor imagers, complementary metaloxide semiconductor images, **56,** 41–42
Seminiferous cord, formation *in vitro* with testicular cells, **57,** 101–103, 105
Semiquantitative assay, for β-galactosidase, **67,** 81
Senescence, **62,** 17
Sensitivity
and background fluorescence, **63,** 113–114
applied to fluorescence measurements, **63,** 310–311
fluorescence, **63,** 313–317
LSC, evaluation, **63,** 63
measurement, flow cytometers, **63,** 127–129
proteins, to fixation/permeabilization, **63,** 291
protoplasts, to changes in osmotica, **63,** 510
Sensors
multiangle scatter, **63,** 7
signal, simultaneously digitized, **63,** 57–58
Sensory cells, live, AFM studies, **68,** 38
Sensory system, recording from, in *Drosophila*, 236–237
Sequence analysis, centrosome-associated protein cDNA, **67,** 129–130
Sequence-tagged sites
construction, **60,** 267, 271
identification in positional cloning, **60,** 274
polymorphism analysis, **60,** 267, 269
Sequencing chromatogram
Gln^{133}, **66,** 127
Ser-65, green fluorescent protein
photobleaching of site-directed mutants, **58,** 33
spectral properties of site-directed mutants, **58,** 23–25, 42, 50, 91, 109, 124, 156, 299, 318
thermal stability of site-directed mutants, **58,** 27–28
Serine proteinase inhibitors, **66,** 250
Sertoli cell, establishment of cell lines
cell preparation
fractionation over discontinuous Percoll gradient, **57,** 96–97
isolation, **57,** 96
clone selection, **57,** 98–99
histology of immortalized cells, **57,** 100–101
immunodetection of simian virus 40 large T antigen, **57,** 99–100
medium for immortalized cell culture, **57,** 99
overview, **57,** 94–95
passaging, **57,** 99
seminiferous cord formation *in vitro* with other testicular cells, **57,** 101–103, 105
transfection
calcium phosphate transfection with simian virus 40 large T antigen gene, **57,** 97–98
principles, **57,** 95–96
Serum, cell culture additive
handling, **57,** 12, 25
heterogeneity, **57,** 25
preparation, **57,** 25
supplementation, **57,** 12–13
toxicity testing, **57,** 54
Serum, starvation, cell-assembled ECM, **69,** 154
Sexual immaturity, **62,** 133, 223–224
Sexual maturity, **62,** 91
Sexual reproduction, *see* Conjugation
Sf9 cells
Bacmid DNA transfection, **67,** 15
γTuSC reconstitution
cell culture, **67,** 12–14
component expression and purification, **67,** 16–18
virus preparation, **67,** 14–16
SFM, *see* Scanning force microscopy
SGLT1, *see* Na^+/D-glucose cotransporter
Shading
digital image processing, **56,** 88–89
high-resolution multimode digital microscopy, **56,** 206
low-light microscopy, **56,** 60
Sheath fluid SF 50, 10–11
Sheath fluid, bacteria, **64,** 546
Sheath stream
center, **63,** 28–29
velocity, **63,** 29–31
SHFR assay, **65,** 112–114
SHG, *see* Second harmonic generation
Shibire mutation, **71,** 204
SHIM, *see* Second harmonic imaging microscopy
Shot noise, **55,** 58–59, 149
Shuttle vectors, **71,** 372
choice of, with AdEasy system, **71,** 391

Shuttle vectors, (*cont.*)
inserting genes of interest into, **71,** 393
Sickle-cell erythropoiesis, **64,** 148–149
Side-scatter (SSC), **66,** 52–54
Signal
targeted GFP mutants, **65,** 375–376
Signal processing
amplification in, **63,** 38–39
analog-to-digital conversion in, **63,** 40
digital, pulse characterization by, **63,** 159–161
in flow cytometry, **63,** 108–110
Signal regulation
acid sphingomyelinase, **66,** 149–150
Signal-to-noise ratio
aequorin, **65,** 360
charge-coupled device cameras, high-resolution multimode digital microscopy, **56,** 206–207
confocal microscopy, **63,** 94
intracellular ion ratio imaging, **56,** 243
low-light electronic cameras
background noise, **56,** 27–28
calculation, **56,** 27–28, 54–56
camera noise, **56,** 48–49
photon counting, **56,** 54–56
readout noise, **56,** 26
sensitivity, **56,** 47–49
shot noise, **56,** 27, 48, 95–96
signal, **56,** 48
temporal averaging, **56,** 95–96
titration experiments, **63,** 283–284
Signal-to-noise ratio, confocal microscopy, **70,** 253–254
Signaling mechanisms, **71,** 113
B35 cell lines in study of, **71,** 288
in hippocampal and cortical neurons, **71,** 112
studies of, **71,** 9
Signaling pathways
cell–matrix adhesion assays
adhesion assay, **69,** 364–365
applications, **69,** 361
cell lysates, **69,** 366–370
cell preparation, **69,** 363–364
coimmunoprecipitation, **69,** 375–376
controls, **69,** 365–366
drug inhibitors, **69,** 376–378
GTPases, **69,** 378
immunoprecipitation, **69,** 370–372
kinase assays, **69,** 375
matrix-coated plates, **69,** 362–363
molecular genetics, **69,** 379–382
phospho-specific antibodies, **69,** 374–375
Western immunoblotting, **69,** 372–374
ECM networks, **69,** 14–15
Signaling regulation
ceramidase, **66,** 159–160
ceramide synthase, **66,** 155
sphingomyelin synthase, **66,** 157
Signalling, receptor mediated, **71,** 12
Signals
absorption, flow cytometric, **63,** 112
cross-talk, sources, **63,** 293–294
emitted by cell passing through laser beam, **63,** 31–32
extinction and absorption, **63,** 112
fluorescence, **63,** 113
fluorescence emission, **63,** 140
homodyning and heterodyning, **63,** 146
light
continuum to flow cytometric data file, **63,** 37–41
from particles: collection, **63,** 33–37
optical, converting to current, **63,** 118–123
sensor, simultaneously digitized, **63,** 57–58
Signals, from repeated-PRINS, **64,** 64, 67
Silane, treatment of DNA on mica, **68,** 360–361
silent heart, random amplified polymorphic DNAs, **60,** 174
Silicon intensified target cameras, features, **56,** 40–41
Silicone aquarium sealant, **71,** 119
Silver grains, confocal applications, **70,** 272–273
Simian virus 40
immortalization of cells with large T antigen gene, **57,** 70–71, 77–78, 97–98, 105–106
immunodetection of large T antigen, **57,** 99–100
Simple sequence length polymorphisms
advantages and disadvantages as polymorphic markers, **60,** 170–171, 177
centromeric markers, **60,** 361–363
gel electrophoresis of products
agarose gel electrophoresis, **60,** 182–183
polyacrylamide gel electrophoresis, **60,** 182
markers in microsatellite genetic linkage map, **60,** 181
polymerase chain reaction, **60,** 182
positional cloning project initiation, **60,** 263–264
primers, **60,** 181
troubleshooting of synthesis and analysis, **60,** 183
SINEs, zebrafish, **60,** 106–107
Singe-cell measurement
intracellular calcium
prostate adenocarcinoma, **66,** 241–242
Single-beam gradient laser trap

dielectric sphere force calculations, **55,** 1–25
gradient trap force, **55,** 9–16
arbitrary trap location, **55,** 11–16, 23–25
Y axis trap focus, **55,** 10–11, 22–23
Z axis trap focus, **55,** 9–10
index of refraction effects, **55,** 19–20
mode profile effects, **55,** 16–20
overview, **55,** 1–4, 20–21
ray forces, **55,** 4–8, 21–22
micromechanical limitations
Brownian motion
hydrodynamic drag, **55,** 139–140
power spectrum, **55,** 136–138
trap calibration, **55,** 138–139
instrumental noise sources, **55,** 147–154
amplifier noise, **55,** 59–60, 151–152
backreflections, **55,** 60, 152–153
beam-pointing, **55,** 59, 62, 153
electronics, **55,** 58–62, 147–152
Johnson noise, **55,** 149–151
power fluctuations, **55,** 58, 152
shot noise, **55,** 58–59, 149
vibration, **55,** 58, 154
micromechanical limitations
dynamic probe interactions, **55,** 146–147
force clamp experiments, **55,** 143–146
position clamp experiments, **55,** 140–143
overview, **55,** 129–132, 154
spectral data analysis
calculation, **55,** 134–135
interpretation, **55,** 132–134
Single-cell gel electrophoresis
development, **64,** 236–237
in alkaline conditions, **64,** 254
Single-cell injection, of *Xenopus* cultures, **71,** 138–139
Single cell isolation, **62,** 198, 231–237
Single cells
cytochrome *c* release, **66,** 372–375
Single-fiber polymerase chain reaction, **65,** 322–325
deleted-mtDNA, **65,** 325
DNA, **65,** 328–329
Single-label imaging, **70,** 55–56
Single-molecule force, **68,** 292
Single neurons, extracting in *Heliosoma*, **71,** 164
Single nucleotide polymorphism (SNP) mapping, **71,** 200
Single-particle tracking, fast 3D, PFM studies, **68,** 210
Single-particle tracking, laser tweezer plasma membrane protein movement studies
elastic intercompartmental barriers, **55,** 185–188
future research directions, **55,** 192
gold labeling particle preparation, **55,** 176–178
protein motion, **55,** 176
video microscopy, **55,** 178–179
Single strand conformation polymorphisms
advantages and disadvantages as polymorphic markers, **60,** 170–171
amplified fragment length polymorphism conversion, **60,** 208–209
applications, **60,** 190
gene mapping of zebrafish analysis of patterns, **60,** 190
electrophoresis of products, **60,** 188–189, 191–192, 272
overview, **60,** 186–187
polymerase chain reaction, **60,** 187–188, 191, 272
primers
design, **60,** 187–189
labeling, **60,** 190–191
linkage studies, **60,** 186
principle of analysis, **60,** 186
Single-stranded DNA, SFM studies, **68,** 388
in situ hybridization, see also Fluorescence *in situ* hybridization
haploid screens, **60,** 55–56, 61, 68
validation of candidate genes, **60,** 250
Sizing, particle, **63,** 328–329
SJD strain, gene mapping, **60,** 358–359
Skeletal muscle mitochondria, **65,** 19–20
complex I
membrane disruption, **65,** 102
Skeletal muscle, video-enhanced differential interference contrast light microscopy performance test, **56,** 181–183
Skin fibroblasts
permeabilized human
ATP synthesis rates, **65,** 140
SlAX 12 NCO adapter vector, **71,** 369–370
selecting and modifying for insertion, **71,** 372–374
Sleeping Beauty, see Tc-like transposable elements
Slice cultures
biolistic transfection of, **71,** 354
in Purkinje neuron studies, **71,** 90
Slide blocking solution, in QDFM, **64,** 37
Slide, generation for data presentation, **58,** 363
Slides
chamber, for immunophenotyping, **64,** 320–321
equilibrated, **64,** 239

Slides (*cont.*)
prepared with 3-aminopropyltriethoxy silane, **64,** 38
pretreatment
and FISH, **64,** 112
for QDFM, **64,** 46–47
three-layer agarose, **64,** 255
Slides, rerun and restained, **63,** 54
Slime mold, *see Dictyostelium discoideum*
Slow axonal transport, mechanism of, **71,** 306
SlowFade antifade, **71,** 187
Slow growth, *see* Cell culture, Long-term storage
Slow redistribution potentiometric indicators, **70,** 432–434
Slow-scan confocal laser scanning microscope
Bio-Rad Laboratories, **70,** 26–27
Carl Zeiss, **70,** 27–29
Leica microsystems, **70,** 31–33
Nikon, **70,** 33–36
Olympus, **70,** 37–39
overview, **70,** 25–26
SM. *see* Sphingomyelin
Small nuclear ribonucleoprotein particle (snRNP), *see also* Protein import, Nucleus, RNA transport, nucleus
fluorescence labeling, **53,** 526–527
immunoaffinity chromatography, **53,** 524–526
import pathway, **53,** 519–520, 546
Small unilamellar vesicles, fusion, AFM studies, **68,** 232–234
SMI, *see* Spindle mitotic index
SMP, **65,** 9
rat liver mitochondria, **65,** 16
respiratory chain activities, **65,** 11
SNAFL, **65,** 372
Snail dissection, **71,** 158
Snails *(Aplysia)*, **71,** 3
Snails *(Heliosoma)*, **71,** 157–169
SNARF1, **65,** 372
Snell's law, **70,** 151
SNOM, *see* Scanning nearfield optical microscopy
snRNP, *see* Small nuclear ribonucleoprotein particle
Society for Developmental Biology (http://sdb.bio.purdue.edu/), **71,** 85
Sodium azide, phosphate-buffered saline with, **63,** 249
Sodium carbonate extraction
yeast mitochondria, **65,** 47
Sodium dodecyl sulfate-polyacrylamide gel electrophoresis (SDS-PAGE)
mitochondrial translation products, **65,** 435–436
separation
plant mitochondria, **65,** 66
yeast actin purification, **65,** 163
Soft agar assay
transformation assays, **66,** 217f, 218
Soft samples, AFM force curves, **68,** 70
Software
use of nominal channels per decade, **63,** 307
WinCyte, for LSC, **63,** 58–62
Software, high-resolution multimode digital microscopy, **56,** 209
Solid-phase binding, basement membrane self-assembly, **69,** 128–129
Solid-state sizing, SFM of DNA, **68,** 370
Solid tumors
growth, **64,** 225
malignancies
dual-color staining, **64,** 396–397
PCR analysis, **64,** 413–414
resistance, **64,** 227
Solubility, antigen, **63,** 261
Soluble
mitochondrial protein, **65,** 47
Soluble extracellular matrix
antigen identity, **69,** 41–42
reporter cell response, **69,** 36–37
Soluble Fc fusion constructs
tumor necrosis factor-related ligands, **66,** 512
Soluble FLAG-tagged LIGHT, **66,** 504
Soluble ligands
tumor necrosis factor-related ligands, **66,** 518–519
Soluble proteins
caspase purification, **66,** 10–12
Soluting binding analysis, basement membrane self-assembly, **69,** 125
Solution nucleation assay, for γTuSC and γTuRC, **67,** 19
Solutions
for centrosome overproduction, **67,** 257
for centrosome protein analysis, **67,** 262
for *Drosophila* embryo extract preparation, **67,** 4
for egg extraction, **67,** 331
for embryo extract preparation, **67,** 7
for γTuSC and γTuRC, **67,** 18–19
for hydroxyurea-arrested CHO cell extract preparation, **67,** 260
for immunofluorescence staining of pericentrin, **67,** 55–56

for immunoperoxidase labeling, **67,** 56–57
for MT nucleation, **67,** 331–333
for MT nucleation reconstitution, **67,** 142–143
for pericentrin enrichment, **67,** 63–64
for pericentrin MT-dependent sedimentation, **67,** 65–66
for S-phase extract preparation, **67,** 270
for yeast SPB MT nucleation, **67,** 169–170
for yeast SPB permeabilization, **67,** 167
Solutions and reagents, protein and gene expression, analysis, **59,** 83–84
Solutions, DNA
adsorption, charged molecule role, **68,** 364
mica, dehydration, **68,** 360
Solutions, for spinal and motor neuron cultures, **71,** 82–85
SFM studies, **68,** 371–374
Somatic cell lines
β1 integrin null cells, **69,** 318–319
fibroblasts from mutant mouse, **69,** 314
immortalized keratinocytes, **69,** 313–314
kidney epithelial cell lines, **69,** 311–313
somitabun, developmental role, **60,** 3
Somites, cell migration
appropriate migration, **69,** 347–348
embryo media, **69,** 343
embryo slicing and embedding, **69,** 345–346
embryo staging, **69,** 343–344
imaging, **69,** 346
microinjection, **69,** 345
microscopy dish preparation, **69,** 346
migratory cycle, **69,** 348
overview, **69,** 342–343
Rac, **69,** 348
slice cultures, **69,** 347
slice environment, **69,** 348
sonic you, mutation effects in development, **60,** 31–33
Sorting
cell, from multicell culture systems, **64,** 214–215
plant chromosomes
after univariate analysis, **64,** 17
and flow karyotyping, **64,** 19–21
Source properties, **71,** 11–12
Southern analysis, telomere length, **64,** 71–72, 91
Southern blot analysis, meiotic recombination intermediates in *Saccharomyces cerevisiae*
cell growth and sporulation, **53,** 265–266
DNA preparation, **53,** 266
double-strand break detection, **53,** 264–265
Southern blot analysis, mouse/zebrafish somatic cell hybrid, **60,** 314, 316–317
Southern hybridization
mtDNA, **65,** 417–418
Southwestern blotting assay, base-unpairing region binding protein, **53,** 336–338
spadetail, mutation effects in development, **60,** 76
Spanner wrench, **70,** 351
Spastic paraplegia, **71,** 291
Spatial filters, **70,** 9
Spatial resolution
averaging filters, **56,** 100–101
digital image processing, noise reduction, **56,** 96–101
low-light electronic cameras, **56,** 28–31, 34, 59–69
quantitative video microscopy, **56,** 119–125
aspect ration, **56,** 120–121
dewarping algorithms, **56,** 124–125
image rotation, **56,** 122–123
image translation, **56,** 121–122
image warping, **56,** 123–125
resolution maximization, **56,** 119–120
video camera–microscope mating, **56,** 147–151
Spatial sampling effects, **70,** 161–163
SPB, *see* Spindle pole body
SPBs, *see* Spindle pole bodies
Spc72p, **67,** 89–92
Species
active *vs.* inactive centrosomes, **67,** 218–219
centrosome parthenogenesis, **67,** 217–218
criteria for choosing, **71,** 9–10
quantity of neuron yields, **71,** 11
Species of *Tetrahymena*, **62,** 5
Specimen preparation, confocal microscopy, **70,** 52, 61–63
Specimen stage
light microscopy, **56,** 139
video-enhanced differential interference contrast light microscopy, **56,** 174
Specimen viability, multiphoton excitation imaging, **70,** 135–137
Specimens
for LSCM, **64,** 317–318
fresh cytological, Ki-67 analysis in, **64,** 431–432
handling
biosafety, **64,** 569
for immunophenotyping study, **64,** 352
hematologic, immunophenotyping, **64,** 319–321, 336–340
small, for immunophenotyping, **64,** 314

Specimens, *see also* Samples
handling and processing, **63,** 262
washing, controversy, **63,** 232–234
Spectral compensation
and optical filtration, **64,** 295–307
four-color, **64,** 303–307
set to correct spectral overlap, **64,** 571
two- or three-color, **64,** 301–303
Spectral data analysis
overview, **55,** 129–132, 154
power spectrum calculation, **55,** 134–135
power spectrum interpretation, **55,** 132–134
Spectral imaging, *see* Ratio imaging
Spectral response, photomultiplier tube, **63,** 120–121
Spectral shift analysis, Hoechst 33342–DNA binding, **64,** 185–187
Spectrin/fodrin, **66,** 259
Spectrin, AFM studies, **68,** 313
Spectrofluorimetric measurements
calcium imaging, **66,** 237–239
Spectrofluorimetry, DNA–drug interactions, **64,** 183–184
Spectrometry
plant senescence, **66,** 449
Sperm
aster formation assay, **67,** 316
pronucleus assembly
cell-free systems, *see* Cell-free nuclear assembly
overview, **53,** 369–370, 418–420
Xenopus, Permeabilization, **67,** 270–271, 280–281
Xenopus sperm permeabilization
buffers, **53,** 371
lysolecithin treatment, **53,** 373
recovery from testes, **53,** 371–373
storage, **53,** 373–374
Sperm sample, preparation for PRINS, **64,** 58
Sperm, *see* Testicular germ cell
Sperm, zebrafish
collection
equipment, **60,** 366–367
gynogenesis-based screening, **60,** 13–14
reagents, **60,** 367–368
squeezing, **60,** 368
testes collection, **60,** 369
ethylnitrosourea mutagenesis of males, **60,** 76
fertilization, *in vitro*, **60,** 15, 44, 46–47, 53, 370–372
freezing
sperm, **60,** 365, 369–370, 372
testes, **60,** 370
gamma-ray mutagenesis, **60,** 62–64
testes dissection, **60,** 371–372
ultraviolet inactivation, **60,** 14, 365–366
Sperm, zebrafish, *in vitro* fertilization, **59,** 51–52
Spermatocytes, laser micromanipulation, **55,** 87–88
Spherical aberrations
confocal microscopy, **70,** 152–153
excitation/emission wavelength effects, **70,** 235
intact tissue microscopy problems, **70,** 303
objective lens, **70,** 261–262, 264–266
Spherical aberrations, ratio imaging, **56,** 226–227
Spheroids
alteration of damage induction in, **64,** 221–222
cells resistant to chemotherapeutic agents, **64,** 218–219
culture system, **64,** 213–214
drug activity patterns in, **64,** 222–223
for multifraction studies, **64,** 226–227
multidimensional complexity, **64,** 212
slow drug penetration in, **64,** 224
Spheroplasting, **65,** 271–272
Spheroplasts
membrane marker proteins, **65,** 171
Sphinganine-*N*-acyltransferase, *see* Ceramide synthase
Sphingolipids, **62,** 37, 48
Sphingomyelin (SM), **66,** 136
enzyme assays, **66,** 149–161
labeling
cell culture, **66,** 138
mass levels
cupric sulfate charring, **66,** 143–144
determination, **66,** 142–144
phosphate analysis, **66,** 142–143
metabolic labeling, **66,** 136–142
Sphingomyelin synthase, **66,** 157–159
materials, **66,** 157–158
molecular probes, **66,** 158–159
principle, **66,** 158
signaling regulation, **66,** 157
Sphingomyelinase
neutral, **66,** 152–161
Spin infection, **66,** 209
Spinal cord medium, **71,** 84
Spinal cord neurons
experimental uses of, **71,** 80–82
plating, **71,** 78–79
Spinal motor neurons, **71,** 3
chick embryo access procedures, **71,** 71–72
chick embryo sources for, **71,** 67–68

dissection of spinal cords for, **71,** 72–75
enrichment by density gradient centrifugation, **71,** 79–80
growing and working with, **71,** 67–85
incubation procedure, **71,** 70
isolation of chick embryos for, **71,** 72
isolation of ventral halves for, **71,** 73–77
materials needed for culture, **71,** 70
plating, **71,** 78–79
Spinal neurons, **71,** 3
Spindle birefringence
individual spindle components
kinetochore fibers, **61,** 197
slow axis orientations and specimen retardances
using polarizing microscope, **61,** 197
using Pol-scope, **61,** 197
time lapse movies, **61,** 197
molecular origin
anisometric shape, **61,** 191
form birefringence
dielectric constant equation, **61,** 192
imbibition measurements, **61,** 193–194
Pisaster ochraceus spindle birefringence, **61,** 193
positive dielectric anisotropy, **61,** 192
qualitative model, **61,** 191–192
specific birefringence, **61,** 192–193
intrinsic birefringence
aromatic side groups, **61,** 195
equation, **61,** 194
Pisaster spindle calculation, **61,** 194
optimum cell types
Chaetopterus pergamentaceous, **61,** 203
echinoderm zygotes, **61,** 199
fertilization envelope removal, **61,** 199, 201
observable properties, **61,** 197–199
plant tissue cells
Dictyostelium discoideum, **61,** 202
Haemanthus katherinae, **61,** 201
Lilium longiflorum, **61,** 202–203
poor yield types, **61,** 201
selection criteria, **61,** 204–205
tissue culture cells, *Taricha granulosa,* **61,** 201
parallel microtubule arrays
average birefringence derivation, **61,** 196
number density estimate procedure, **61,** 196
single and bundled microtubules
measured peak retardances, **61,** 195
retardance cross section for single microtubule, **61,** 195
single fiber birefringence theory, **61,** 196
spindle fiber dynamics, **61,** 190–191
Spindle component antibodies
factors controlling antibody production
antigenicity of mitotic spindle components, **61,** 243
quality of isolated spindles, **61,** 243–244
hybridoma technology advantages, **61,** 234
importance of probe specificity, **61,** 243
methods
cell freezing, **61,** 241
cell fusion
materials, **61,** 238
procedure, **61,** 238–239
classification of clones, **61,** 242
expansion, **61,** 241
immunization of mice, **61,** 236–237
immunogen preparation
materials, **61,** 235
procedure, **61,** 235–236
large-scale antibody production
ascites, **61,** 242
hybridoma supernatant, **61,** 242
myeloma cell culture
materials, **61,** 237
procedure, **61,** 237
preparation of immunized mouse spleen cells
materials, **61,** 237
procedure, **61,** 238
purification, **61,** 243
screening
immunoblot screening, **61,** 240
immunofluorescence screening, **61,** 239
subcloning by limiting dilution
materials, **61,** 240
procedure, **61,** 240–241
Spindle mitotic index (SMI), **61,** 358
Spindle pole bodies
assembly
assays, **67,** 243–245
associated proteins, **67,** 247–251
observation methods, **67,** 251–252
centrosome assay, **67,** 300–301
centrosome-independent assembly
associated reaction, **67,** 246–247
DNA bead preparation, **67,** 245–246
centrosome reproduction assay, **67,** 294–295
genetic analysis
dosage-dependent suppressor screen, **67,** 97–100
extragenic suppressor screen, **67,** 100–102
synthetic dosage lethality, **67,** 108

Spindle pole bodies (*cont.*)
synthetic lethal screen, **67,** 103–107
mitotic, MT nucleation at, **67,** 171
NuMA role, **67,** 224–225
precipitation
antibody couplings, **67,** 84
cell lysis, **67,** 85–86
controls, **67,** 89
gene fusion construction, **67,** 82–84
HA- and Myc-tagged proteins, **67,** 86–87
protein A fusion proteins, **67,** 87–88
protein detection, **67,** 88
with polyclonal antibodies, **67,** 87
proteins, *in vitro* binding
controls, **67,** 92
Spc72p to Kar1p, **67,** 89–92
related structures, mitotic, **67,** 298
yeast
γ-tubulin role, **67,** 173
3D modeling, **67,** 37–40
activator identification, **67,** 173–174
interphase activation, **67,** 171–173
MT nucleation, **67,** 169–171
permeabilization, **67,** 167–169
two-hybrid system, **67,** 73–74
Spindle pole body
functions in yeast, **58,** 96
green fluorescent protein fusion protein studies
γ-tubulin, **58,** 99–100
Nuf2, **58,** 100–101
photobleaching and dynamics, **58,** 103
Spindle poles, *Xenopus* oocytes, **70,** 412
Spindle, *Drosophila*
Ncd-green fluorescent protein fusion protein
advantages over direct cytoskeletal protein fusion, **58,** 154–155, 162
confocal microscopy imaging
embryo preparation, **58,** 156–158
filters, **58,** 158–159
oocyte preparation, **58,** 156–157
time-lapse imaging, **58,** 158
construct design, **58,** 155–156
mutant analysis, **58,** 160
photoactivation applications, **58,** 160–161
subcellular localization, **58,** 159–160
transgene expression, **58,** 154
overview of dynamics, **58,** 153–154
Spinning disk confocal microscopy
monoscanning confocal microscope, **70,** 47–49
overview, **70,** 43–45
tandem scanning confocal microscope, **70,** 45–47
Spinocerebellar ataxia, associated with Purkinje neurons, **71,** 91
Spirogyra, fluorescence lifetime imaging, **56,** 290–291
Spirogyra, cytoplasmic streaming, **55,** 199–202
Spisula solidissima
MT nucleation
cosedimentation assay, **67,** 158–161
disassembly and reassembly, **67,** 152–153
immobilized centrosome recovery assay, **67,** 153–155
IVEM tomography, **67,** 155–156
linear sucrose gradient centrifugation, **67,** 157–158
oocyte lysate preparation, **67,** 149
Spisula solidissima oocyte centrosome isolation
activated lysate centrosome isolation
functional reconstitution assay
hemacytometry, **61,** 47
peak centrosome-containing sucrose fraction, **61,** 47, 49
polarized light microscopy of aster formation, **61,** 47
materials and equipment, **61,** 45
sucrose-density gradient centrifugation, **61,** 45–47
clam dissection, **61,** 37–38
electron microscopy
analysis, **61,** 53
colloidal gold localization studies, **61,** 55
intermediate voltage EM tomography and ring structures, **61,** 53
procedures
isolated centrosome preparation, **61,** 53
reconstituted aster preparation *in vitro*, **61,** 53
favored model properties, **61,** 36
immunofluorescence of centrosomes and asters
asters reconstituted *in vitro*, **61,** 52
clarified *Spisula* oocyte lysate
materials, **61,** 50–51
procedure, **61,** 51–52
Elvanol mounting medium preparation
materials, **61,** 49–50
procedure, **61,** 50
purpose, **61,** 49
obtaining organisms, **61,** 37
oocyte activation test
KCl-induced germinal vesicle breakdown (GVBD), **61,** 41
materials, **61,** 40

oocyte preparation, **61,** 37–40
preparation of microtubule protein
sea urchin tubule source
advantages, **61,** 43
materials and equipment, **61,** 44
procedure, **61,** 44
Spisula oocyte source
procedure, **61,** 44–45
preparation of oocyte lysates
materials and equipment, **61,** 41–42
procedure
aster buffer, **61,** 42
aster formation assay, **61,** 43
Eppendorf Pipetman lysis, **61,** 42
KCl activation, **61,** 42
lysate clarification, **61,** 43
vitelline layer removal, **61,** 42
vortex lysis, **61,** 42
storage, **61,** 55
uses, **61,** 55
yield, **61,** 54–55
Spongiome, **62,** 51
Spores, bacterial, **64,** 546
SPR, *see* Surface plasmon resonance
Sprague-Dawley rats, **71,** 113
Squamous cell carcinomas, clinical studies of proliferation, **64,** 456–461
Squamous cheek cells, video-enhanced differential interference contrast light microscopy performance test, **56,** 180–181
Squid giant axon, diameters of, **71,** 9
SSC, **66,** 52–54
SSCPs, *see* Single strand conformation polymorphisms
SSLPs, *see* Simple sequence length polymorphisms
Stability, illumination, for fluorescent microscopy, **64,** 75–76
Stable transfectants
in B35 cells, **71,** 295–296
isolation of, **71,** 281
Stage drift, **71,** 320
Stage scanning confocal microscopy, **70,** 24–25
Stage, *see* Specimen stage; Substage
Staining
antibodies, FLIM, **69,** 269
blood cells, **63,** 212
cell-assembled ECM, **69,** 153
cell surface marker: CFSE labeling method, **63,** 381–382
costaining with PECY5-CD8, **64,** 303
cultured cells, **63,** 361–362
DNA, bacteria, **64,** 541–542
double indirect, **63,** 294
dual, **63,** 12
fixed bacteria, **64,** 543–544
for analysis of RNA synthesis, **64,** 132
for APO2.7 expression, **63,** 490–491
for cyclins, **64,** 432–433
for evaluating drug retention and efflux, **64,** 197–198
for flow cytometry, **63,** 137–138
for viability status with antitubulin antibody, **63,** 491–492
immunochemical, cell preparation and, **63,** 422–423
immunofluorescence, and erythrocyte lysis, **64,** 345–346
indirect, blocking for, **63,** 232
intracellular, **63,** 242–243
CFSE method, **63,** 383
mitotic cells, **63,** 345–346
nuclear isolation and, **64,** 271–272
nuclear matrix isolation and, **64,** 273
pericentrin, for centrosome defects, **67,** 67–68
protocols for mitochondria analysis, **64,** 125
simultaneous
APO2.7 and cell surface antigens, **63,** 263
Ki-67 protein and cell surface antigen, **63,** 407–408
T cell receptor z and cell surface antigens, **63,** 263–264
solid tumors and tissues, **63,** 362–364
tissue culture cells, with antibodies, **67,** 307–308
vital *E. coli,* **64,** 544–546
with JC-1, **63,** 472, 477–478
with one to four antibodies, **63,** 239–242
Xenopus oocyte keratin filaments, **70,** 412
Xenopus oocyte microtubules, **70,** 411
Xenopus oocytes and egg chromosomes, **70,** 393–394
Stalk cells, **66,** 478
calcofluor, **66,** 484
differentiation
Dictyostelium discoideum, **66,** 477–478
double fluorescein diacetate/propidium iodide staining, **66,** 483
trypan blue, **66,** 484
Stamen hair cell division studies
advantages, **61,** 419
culture methods
cutinase, **61,** 420
dissection, **61,** 420
findings
microinjection

Stamen hair cell division studies (*cont.*)
calcium levels, **61,** 421, 423
fluorescent calmodulin, **61,** 424
heterologous tubulin, **61,** 421
mitotic microtubule kinetics, **61,** 423
phalloidin, **61,** 423
profilin, **61,** 423
pharmacologic studies
BAPTA, **61,** 421
caffeine, **61,** 421, 423
calcium antagonists, **61,** 421
kinases, **61,** 421
Standardization
common cytometry measurements, **63,** 312–330
flow cytometric, **63,** 300–302
light scatter, **63,** 327–328
quantitative fluorescence cytometry, **63,** 311–312
Staphylococcal α-hemolysin, AFM studies, **68,** 270
Star strains, **62,** 153–155, 521–522
mating types, **62,** 232
table, **62,** 190–191
Starvation media, *see* Culture media
Starvation-mediated development, **62,** 88–98
Staurosporin
AIF, **65,** 153
STEM, *see* Scanning transmission electron microscopy
Step size, confocal microscopy, **70,** 54
Sterecon system, **61,** 102–103, 105, 308
Stereo image generation, **70,** 58–59, 113
Stereocilia, **68,** 144–146
Sterile technique, animal cell culture, **57,** 13–14, 51–52
Sterilization procedures, spinal and motor neuron cultures, **71,** 82
Sterols, **62,** 36
Stiffness, polyacrylamide substratum, **69,** 337
Stigmatellin, **65,** 105–106
Still cameras, digital, **56,** 40
STM, *see* Scanning tunneling microscopy
Stochastic interactions, between neurons and target cells, **71,** 145
Stock solutions, for plant flow cytogenetics, **64,** 6–12
Stoelting (www.stoeltingco.com), **71,** 71
StonedB protein, in *Drosophila* vesicle recycling, **71,** 204
Storage
tissues, **67,** 324–325
Xenopus egg extract, **67,** 325
Storage, blood, prior to cell preparation, **63,** 208–209
Storage, *see* Cell culture, long-term storage
Strain, induced, AFM calculations, **68,** 181–182
Strains, *see also* Mutant strains, **62,** 5, 18, 189–191, 219
B2086, **62,** 232
CU427, **62,** 232, 520
CU428, **62,** 232, 502, 519
CU438, **62,** 232
diversity, **62,** 18
heterokaryon, **62,** 270
inbred, **62,** 13, 189
nomenclature, **62,** 4, 188
nullisomic, **62,** 190–191, 265
table, **62,** 232
Streptavidin–biotin unbinding, AFM studies, **68,** 306–307
Streptavidin, protein labeling, **63,** 192
Streptolysin O
mitochondria, **66,** 371–372
Stress-activated kinases (SAPKs), *see* C-Jun N-terminal kinases
Striated rootlets, **62,** 57
Stroke, cortical and hippocampal involvement in, **71,** 112
Strongylocentrotus purpuratus, MT isolation, **67,** 151–152
Structural biology, cryo-AFM
individual molecules, **68,** 248–250
large assembly surface details, **68,** 250–252
STS, *see* Sequence-tagged sites
stumpy, mutation effects in development, **60,** 78
Subcellular delivery, of PTP cargo, **71,** 329
Subcellular fractionation, of fly heads, **71,** 206–207
Subdiploid DNA content
DNA fragmentation assay, **66,** 87–90
Subfractionation, **65,** 203
Submitochondrial particles
hydrogen peroxide, **65,** 82
membrane potential measurements, **65,** 85
Submitochondrial particles (SMP), **65,** 9
rat liver mitochondria, **65,** 16
Substage, light microscopy, **56,** 138–139
Substance P immunoreactive sensory neurons, **71,** 140
Substitution, maximum, in protein labeling, **63,** 190
Substrate oxidation

beef heart mitochondria, **65,** 7
Substrate
for chick ciliary ganglion neurons, **71,** 37
for peripheral neurons, **71,** 19–22
Substrate coatings, **71,** 21
for spinal and motor neuron cultures, **71,** 84
Substratum-bound molecules, as guidance cues, **71,** 2–3
Substrates, AFM studies, **68,** 215
Subtilisin, **66,** 5
Subventricular zone, migration to olfactory bulb
brain slice preparation, **69,** 349
DCC, **69,** 351–353
neuronal precursor cells, **69,** 351
overview, **69,** 349
Succinate, **65,** 129, 137
Succinate-cytochrome *c* reductase assay, **65,** 113–114
Succinate dehydrogenase (SDH), **65,** 314, 326
Succinate-hexacyanoferrate reductase (SHFR) assay, **65,** 112–114
N-Succinnimidyl-3-(S-acetylthio) propionate, AFM tips, **68,** 119
Succinate:ubiquinone oxidoreductase activity, **65,** 103
Sucrose gradients
density gradient purification, **65,** 57
Sucrose gradients, for embryo extract preparation, **67,** 8–10
Sulfation, heparan sulfate, chlorate treatment, **69,** 90–91
Sulfite oxidase, **66,** 366
Super fluor, **70,** 258
Superior cervial ganglia, culturing neurons from, **71,** 306–310
Superose 6 column
apoptosome fractionation, **66,** 24
Superose 12 column
gel-filtration chromatography
Apaf-1 purification, **66,** 22–23
Superoxide anion concentration
lucigenin-based assay, **66,** 312–313
Superoxide anion production assay
intact cells, **66,** 311
purified plasma membrane, **66,** 310–311
Suppliers, of neuron species, **71,** 13
Supported lipid bilayers, AFM studies
Langmuir trough, **68,** 234–235
Langmuir trough and small Teflon wells, **68,** 235–236
small vesicle fusion, **68,** 232–234
Suppressor screen, for SPBs
dosage-dependent screen, **67,** 97–100
extragenic screen, **67,** 100–102
SURF-1 gene
mutations, **65,** 136
Surface chemistry, AFM studies
basic studies, **68,** 215
cleaning, **68,** 117–118
crosslinker binding, **68,** 119
esterification, **68,** 118
large assemblies, cryo-AFM studies, **68,** 250–252
ligand binding, **68,** 119
ligand density and functionality, **68,** 119–120
probe surfaces, **68,** 120–121
Surface plasmon resonance
tumor necrosis factor-related ligands, **66,** 517
Surface plasmon resonance, integrin ligand binding, **69,** 254–256
Surface tension, affecting success of biolistics, **71,** 364
Surfaces
glass, preparation, **69,** 327
gold-coated, **69,** 393, 397
micropatterned culture, *see* Micropatterned culture surfaces
SFM of supercoiled DNA, **68,** 385–388
treatments for DNA adsorption, **68,** 365
Surgical oncology, ploidy and proliferation in, **64,** 447–448
Surrogate
appropriate cellular, **64,** 295
measure of dynamic proliferation: labeling index as, **64,** 473
Survivability
enhancing, of hippocampal and cortical neuronal cultures, **71,** 124
of *Xenopus* cultures, **71,** 138–139
Suspension adaptation
approaches, **57,** 224
culture, **57,** 225–226
medium, **57,** 224
scaleup, **57,** 226
spinners, **57,** 224–225
Suspension cells
cell-free systems, **66,** 169–170
transmitochondrial hybrid generation, **65,** 404–405
Suspension, infection of neuronal tissues in, **71,** 411
Sutter P-97 Flaming-Brown micropipette puller, **71,** 311
SUVs, *see* Small unilamellar vesicles

SV40, *see* Simian virus 40
SWISS-PROT, **65,** 179, 182
Sympathetic explant culture, **71,** 29
Sympathetic ganglion, rat, dissection of, **71,** 26
Sympathetic neuronal culture, maintenance of, **71,** 23
Synapses, production of, **71,** 10
Synaptic interactions
  cerebellum as study model for, **71,** 91
  enhanced via cocultures, **71,** 145
  in *Drosophila* studies, **71,** 203
  in hippocampal and cortical neurons, **71,** 111, 112
Synaptic vesicles, **71,** 205
Synaptogenesis
  enhancing, **71,** 124
  in *Drosophila* studies, **71,** 196
  in hippocampal and cortical neurons, **71,** 112
Synaptonemal complex (SC)
  dimensions, **53,** 236–237
  electron microscopy
    negative staining, **53,** 240–241
    Saccharomyces cerevisiae complexes, **53,** 273–274
    serial sectioning, **53,** 240
    tomography, **53,** 240
  electron microscopy *in situ* hybridization, **53,** 253
  fluorescence *in situ* hybridization
    detection systems, **53,** 277–278
    fluorescence microscopy and imaging, **53,** 252, 278–280
    hybridization conditions, **53,** 277
    materials, **53,** 250–251
    probe labeling, **53,** 276–277
    proteinase K treatment, **53,** 251–252
    sample preparation, **53,** 250
  immunostaining of *Saccharomyces cerevisiae* proteins
    antibody specificity, **53,** 274–275
    epitope tagging, **53,** 275–276
    sample preparation, **53,** 274
    staining of slides, **53,** 275
  lateral elements, **53,** 237
  preparation of meiotic cells
    cryogenic preservation of seminiferous tubules, **53,** 238
    ovarian material
      birds, **53,** 239–240
      mammals, **53,** 239
    Saccharomyces cerevisiae, **53,** 269–270
    testicular biopsy, **53,** 238–239
  recombination nodule association, **53,** 236, 247
  whole-mount techniques
    drying-down technique, **53,** 247
    microspreading nuclei on liquid surfaces
      advantages and limitations, **53,** 246
      fixation, **53,** 244
      materials, **53,** 243–244
      principle, **53,** 242–243
      solution preparation, **53,** 244–245
      staining for electron microscopy, **53,** 243
      technique, **53,** 245–246
    spreading of *Saccharomyces cerevisiae* nuclei with detergent
      cell growth, **53,** 270
      detergent, **53,** 271
      fixative, **53,** 271–272
      silver staining, **53,** 272–273
Synaptosomes
  transmitochondrial cybrid generation, **65,** 406
Synchronization, cell cycle
  centrifugal elutriation, **57,** 244–246
  G0 arrest, release, **57,** 237, 239
  growth-dependence *vs.* cell cycle regulation in measured events, **57,** 247–248
  heterogeneity of synchronized populations, **57,** 246–247
  M phase block, release, **57,** 239–240
  mitotic detachment, **57,** 241–242, 244
  S phase block, release, **57,** 240–241
Synchronization, cell cycle, plant chromosomes, **64,** 19, 21
Synchronized cells, centrosome activity, **67,** 215–217
Synchrony, **62,** 77, 78, 233
Synchrony experiments, BrdUrd-labeling in, **63,** 358
Synclone, **62,** 133
Syndecan-4, null fibroblasts, **69,** 321
Syngens, **62,** 10, 21, 188
Synkaryon, **62,** 9, 132, 228
Synthetic dosage lethality, for SPBs, **67,** 108
Synthetic fluorescent dyes
  calcium-specific probes, **66,** 234–236
Synthetic lethal screen, for SPBs, **67,** 103–107
Synthetic matrix
  cell-assembled ECM
    advantages, **69,** 152
    cells and proteins, **69,** 152–153
    disadvantages, **69,** 152
    DOC-soluble and -insoluble material, **69,** 154
    immunofluorescence staining, **69,** 153
    serum starvation, **69,** 154

time course, **69,** 154
troubleshooting, **69,** 155
provisional matrix
advantages, **69,** 147
angiogenesis, **69,** 156
cell migration, **69,** 156
contraction, **69,** 150
degradation, **69,** 156
disadvantages, **69,** 147
fibrinogen preparation, **69,** 148
immunofluorescence, **69,** 150–151
molecular event model, **69,** 155
other protein inclusion, **69,** 148–150
synthesis *in vitro*, **69,** 148
tissue cultures, **69,** 156–157
tissue engineering, **69,** 156
tissue injury, **69,** 155
troubleshooting, **69,** 151
Syringe-driven mixing, **63,** 174–175
System limitations, *see* Micromechanical limitations
SYTO 13, Gram stain, **64,** 525–526
SYTO, 18, **65,** 339
SYTOX, **62,** 297–310

T

T antigen, scanning transmission electron microscopy, **53,** 159
T cell
human peripheral blood lymphocytes, isolation and culture, **53,** 313
nucleolus activation model, **53,** 312–313
phytohemagglutinin activation, **53,** 312–313
T-cell receptor
as clone-specific markers, **64,** 404–405
Vb antigen analysis, **64,** 583
T cell receptor z, and cell surface antigens: simultaneous staining, **63,** 263–264
T-cell, immortalization by oncogene transfection
bone marrow extract preparation, **57,** 79
characterization of cells, **57,** 81–83
efficiency, **57,** 81
electroporation, **57,** 80–81
oncogene expression plasmid construction, **57,** 79
peripheral blood mononuclear cell preparation, **57,** 79–80
yolk lipoprotein preparation, **57,** 79
T lymphocytes
cultures, **63,** 392
MHC class II expression on, **63,** 451–452
T system, device for mixing, **63,** 172–174
T. pyriformis, see *Tetrahymena pyriformis*
T8993G mutant cybrid cells, **65,** 141–143
Tac-α5, expression cloning, **69,** 219
Tag Image File Format, **58,** 95, 179, 191–192, 361
Tag sequences, integrins, **69,** 232
Tail moment threshold, **64,** 244–245
Tandem conjugate dyes, **63,** 201
Tandem scanning confocal microscope, **70,** 45–47
Tapping mode atomic force microscopy
basic operation, **68,** 35
benefits, **68,** 338–339
overview, **68,** 261–263
Tapping recording mode photonic force microscopy, **68,** 209–210
Taq polymerase, for PRINS, **64,** 56–57
Target cells
retroviral transduction, **66,** 209–210
Target enzymes, availability: quality control, **64,** 180–182
Target interactions
altered in mutations, **71,** 224–228
in chick embryos, **71,** 68
study of Tibia-1 pathway, **71,** 172
Target organisms, flow cytometric identification, **64,** 518–520
Targeted gene disruptions, **71,** 196
Targeted green fluorescent protein mutants
recombinant expression, **65,** 375
signal, **65,** 375–376
YFP calibration, **65,** 376
Targeted probes
aequorin, **65,** 358–360
TargetP, **65,** 180–181
Taricha granulosa lung cell cultures, **61,** 160, 201
Tau–green fluorescent protein, fusion protein analysis in *Drosophila* nervous system development, **58,** 170–173
Tax, immortalization of cells, **57,** 73–74
Taxol, *see* Drug role in analyzing mitosis
Tc-like transposable elements
functional domains, **60,** 110–112
integration mechanism, **60,** 109–110
Sleeping Beauty, **60,** 122, 124
subfamilies in zebrafish genome, **60,** 112
transformation vectors, **60,** 121–122, 124
TCBP-25 and 23, 455–464
TDFM, *see* Transverse dynamic force microscopy
TDT-mediated dTUP-biotin nick-end labeling. *see* TUNEL

Technic, Inc., **71,** 355
Teflon, small wells, AFM studies, **68,** 235–236
Telomerase, **62,** 141, 395–412
  assay, **62,** 403, 404
  mutant phenotypes, **62,** 74–75, 423–424
  purification, **62,** 404–412
  purification, small scale, **62,** 429–430
  RNA, **62,** 398–399
  RNA and mutagenesis, **62,** 417–431
Telomere length measurement
  biological studies, **64,** 93
  calibration of results, **64,** 85–86
  chromosome segmentation for, **64,** 82–85
  fluorescence labeling of sample, **64,** 72–73
  fluorescence microscopy imaging system, **64,** 73–77
  human cells, Southern analysis, **64,** 91
  presentation of results to user, **64,** 86–88
  sample preparation for, **64,** 76, 79–80
  system description, **64,** 77–79
  TFL-TELO algorithm, **64,** 77
Telomere, length in immortalized cells, **57,** 84
Telomeres, **62,** 139–141
  cloning, **62,** 425–427
  integrated fluorescence intensity, **64,** 73
    algorithm, **64,** 88–90
    validation of method, **64,** 92
  quantitative FISH and image analysis, **64,** 72
  segmentation, **64,** 80–82
  Southern analysis, **64,** 71–72
  structure–function, **64,** 70–71
Telomeres, zebrafish, **60,** 104
TEM, *see* Transmission electron microscopy
TEM, *see* Transmission electron microscopy, 3D
Temperature
  effect on blood prior to cell preparation, **63,** 209
  for cell adhesion assay, **63,** 610
Temperature
  and choice of neuronal culture, **71,** 12
  of cell suspension, as electroporation parameter, **71,** 347
Temperature-sensitive paralytics, **71,** 204
Template DNA
  preprotein synthesis, **65,** 191–193
    PCR, **65,** 191–193
Temporal averaging, noise reduction, **56,** 95–96
Temporal coupling, egg activation and centrosome injection, **67,** 215
Temporal imaging
  fluorescence microscopy, **56,** 277–302
    lifetime-resolved camera, **56,** 287–292
      applications, **56,** 290–293
      data acquisition, **56,** 289–291
      image intensifier, **56,** 288–289
      light source, **56,** 289
      optics, **56,** 289
    methods, **56,** 279–287
      frequency domain measurement, **56,** 279–281
      photobleaching effects, **56,** 190, 285–287
      simultaneous multiple-lifetime component measurement, **56,** 281–285
      time domain measurement, **56,** 279–281
    overview, **56,** 277–279, 301–302
    pump–probe microscopy, **56,** 298–301
      applications, **56,** 300–301
      instrumentation, **56,** 299–300
    two-photon microscopy, **56,** 293–298
      applications, **56,** 297
      data acquisition, **56,** 296–297
      detector, **56,** 296–297
      instrumentation, **56,** 294–297
      light source, **56,** 295
      optics, **56,** 295–296
      scanner, **56,** 295–296
  low-light microscopy, **56,** 60
  quantitative video microscopy, **56,** 133
Tenascin, in *Xenopus* cultures, **71,** 140,146
Tenascin, unfolding, **69,** 175–177
Terminal deoxynucleotidyl transferase apoptosis assay, *see* TUNEL
Terminal restriction fragments, **64,** 71
Testes, *see* Sperm
Testicular cell, *see* Leydig cell, Peritubular cell, Sertoli cell, Testicular germ cell
Testicular germ cell
  establishment of cell lines
    cell preparation
      fractionation over discontinuous Percoll gradient, **57,** 96–97
      isolation, **57,** 96
    characterization of cotransfected immortalized cells, **57,** 107–108
    clone selection, **57,** 98–99
    flow cytometry analysis of DNA, **57,** 106–107
    histology of immortalized cells, **57,** 100–101, 108
    immunodetection of simian virus 40 large T antigen, **57,** 99–100
    medium for immortalized cell culture, **57,** 99
    overview, **57,** 94–95
    passaging, **57,** 99

transfection with simian virus 40 large T antigen gene
calcium phosphate transfection, **57,** 97–98
cotransfection with temperature-sensitive p53 mutant, **57,** 105–106
principles, **57,** 95–96
seminiferous cord formation *in vitro* with other testicular cells, **57,** 101–103, 105
spermatogenesis, **57,** 94
tumor lines, **57,** 94
Testing programs, for flow cytometric immunophenotyping assays, **64,** 576–577
Tet-off and tet-on adenoviral vector systems, **71,** 413
Tetrahymanol, **62,** 36
Tetrahymena piriformis, **65,** 28–29
Tetrahymena pyriformis, **62,** 6, 11–12, 21, 31, 188
Tetrahymena thermophila
CSU-10 studies, **70,** 96–97
indirect immunofluorescence, **70,** 338, 340–344
live fluorescence microscopy equipment, **70,** 347–358
overview, **70,** 338
rDNA, FISH, **70,** 344–347
Tetrahymena
as a model system, **62,** 99–103, 173–176
evolution, **62,** 29
nomenclature, **62,** 21, 188–189
strains, **62,** 4–13, 31
Tetramethylrhhodamine labeling, **61,** 159
Tetramethylrhodamine methyl ester (TMRM)
mitochondrial membrane potential, **65,** 292
Tetramethylrhodamines, **65,** 295
Tetraphenylphosphonium (TTP^+)
electrode, **65,** 288
Tetraphenylphosphonium electrode, **63,** 469
Tetrins, **62,** 54, 58, 309, 442–446
Texas Red, protein labeling, **63,** 193
TF55, scanning transmission electron microscopy, **53,** 160
TFL-TELO algorithm, **64,** 77
TGase, **66,** 112–114
activity assays, **66,** 114–115
TGase 1 enzyme
involucrin peaks
HPLC separation, **66,** 126
TGase glutamyl donor substrate proteins
affinity isolation, **66,** 122–123
affinity labeled
identification, **66,** 124–125
C. elegans, **66,** 123–124, 124
DALP-MC, **66,** 120–121
in situ identification
apoptotic cells, **66,** 119–124
X-biotin, **66,** 123–124
TGase-interacting proteins
immunoprecipitation
apoptotic cells, **66,** 119
TGase-mediated glutamine deamidation
peptide sequencing, **66,** 130–131
TGase reactive glutamine residues
protein sequencing, **66,** 125–126
TGase substrate proteins
cross-links
location, **66,** 126–129
TGF-β, *see* Transforming growth factor β
TGF-β1, *see* Transforming growth factor β1
Therapy
cancer, cytometric assays in, **64,** 475–476
cell production rates and, **64,** 469–470
experimental, and cell proliferation, **64,** 474
Thermodynamics
equilibrium, DNA on mica, **68,** 363–364
vascular endothelial cadherin, **68,** 129–130
Thickening agents, for peripheral neuron cultures, **71,** 24
Thick tissue
agarose embedding, **70,** 330
antibody penetration, **70,** 315–319
fixation, **70,** 308–313, 329
fluorochromes, **70,** 319–321
mounting sections, **70,** 331–332
preparation, future prospects, **70,** 328–329
procedures, **70,** 330–331
Vibratome sectioning, **70,** 314–315, 330
Thin-layer chromatography
metabolic labeling, **66,** 139–140
Thormar effect, **62,** 77
Thr-203, spectral properties of green fluorescent protein site-directed mutants, **58,** 24–26, 317
Three-color laser scanning confocal microscopy, **70,** 206–207, 217–221
Three-dimensional matrix, cell culture
basement membrane, **69,** 407–408
collagen I, **69,** 405–407
Three-dimensional modeling, SPBs and organelles, **67,** 37–40
Three-dimensional plot, and histograms, **63,** 166
Three-dimensional reconstruction
chromatic aberration corrections, **70,** 279–280
CSU-10 digital signal processing, **70,** 113, 115–116
definitions, **70,** 61
image processing, **70,** 65

Three-dimensional reconstruction (*cont.*)
image registration, **70,** 63
image storage, **70,** 63–65
overview, **70,** 60–61
pixel resolution, **70,** 63
specimen preparation, **70,** 61–63
volume visualization, **70,** 65–66
web sites, **70,** 78
Xenopus oocyte cytoskeleton
brightness and contrast extremes, **70,** 402
digital image processing, **70,** 403–404
image brightness optimization, **70,** 400–402
image scale optimization, **70,** 399
objective choice, **70,** 398–399
optical sectioning, **70,** 399–400
presentation, **70,** 404–408
Threshold contours, LSC
computer-drawn, **63,** 58–59
Thrombospondins
TSP1, **69,** 8–9
TSP2, **69,** 9, 321–322
Throughput, enhanced in neural-enriched cultures, **71,** 143–144
Thymic epithelial cells, cell adhesion studies, **63,** 602
Thymidine analogs, advantages of MAbs over, **63,** 356
Thymidine analogs, in clinical studies of cell production rates, **64,** 452–474
Thymidine kinase, **65,** 400–401
Thymidine, cell cycle synchronization, **57,** 240–241
^{3}H-Thymidine labeling index, **64,** 363, 365–366
breast cancer, **64,** 368–369
correlation with chemotherapy, **64,** 373, 375–376
Thymocytes, **66,** 223
Thymus cells, phase-resolved measurements, **63,** 141–143
Tibia-1 pathway
advantages of studying, **71,** 172
axon guidance mechanisms in, **71,** 176–183
blocking antibody/peptide molecules in culture, **71,** 185
ectopic expression in developing limb bud, **71,** 185, 190–191
embryo cultures, **71,** 186–187
embryonic development of, **71,** 173–176
embryonic dissection, **71,** 186
filleting limb buds to access, **71,** 188–189
guidance molecules in, **71,** 177–183
immunochemistry protocols, **71,** 187
laboratory protocols for studying, **71,** 186–191
neuronal anatomy of, **71,** 173–176
neuronal labeling protocol, **71,** 190
peripheral to central nervous system direction, **71,** 174
procedures for analyzing, **71,** 183–186
staging grasshopper embryos for, **71,** 183
TIFF, *see* Tag Image File Format
Tilted flask
quick development assay, **66,** 479
Tilted flask protocol, **66,** 478–479
TIM machinery defects
protein import assaying, **65,** 206–207
Time
for statistical significance of antibiotic effects, **64,** 561
in flow cytometric diagnostic procedure, **64,** 553–554
role in tumor proliferation cytometry, **64,** 451
Time–date generator, video-enhanced differential interference contrast light microscopy, **56,** 178
Time-lapse imaging
fluorescence, **71,** 319
in Tibia-1 pathway studies, **71,** 185, 187
Time-lapse microscopy
data analysis, **63,** 553–554
measurement of cell migration, **63,** 550–554
Time-lapse microscopy, for *Drosophila* centrosome analysis, **67,** 121
Time-point measurements, single and multiple, **63,** 367–369
Time-resolved imaging, **65,** 367
Time-window device
earliest time resolution by, **63,** 175
for large number of measurements, **63,** 172–174
Time-zero device, for very early time point measurements, **63,** 174
Time, *see also* Pseudotime
artifacts dependent on: quality control, **63,** 178–179
list mode feature of LSCM, **63,** 77–78
measurements based on: applications, **63,** 179–181
particles separated by, **63,** 30–31
resolved fluorescence measurements, **63,** 146–147
role in sample mixing and delivery for flow cytometry, **63,** 171–175
Timing
in cell adhesion assay, **63,** 610
integrators and pulse width measurement circuits, **63,** 153–154
peak detector, **63,** 153

Timing of expression, in biolistics, **71,** 364
TIR fluorescence microscopy, *see* Total internal reflection fluorescence microscopy
Tirapazamine, induced DNA damage, **64,** 243
Tissue
  acquisition, **66,** 410
  homeostasis, **66,** 394
  preservation, **66,** 410–412, 411
Tissue culture dishes
  for peripheral neurons, **71,** 19
  in *Xenopus* cultures, **71,** 145
Tissue culture plastic, as substrate, **71,** 11–12
Tissue cultures
  cryostat sections, adhesion assays, **69,** 37–40
  synthetic provisional matrix, **69,** 156–157
Tissue dispersal, methods, **63,** 238
Tissue engineering, synthetic provisional matrix, **69,** 156
Tissue explants, culturing in *Xenopus,* 139
Tissue injury, synthetic provisional matrix, **69,** 155
Tissue sections, Ki-67 analysis in, **64,** 432
Tissue sources, for detection of MRD, **64,** 387
Tissue thickness, as parameter of biolistics, **71,** 355
Tissues
  cell membrane-anchored wild-type integrin extraction, **69,** 225–227
  culture cells, antibody staining, **67,** 307–308
  ECM antigen source, **69,** 42
  HS binding probes, **69,** 101–102
  immunoperoxidase labeling, **67,** 56–57
  instructive ECM and reporter cells, **69,** 47
  integrin extraction, **69,** 233
  procurement, **67,** 324–325
  storage, **67,** 324–325
  tumor, centrosome defects, **67,** 67–68
Titer assays, for adenovirus vectors, **71,** 400–402
Titin
  size, **58,** 252
  Z-band protein, *see* Zeugmatin–green fluorescent protein fusion protein
Titration experiments, in antibody–antigen reactions, **63,** 283–286
TLI, *see* ^{3}H-Thymidine labeling index
TMRE, **65,** 88–90; **66,** 381; **70,** 434–437, 439
  cellular potentials, **65,** 90
  mitochondrial membrane potential, **65,** 292
TNF receptor-associated factor (TRAF), **66,** 501
TNF, *see* Tumor necrosis factor
Tobacco BY-2 culture cells
  cell division studies
    advantages, **61,** 414
    DIC, **61,** 414
    mitotic index, **61,** 415
    multiplication specifications, **61,** 414
  culture methods
    media, **61,** 415
    source, **61,** 415
    synchronization
      aphidicolon, **61,** 415
      hydroxyurea, **61,** 415
      propyzamide, **61,** 415
  findings
    callose synthesis, **61,** 415
    gene introduction, **61,** 416
    KatAp microtubule motor protein, **61,** 416
    phragmoplast microtubule translocation, **61,** 415
    phragmoplastin dynamics, **61,** 416
    TKRP125 protein role, **61,** 416
Toluidine blue, nucleolus staining, **53,** 305–306
Toluidinylnaphthalene sulfonate, **66,** 264
TOM machinery defects
  yeast mutants, **65,** 206
Tomographic electron microscopy (TEM), **65,** 245–254
  application, **65,** 248–249
  computer analysis, **65,** 254
  cryosubstitution, **65,** 250–252
  data, **65,** 252–254
  frozen-hydrated specimens, **65,** 252
  historical development, **65,** 246
  mechanics, **65,** 245–246
  plastic-embedded specimens, **65,** 249–250
  specimen preparation, **65,** 249–252
  three dimensional, **65,** 246–248
Tomography, IVEM, for *Spisula* MT nucleation, **67,** 155–156
Tomography, three-dimensional reconstruction in transmission electron microscopy
  resolution, **53,** 114
  sample preparation and data collection, **53,** 114–115, 120–121, 240
  software, **53,** 113, 115
  tilt angles, **53,** 114
Topaz mutant, characterization, **58,** 25
Topography, in analysis of cell migration, **63,** 553
Topoisomerase inhibitors, **66,** 368
Total internal reflection fluorescence microscopy, single-molecule detection
  criteria of fluorophores, **58,** 50
  data acquisition and analysis, **58,** 71–72
  detectors, **58,** 52–53, 70–71
  dust background, **58,** 54–55

Total internal reflection fluorescence microscopy, single-molecule detection (*cont.*)
filters, **58,** 53–54, 68–69
illumination
geometries, **58,** 53–54
lasers, **58,** 67–68
principles, **58,** 51–52
kinesin measurements with green fluorescent protein
advantages, **58,** 64–65
assays, *in vitro* and *in vivo*, **58,** 65–67
overview, **58,** 49–51
processivity, **58,** 65
objective lens, **58,** 68–69
optical layout, **58,** 55–57, 68–69
S65T green fluorescent protein characteristics
comparison to Cy3 emission, **58,** 55, 58–62
intensity, **58,** 60–61, 63–64
noise, **58,** 61
proton transfer, **58,** 63
reversible photobleaching, **58,** 61–64
vibration isolation, **58,** 69–70
Total tissue homogenates
protein complex isolation, **65,** 240–241
Total value, LSCM list mode feature, **63,** 62–63
Touch preparation
vs. cytocentrifugation
LSC, **66,** 75
Touch preparations, MT nucleation assay, **67,** 327–328
Toxicity, two-hybrid vector test, **67,** 79
Toxicity, reduced in biolistic transfection, **71,** 354
Toxicological assays, **71,** 287–288
chick forebrain neurons in, **71,** 53
of protein transduction domains and peptides, **71,** 328–329
Toxins
mitochondria, **66,** 386
TPEF, *see* Two-photon excitation fluorescence
TPEM, *see* Two-photon excitation microscopy
TPP, *see* Tetraphenylphosphonium electrode
Tr1 cell, in Tibia-1 pathway studies, **71,** 174
Tracking mechanics
head position, **55,** 165
single-particles
elastic intercompartmental barriers, **55,** 185–188
future research directions, **55,** 192
gold labeling particle preparation, **55,** 176–178
protein motion, **55,** 176
video microscopy, **55,** 178–179
Traction forces
image collection, **69,** 333–334
microscope setup, **69,** 332–333
polyacrylamide substratum
deformation analysis, **69,** 334–335
deformation data, **69,** 334
preparation, **69,** 332
TRAF, **66,** 501
trans-interaction force, vascular endothelial cadherin, **68,** 127–128
Transbilayer lipid movement assays, **66,** 355–358
data analysis, **66,** 357
NBD-phospholipid transport measurements, **66,** 356–357
Transcription analysis, *see Xenopus*
Transcription control, **71,** 201
Transcription-dependent mtDNA replication, **65,** 414–415
Transcription factors, LSCM analysis, **64,** 437–438
Transcription foci, nuclear detection, **64,** 136–137
Transcription gene products, localization, **59,** 78–82
Transcription, promoters, **62,** 149
Transduction channel, OHC, gating, **68,** 161–165
Transfected c-Jun N-terminal kinases (JNKs)
JNK activation
in vitro kinase assay, **66,** 191–192
Transfection
affecting choice of neuronal culture, **71,** 2
biolistic, **71,** 354–367
by nuclear injection, **71,** 310–314
chick ciliary neurons, with DNA plasmids, **71,** 39
expression cloning, **69,** 215–216
FLIM, **69,** 267
focused on single or small groups of cells, **71,** 367
in B35 cell lines, **71,** 295–297
integrin-mediated signaling pathways, **69,** 379–381
low efficiencies in PC12 cell lines, **71,** 278
mediated by Chariot PTP, **71,** 328–329
mammalian cells, **67,** 187
of B35 neuroblastoma cells, **71,** 287
with GFP, **71,** 293
of PC12 cell lines, **71,** 268
of primary central and peripheral nervous system neurons by electroporation, **71,** 339–349
peripheral neurons and, **71,** 18
recombinant laminin generation, **69,** 121–123

secretion assays with human growth hormone, **71,** 271–283
Sf9 cells, with Bacmid DNA, **67,** 15
transient electrical permeabilization, **71,** 340
transient, for centrosome-associated protein expression, **67,** 132–133
vs. infection methods, **71,** 370
Transfection harvests, expanding, **71,** 399–400
Transfection protocols, **61,** 124–126
Transfection, *see* Gene delivery
Transfections
Bcl-2 family, **66,** 36–38
Transformation
antisense ribosome, **62,** 168, 533–546
co-transformation, **62,** 171, 495, 554–555
comparison among methods, **62,** 471, 487–488
drug selection, **62,** 482
electroporation, **62,** 168–169, 421, 485–500
electroporation efficiency, **62,** 487
essential genes, **62,** 513–530, 542–544
flanking selectable marker, **62,** 552
gene disruption, **62,** 284, 492–493
gene replacement, **62,** 170–172, 492, 494, 549
knockout heterokaryon, **62,** 513–530, 555
macronucleus, **62,** 163–172
microinjection, **62,** 163–165, 469–482
micronucleus, **62,** 172–174, 501–511
particle bombardment, **62,** 169, 172–173, 501–510
preparation of DNA, **62,** 488–489
random mutagenesis, **62,** 523
replacement with a tagged gene, **62,** 552
reversion, **62,** 422
rRNA gene, **62,** 469–470, 536
selectable marker, *see* Drug selection, Selectable markers
selection, *see* Drug selection
summary of methods, **62,** 163–173
vectors, **62,** 165–168
vegetative cells, **62,** 503
Transformation assays
focus assay, **66,** 216–218, 217
nonimmortal cells, **66,** 216–219
nude mice transplantation, **66,** 219
soft agar assay, **66,** 217, 218
Transformation, yeast, for two-hybrid assay, **67,** 79–80
Transformed yeast
caspases, **66,** 459
Transforming growth factor β1, TSP1, **69,** 9–10
Transforming growth factor β, effect on proliferation and hemoglobin profiles, **64,** 145–147
Transforms
digital image processing, **56,** 105–108
quantitative video microscopy, **56,** 125–126
Transgenes
expression not regulated with adenoviruses, **71,** 413
expression via replication-deficient adenovirus, **71,** 388–413
expression via retroviral infection, **71,** 369–385
in *Drosophila* studies, **71,** 220–222
stable expression of, **71,** 8
upper limit on usable size, **71,** 371
use in *Xenopus* cultures, **71,** 130
Transgenic animal, *see also* Embryonic stem cell
aggregation
embryo, preparation, **57,** 290
embryonic stem cell preparation, **57,** 290
embryonic stem cell–embryo aggregation, **57,** 291
plate preparation, **57,** 289
chimera production, overview, **57,** 289
embryonic stem cell applications in transgenic technology, **57,** 280–281, 292
germline transmission, **57,** 291
laboratory requirements for production, **57,** 281
Transgenic flies, **71,** 201
Transgenic line screening
plant mitochondria, **65,** 57–59
Transgenic models, antigen receptor, **63,** 395–396
Transgenic zebrafish
applications, **60,** 133–134
developmental mutant screening, **60,** 38
generation wth transposons, **60,** 119, 121–122, 124
green fluorescent protein
promoter analysis in living embryos
fluorescence imaging, **60,** 146
polymerase chain reaction-based mutagenesis of constructs, **60,** 145–146
promoter/enhancer construct generation, **60,** 144–147
reporter gene identification of founder fish, **60,** 143
large-insert clones in generation, **60,** 251
polymerase chain reaction, identification of offspring, **60,** 89
production
equipment, **60,** 134–135

Transgenic zebrafish (*cont.*)
founder fish identification by polymerase chain reaction, **60,** 141–142
homozygous transgenic fish, **60,** 143–144
microinjection of DNA
apparatus setup, **60,** 138
egg collection and preparation, **60,** 139–140
fish maintenance, **60,** 136–137
needle preparation, **60,** 139
plasmid preparation, **60,** 136
plate preparation, **60,** 137–138
technique, **60,** 140–141
reagents, buffers, and kits, **60,** 135–136
Transglutaminase (TGase), **66,** 112–114
activity assays, **66,** 114–115
Transient fusion, depressions, AFM studies, **68,** 43–47
Transient transfection, for centrosome-associated protein expression, **67,** 132–133
trans-interaction force, vascular endothecial cadherin, **68,** 127–128
Transition tissues, proliferation in, **64,** 467
Translation gene products, localization, **59,** 78–82
Translational studies
first generation, with proliferation markers, **64,** 377–378
human tumors, role of proliferation indices, **64,** 366–367
proliferation markers
and prognosis, **64,** 367–372
and response to systemic treatments, **64,** 372–377
Translocase activity
assays, **66,** 356
Translocation, t(9;22), **64,** 335
Transmembrane potential, increse during electroporation, **71,** 341
Transmission electron microscope, single-beam gradient laser trap, dielectric sphere force calculations
$_{00}$-mode profile effects, **55,** 16–17
$_{01}$-do-nut-mode profile effects, **55,** 17–19
Transmission electron microscopy
bacteriophage T5, STM comparison, **68,** 6–7
cells grown on filters or matrices, **57,** 304–305
monolayer cultures
dehydration, **57,** 301
embedding in plastic, **57,** 303
Epon separation from culture dish, **57,** 303
glutaraldehyde fixation, **57,** 298, 300
mounting of cells, **57,** 303–304
osmium tetroxide postfixation, **57,** 300–301
substrates for growth, **57,** 298, 304–305
uranyl acetate staining, **57,** 301, 304
suspension cultures
fixation, **57,** 305–307
harvesting, **57,** 305, 307–308
staining, **57,** 307
transient fusions at depressions, **68,** 46
Transmission electron microscopy (TEM)
cell-free nuclear assembly monitoring, **53,** 402, 448–449
chromatin and nucleosomes
in situ
buffer system selection, **53,** 169
cryoultramicrotomy, **53,** 175
data collection, **53,** 175–176
low-temperature embedding and ultraviolet polymerization, **53,** 172
Lowicryl resin handling and sectioning, **53,** 170–173
overview, **53,** 168
progressive lowering of temperature embedding, **53,** 169–171
rapid freezing and freeze substitution, **53,** 169–172
staining of sections, **53,** 173–175
in vitro
adhesion to substrate, **53,** 178–179
cryoelectron microscopy, **53,** 180–183
fixation, **53,** 177–178
heavy metal shadowing, **53,** 178
negative staining, **53,** 180
nuclei isolation, **53,** 176
overview, **53,** 168
positive staining, **53,** 178–180
soluble chromatin preparation, **53,** 176–177
fluorescence microscopy interfacing, **53,** 101–103, 110
image analysis
gray-scale mapping, **53,** 109–110
software, **53,** 109
lamin staining, **53,** 102
monitoring of lamin aggregation and polymerization, **53,** 405
nucleoporin localization with immunogold electron microscopy
cell permeabilization
digitonin, **53,** 289–290
freeze-thawing, **53,** 290–291

mechanical perforation, **53,** 290
immunolabeling
cryosections, preparation and processing, **53,** 294
oocytes, *in vitro* labeling in amphibians, **53,** 293–294
permeabilized cells and isolated nuclei, **53,** 293
plastic sections, preparation and processing, **53,** 295
mAb414 studies, **53,** 297
mapping strategies, **53,** 288–289, 296
nucleus and nuclear envelope isolation
oocytes, amphibian, **53,** 292–293
rat liver, **53,** 291–292
tissue culture cells, **53,** 292
precision of localization, **53,** 297–299
principle, **53,** 288
transmission electron microscopy, **53,** 296
recombination nodule
sample preparation from avian oocytes, **53,** 248–249
staining
mammalian nodules, **53,** 248
plant nodules, **53,** 249
specimen preparation
embedding, **53,** 104–105
fixation, **53,** 103–104
whole mount, **53,** 104–105
staining of nuclear components
DNA, **53,** 106–109
green fluorescent protein–lac repressor fusion protein staining of chromosomes, **53,** 108–109
heavy metals, **53,** 105–106
immunostaining, **53,** 107–108, 117–119
RNA, **53,** 107
synaptonemal complex
negative staining, **53,** 240–241
Saccharomyces cerevisiae complexes, **53,** 273–274
serial sectioning, **53,** 240
tomography, **53,** 240
three-dimensional reconstruction
overview, **53,** 101
serial sectioning
alignment of sections, **53,** 112–113
reproducibility of preparative steps, **53,** 112
sectioning technique, **53,** 111, 119–120
stereoviews, **53,** 110–111
tomography
resolution, **53,** 114
sample preparation and data collection, **53,** 114–115, 120–121, 175
software, **53,** 113, 115
tilt angles, **53,** 114
transcriptionally active gene visualization in *Xenopus oocytes*
grid
contrast enhancement, **53,** 488–489
deposition of dispersed nuclear contents, **53,** 486, 488
materials, **53,** 478–481
microinjection of plasmids, **53,** 483–485
microscopy parameters, **53,** 489–490
Miller chromatin spreading, **53,** 485–486, 490–491
oocyte removal, **53,** 481–482
preparation for microinjection, **53,** 482–483
principle, **53,** 474–476
promoter selection for plasmid genes, **53,** 491–492
regulatory sequence analysis, **53,** 493–495
ribonucleoprotein fibril mapping, **53,** 492–493
solution preparation, **53,** 477–478
trans-acting factor analysis, **53,** 493–494
Transmission electron microscopy, 3D
analysis and display of 3D reconstructions
continuous-density volume rendering
surface rendering, **61,** 105
volume rendering, **61,** 105
contour modeling and rendering
marching cubes algorithm software, **61,** 104
MOVIE.BYU software, **61,** 104
rendering parameters, **61,** 104
tiling, **61,** 103–104
segmentation
automatic, **61,** 102
definition, **61,** 102
interactive, **61,** 102
stacked contour uses, **61,** 103
Sterecom software and 3D tracing, **61,** 102–103
Tinkerbell interactive volume editor, **61,** 103
slice viewing
3D window cut advantages, **61,** 100
ideal viewing strategy, **61,** 100
serial section image processing, **61,** 100
tomographic image processing, **61,** 100
voxel thickness, **61,** 100
electron tomography

Transmission electron microscopy, 3D (*cont.*)
alignment methods
colloidal gold, **61,** 96
double-tilt data set, **61,** 96
markerless schemes, **61,** 96
rationale, **61,** 96
computing 3D reconstruction
back projection, **61,** 97
test volume, **61,** 97–98
weighting, **61,** 97
definition, **61,** 82–83
digitization, camera selection and specifications, **61,** 95
double-tilt scheme
angular intervals, **61,** 93
charge coupled device (CCD) camera, **61,** 93, 94, 95
colloidal gold distribution, **61,** 92
definition, **61,** 90
goniometer adjustment, **61,** 93
magnification ranges, **61,** 93
occlusion avoidance, **61,** 92
stage eccentricity adjustment, **61,** 93–94
electron dose minimization
plastic section shrinkage, **61,** 94
specimen thickness stabilization, **61,** 94
strategies, **61,** 94–95
instrumentation requirements
reference institutions, **61,** 89
tilt stage specifications, **61,** 88–89
pixel size
determination, **61,** 95
direct digital recording, **61,** 95–96
reconstruction advantages over stereo pairs, **61,** 83
serial section tomography
mass loss, **61,** 86–87
surface collapse, **61,** 87
specimen geometry limitations
anisotropic resolution, **61,** 89
conical tilting use, **61,** 89–90
elongation factor, **61,** 89
image degradation, **61,** 89
slab geometry, **61,** 89
tilt image angular range, **61,** 89, 93
specimen preparation
BODIPY-ceramide photoconversion, **61,** 87
colloidal gold size selection, **61,** 88
fiducial markers, **61,** 88
freeze substitution, **61,** 88
rapid-freezing methods, **61,** 87
selective staining, **61,** 87
tilt angles, **61,** 82–83, 85–86
tilt series normalization
correlation of gray-scale values to mass density, **61,** 97
importance in unstained specimens, **61,** 97
negative log value computation, **61,** 97
tomography *vs.* serial section analysis
accelerating voltage requirements, **61,** 86, 89
resolution limit *vs.* detection limit, **61,** 86
resolution requirement, **61,** 84
serial section depth dimension limits, **61,** 84
tomographic resolution equation, **61,** 85
tomography depth dimension limits, **61,** 84
tomography disadvantages, **61,** 86
x–y resolution, **61,** 84–85
serial section reconstruction
aligning and stacking
continuous-density volume construction, **61,** 99
contour matching, **61,** 99
fiducial markers, **61,** 99
sectioning
serial tomographic reconstruction, **61,** 98
stereo viewing, **61,** 98–99
z resolution, **61,** 98
uses, **61,** 98
software packages
SPIDER, **61,** 105
Sterecom, **61,** 102–103, 105
Transmission electron microscopy, ultrastructural analysis with, **63,** 495
Transmission EM
embryo fixation
Drosophila, **66,** 330
Transmitochondrial animal cell lines
preparation, **65,** 398
Transmitochondrial animal cell technology, **65,** 397–410
heteroplasmy, **65,** 409–410
p^0 cell generation, **65,** 399–407
transmitochondrial cybrid generation, **65,** 401–407
transmitochondrial hybrid generation, **65,** 407–409
Transmitochondrial cybrid generation
adherent cells, **65,** 402–404
cultured cells, **65,** 402–405

p^0 cells, **65,** 401–406
p^+ cells, **65,** 406–407
platelets, **65,** 405–406
suspension cells, **65,** 404–405
synaptosomes, **65,** 406
Transmitochondrial hybrid cells
microcell-mediated chromosome, **65,** 407–409
mtDNA transfer, **65,** 407–409
Transparency, generation for data presentation, **58,** 363
Transport, blood, prior to cell preparation, **63,** 208–209
Transposons
DNA transposons
frequency in genome, **60,** 105
miniature inverted-repeat transposable elements, **60,** 112–113
Tc1-like transposable elements, **60,** 109–112
retrotransposons
classification, **60,** 106
DANA, **60,** 107–108
frequency in genome, **60,** 105
LINEs, **60,** 108
SINEs, **60,** 106–107
unclassified and composite elements, **60,** 108
Transverse dynamic force microscopy
application, **68,** 353–354
overview, **68,** 352–353
Transverse dynamic force spectroscopy
application, **68,** 353–354
overview, **68,** 351–353
Transwell migration assay, in B35 cell lines, **71,** 299–300
Trapping, Mabs for instructive ECM identification, **69,** 45–47
Traps, *see* Laser scissors, Laser tweezers
Tricaine, safety in handling, **53,** 372
Trichocysts, **62,** 349
Tricine-sodium dodecyl sulfate-polyacrylamide gel electrophoresis, **65,** 237
Trifuoperazine, **66,** 264
Trigger circuit, **63,** 151–153
Trimerization
tumor necrosis factor-related ligands, **66,** 509–511
TRITC-dextran, **71,** 132
Triton X-100, **70,** 316–319
Triton X-100, for yeast SPB permeabilization, **67,** 169
Triton X-100, proteoglycan extraction, **69,** 55
tRNA, **62,** 50
Trophic factors, in chick embryos, **71,** 68
Trypan blue, **66,** 483–485
stalk cells, **66,** 484
viability assays, **66,** 214–215
Trypsin–Giemsa banding, chromosome analysis, **57,** 209–210
Trypsin, in chick ciliary neuron cultures, **71,** 42
Tryptic involucrin peptides
amino acids, **66,** 129
HPLC elution profile, **66,** 128
TSPs, *see* Thrombospondins
TTP^+
electrode, **65,** 288
Tublin dynamics, zebrafish, **59,** 218
α-Tubulin, **67,** 181–182
β-Tubulin, **67,** 181–182
γ-Tubulin
cDNA and genes, **67,** 196–200
in mammalian cells
in ectopic MT nucleation, **67,** 185–189
in MT nucleation, **67,** 179–185
in SPB activation, **67,** 173
plant
antibodies against, **67,** 200–206
functional studies, **67,** 207
immunofluorescence microscopy, **67,** 206–207
γ-Tubulin-containing complexes, **67,** 6–11
γ-Tubulin ring complex
and γTuSC, isolation, **67,** 7–8
capping assay, **67,** 20–21
direct labeling, **67,** 21–23
from *Drosophila* embryo extracts, **67,** 3–6
functional assays, **67,** 18–19
on-glass nucleation assay, **67,** 20
solution nucleation assay, **67,** 19
γ-Tubulin simplex complex, **67,** 18–19
γ-Tubulin small complex
and γTuRC, **67,** 7–8
capping assay, **67,** 20–21
direct labeling, **67,** 21–23
from *Drosophila* embryo extracts, **67,** 4–6, 10–11
on-glass nucleation assay, **67,** 20
purification, **67,** 8–10
reconstitution in Sf9 cells, **67,** 12–18
solution nucleation assay, **67,** 19
γ-Tubulin–green fluorescent protein, fusion protein studies in yeast, **58,** 99–100
γTuRC, *see* γ-Tubulin ring complex
γTuSC, *see* γ-Tubulin small complex
α-Tubulin–green fluourescent protein, fusion protein studies in yeast, **58,** 97–99

γ-Tubulin–fluourescent protein, fusion protein
studies in yeast, **58,** 99–100
Tubulin, **62,** 60–63, 291–310, 441–446
Cy3-labeled, **67,** 45–46
for *Drosophila* centrosome analysis, **67,** 116
gamma tubulin, **62,** 54, 61
preparation, **67,** 170
Tubulin, and surface antigen MC5:
simultaneous staining, **63,** 264–266
Tubulin GTPase activity, **61,** 281
Tubulins, *Xenopus* oocytes, **70,** 411
Tumor cells
difference from those in tissue
culture, **64,** 211–212
subpopulations, proliferative
conditions, **64,** 360
Tumor cells, early DNA measurements, **63,** 10
Tumor necrosis factor (TNF), **66,** 211
alpha cytotoxicity assay
Bcl-2 family, **66,** 39–40
ligands
adenovirus, **66,** 520–523
agonistic antibodies, **66,** 519
antagonistic antibodies, **66,** 519
apoptosis, **66,** 499–524
apoptosis induction, **66,** 517–520
confirmation, **66,** 519
membrane-bound ligands, **66,** 519–520
soluble ligands, **66,** 518–519
target cell, **66,** 517
cross-linking, **66,** 509–511
Dot blotting, **66,** 508–509
ELISA, **66,** 505–506, 516–517
flow cytometry, **66,** 513–514
flow cytometry-binding, **66,** 514
gel-filtration analysis, **66,** 511
HCMV, **66,** 524
immunoprecipitation, **66,** 515–516
insect cells
baculovirus, **66,** 503–504
mammalian cells, **66,** 504–505
detection and quantification, **66,** 505–509
purification, **66,** 511–512
quantitative Western blotting, **66,** 507–508
reagents, **66,** 503–512
receptor antagonism, **66,** 519
receptor binding characteristics,
66, 512–517
receptor-mediated ligand
precipitation, **66,** 514–516
recombinant soluble forms, **66,** 503
soluble Fc fusion constructs, **66,** 512
surface plasmon resonance, **66,** 517
trimerization, **66,** 509–511
viral infection, **66,** 520–524
TNFR superfamily, **66,** 500–502, 500
Tumor-specific markers, in PCR, **64,** 403–404
Tumor tissues, centrosome defects, **67,** 67–68
Tumors
endocrine therapy, **64,** 376–377
estrogen receptor-stained, **64,** 436
heterogeneity, **64,** 446–447
human
and experimental tumors, **64,** 360–361
classifications, **64,** 446
proliferation
adenocarcinomas, **64,** 462–463
overview of data on, **64,** 470–474
squamous cell carcinomas, **64,** 456–461
time-dependent parameters, **64,** 451
myoepithelial, Humatrix preparation
from, **63,** 570–571
solid
discrimination of apoptotic and necrotic
cells, **63,** 536
sample labeling and fixation, **63,** 360–361
staining procedure, **63,** 362–364
solid, *see* Solid tumors
stain concentration profile, **64,** 217
translational research, **64,** 366–377
Tunable dye lasers, **70,** 189
TUNEL
detection of DNA breaks by flow
cytometry, **63,** 518–519
flow cytometry assays
TUNEL/cell cycle analysis
filters, **57,** 269
fixation of cells, **57,** 268
labeled nucleotides, **57,** 275–276
performance of assay, **57,** 269–272
principle, **57,** 259, 267
staining, **57,** 268–269, 271–272
TUNEL/surface staining, **57,** 257–259
in situ, **63,** 519–520
principle and applications, **57,** 255–256,
257–259
study of apoptosis in plants, **63,** 506
TUNEL (TDT-mediated dTUP-biotin
nick-end labeling), **66,** 322, 324f, 394,
397–402, 399f, 405–406
Catharanthus roseus, **66,** 443
Drosophila
embryos, **66,** 326–327
eye discs, **66,** 327–328

eye discs, **66,** 329
materials, **66,** 398–401
plants, **66,** 443–447
cell cultures, **66,** 445–446
leaves, **66,** 443
reagents, **66,** 446–447
roots, **66,** 444
protocol, **66,** 401–402
vs. poly(ADP-ribose) polymerase, **66,** 99–100
TUNEL (TDT-mediated dTUP-biotin nick-end labeling) assay
commercial kits, **66,** 92
data analysis, **66,** 92–93
DNA fragmentation, **66,** 90–93
flow cytometry, **66,** 91
LSC, **66,** 91–92
materials, **66,** 90–91
Turbo Sort, **66,** 64
Two-color laser scanning confocal microscopy
argon ion lasers, **70,** 193–195, 201–204, 212–213
helium–neon lasers, **70,** 204
krypton–argon ion laser, **70,** 204–206, 225
Two-dimensional gel electrophoresis
Arabidopsis mitochondrial protein, **65,** 69
Two-hybrid analysis
Saccharomyces cerevisiae, **65,** 219
Two-hybrid system
for β-galactosidase, **67,** 81–82
functionality tests, **67,** 78–79
fusion protein expression test, **67,** 78
gene cloning into, **67,** 75–76
growth media, **67,** 79
nucleotide sequence confirmation, **67,** 76–77
self-activation test, **67,** 78
toxicity test, **67,** 79
yeast
cell mating, **67,** 80–81
SPB components, **67,** 73–74
strains, **67,** 79
transformation, **67,** 79–80
Two-photon excitation fluorescence, **70,** 441–445, 447–448
Two-photon excitation microscopy
advantages and limitations, **58,** 41
lasers, **58,** 40–42
principle, **58,** 40
Two-photon microscopy, temporal imaging, **56,** 293–298
applications, **56,** 297
data acquisition, **56,** 296–297
detector, **56,** 296–297
instrumentation, **56,** 294–297
light source, **56,** 295
optics, **56,** 295–296
scanner, **56,** 295–296
Type 5 adenovirus, structure, modification, and replication of, **71,** 389
Tyr-66 mutation of green fluorescent protein, **58,** 22–24, 299–300, 318
Tyr-203, reversible photobleaching of green fluorescent protein mutants, **58,** 16, 26
Tyr-66, spectral properties of green fluorescent protein site-directed mutants, **58,** 22–24, 299–300, 318
Tyr-154, spectral properties of green fluorescent protein site-directed mutants, **58,** 300
Tyrosine phosphorylated proteins, **63,** 617

U

Ubiquinol-cytochrome c reductase
complex II assays, **65,** 79–80
Ubiquinol:ferricytochrome c oxidoreductase, **65,** 106
Ubiquitin
web site, **66,** 249
Ultracentrifuge, **71,** 138
Ultraviolet absorption, early measurements, **63,** 4
Ultraviolet imaging, **70,** 269–271
Ultraviolet light, micropatterned culture surface, **69,** 391–392
Ultraviolet photolysis, in live-cell imaging studies, **71,** 152–154
Ultrawide low-magnification confocal imaging, **70,** 282, 287
Unbinding force, ligand–receptor recognition
distribution, **68,** 123
dynamic force microscopy, **68,** 124–125
force–distance cycle, **68,** 121–123
(Undec-1-en-1-yl)tri(ethylene glycol), **69,** 395–396
Unisomic, **62,** 191, 228
Universal Imaging Corporation, **71,** 319
Universal precautions, preparation of cells from blood, **63,** 207–208
unplugged, mutation effects in development, **60,** 31–33, 79
Unsharp masking, CSU-10 digital signal processing, **70,** 108–110, 112–113

Untreated glass, as substrate in *Xenopus* cultures, **71,** 146
Uranyl acetate, cell staining for transmission electron microscopy, **57,** 301, 304
User-interaction features, telomere length measurement, **64,** 86–88
Uterine cervix, squamous carcinomas, **64,** 459
UV imaging, *see* Ultraviolet imaging
UV microbeam irradiation, **61,** 418

V

V163A green fluorescent protein
 intensity of fluorescence emission, **58,** 318–319
 thermal stability and folding kinetics, **58,** 27, 109, 142, 329
Vacuoles
 alveolar, **62,** 34–35, 38–40
 contractile, **62,** 51–52, 86–87, 314
 food, **62,** 44, 98
valentino, gamma-ray mutagenesis, **60,** 65–66, 68
Vanadate-mediated photolysis of dynein, **62,** 434–439
Varl, **65,** 432
vas, mutation effects in development, **60,** 61
Vascular endothelial cells, cadherin, dynamic force microscopy
 binding strength, **68,** 128–129
 Ca^{2+} dependence, **68,** 127–128
 conformation, **68,** 126
 overview, **68,** 125–126
 thermodynamics and structural parameters, **68,** 129–130
 trans association, **68,** 127–128
Vascular gene expression, **59,** 313–333
Vascular smooth muscle cells
 propagation, **63,** 573–574
 spindle-shaped morphology, **63,** 576
VE cells, *see* Vascular endothelial cells
Vectors
 autonomously replicating, **62,** 495–496, 555
 recombinant caspase expression, **66,** 7
 retroviral, **66,** 203, 204
 table, **62,** 164
Velocity
 fluid stream, **63,** 45,47
 sheath stream, **63,** 29–31
Velocity gradient sedimentation, for pericentrin enrichment, **67,** 64
Ventral spinal cord (chick), isolation of, **71,** 76
Ventralized mutants, zebrafish, **59,** 160–162
Vertebrate axon guidance mechanisms, similarity with grasshopper Tibia-1 pathway, **71,** 172
Vertebrate growth cones, compared with *Heliosoma* growth cones, **71,** 159
Vertebrate retinal ganglion cells, **71,** 3
Very-high-speed full-frame confocal imaging, **70,** 104–105
Vesicle
 binding to decondensed chromatin, **53,** 433–434
 fusion in male pronuclei formation
  dextran exclusion assay, **53,** 435–436
  fluorescence assay, **53,** 435
  incubation conditions, **53,** 434
 isolation and characterization from sea urchin extracts, **53,** 426–429
 lamin immunodepletion studies, **53,** 443
 membrane protein analysis
  N-ethylmaleimide treatment of vesicles, **53,** 444–445
  extraction of proteins, **53,** 445–446
  nuclear envelope preparation, **53,** 443–444
  trypsinization of vesicles, **53,** 444–445
Vesicle dynamics, **71,** 226–227
Vesicular stomatitis virus (VSV-G) retroviral vectors, **66,** 203
Viability, **62,** 16, 211–212
 blood cells, **63,** 211–212
 cell, measuring, **63,** 244
Viability assays
 crystal violet, **66,** 215–216
 trypan blue, **66,** 214–215
Viability staining, **63,** 265
 with antitubulin antibody, **63,** 491–492
 with trypan blue, **63,** 495
Vibration, instrumental noise, **55,** 58, 154
Vibration isolation
 high-resolution multimode digital microscopy, **56,** 199
 subtraction enhancement, **56,** 80
 video-enhanced differential interference contrast light microscopy, **56,** 176
Vibratome technique
 drawbacks, **70,** 327
 future prospects, **70,** 328
 overview, **70,** 314–315, 330
 retinal cell biology, **70,** 326
Video cameras, *see also specific types*
 image processing, *see* Digital image processing
 low-light electronic cameras, **56,** 19–43
  detector features, **56,** 35–42
   color video cameras, **56,** 39–40

complementary metal-oxide
semiconductor imagers, **56,** 41–42
fast-scan cameras, **56,** 37
intensified cameras, **56,** 37–39
silicon intensified target cameras, **56,** 40–41
slow-scan cameras, **56,** 36–37
imaging device characterization parameters
camera noise, **56,** 26–28
controllers, **56,** 72–73
dynamic range, **56,** 32–33, 58–59
fill factor, **56,** 24–26
frame rate, **56,** 33–34, 62
gain, **56,** 33
imaging rate, **56,** 34–35
linearity, **56,** 31–32
quantum efficiency, **56,** 23–26
sensitivity, **56,** 23–26, 34–35
signal-to-noise ratio, **56,** 26–28
spatial resolution, **56,** 28–31, 34, 59–60
spectral response, **56,** 23
time response, **56,** 32
uniformity, **56,** 31–32
overview, **56,** 19–22, 42–43
applications, **56,** 19–20
image quality, **56,** 20–21
low-light detection methods, **56,** 21–22
microscope mating
image resolution, **56,** 147–151
light microscope adapters, **56,** 140
optical considerations, **56,** 151–152
Video-enhanced differential interference contrast
light microscopy, **56,** 153–183
image formation, **56,** 155–161
bias retardation, **56,** 160–161
image features, **56,** 158–160
microscope design, **56,** 155–158
microscope selection, **56,** 168–179
analyzer, **56,** 171–173
beam-splitting prisms, **56,** 174
camera detector, **56,** 175–176
compensation methods, **56,** 174
condenser, **56,** 173–174
considerations, **56,** 168–169
filters, **56,** 171
illumination, **56,** 169–171
magnification, **56,** 175–176
objective, **56,** 173–174
polarizer, **56,** 171–173
specimen chambers, **56,** 175
specimen stage, **56,** 174
vibration isolation, **56,** 176
overview, **56,** 153–155, 161–167
contrast enhancement
analog method, **56,** 162–165
digital method, **56,** 165–167
video basics, **56,** 161–162
performance test specimens, **56,** 179–183
diatoms, **56,** 179–180
embedded skeletal muscle, **56,** 181–183
squamous cheek cells, **56,** 180–181
vertebrate cell monolayer study, perfusion
chamber, **56,** 253–271
assembly, **56,** 270–271
chamber design, **56,** 256–260, 265–266
construction, **56,** 260–265, 267–269
historical perspectives, **56,** 256
overview, **56,** 253–256
temperature regulation, **56,** 269–270
video components
analog processor, **56,** 176–177
camera, **56,** 176–177
considerations, **56,** 168–169
digital image processor, **56,** 177–178
digital storage, **56,** 179
peripherals, **56,** 178–179
time–date generator, **56,** 178
Video microscopy
basic adjustments, **56,** 1–17
black-and-white video signal, **56,** 2–4
camera adjustments
black level, **56,** 10–12
coordinate monitor adjustments, **56,** 14–17
gain, **56,** 5–9
gamma, **56,** 6
linearity, **56,** 6
specimen illumination intensity, **56,** 9–10
monitor controls, **56,** 12–17
brightness, **56,** 13
contrast, **56,** 13–14
coordinate camera adjustments, **56,** 14–17
monitor response, **56,** 14
overview, **56,** 1–2
cameras–microscope mating
image resolution, **56,** 147–151
light microscope adapters, **56,** 140
optical considerations, **56,** 151–152
digital image processing, **56,** 63–90
analog verses digital processing, **56,** 70–73
analysis, **56,** 83–90
brightness, **56,** 86–90
image accuracy, **56,** 83–86
image processor, **56,** 89–90
optical system characterization, **56,** 87–88
shading correction, **56,** 88–89

Video microscopy (*cont.*)
fundamentals, **56,** 63–70
conventional microscopy compared, **56,** 65–66
image sampling, **56,** 68–69
imaging process, **56,** 66–67
overview, **56,** 63–64
pixels, **56,** 67–68, 82, 92–94
small object visualization, **56,** 69–70
voxels, **56,** 67–68
mechanics, **56,** 73–83
contrast enhancement, **56,** 77–79
detail enhancement, **56,** 82–83
image averaging, **56,** 81
image integration, **56,** 81–82
image subtraction, **56,** 79–80
processor components, **56,** 74–76
enhanced differential interference microscopy, *see* Video-enhanced differential interference contrast light microscopy
laser tweezer plasma membrane protein movement studies, single-particle tracking, **55,** 178–179
laser tweezers, recording system, **55,** 38–39
light microscope adapters, **56,** 140
quantitative imaging, **56,** 117–133
image definition, **56,** 118–119
point spread function, **56,** 126–127
quantitative information, **56,** 119
resolution limits, **56,** 127–132
dynamic range, **56,** 128–130
flat field correction, **56,** 131–132
intensity corrections, **56,** 127
manual camera control, **56,** 127–128
nonlinearity correction, **56,** 130
spatial correction applications, **56,** 119–125
aspect ratio, **56,** 120–121
dewarping algorithms, **56,** 124–125
image rotation, **56,** 122–123
image translation, **56,** 121–122
image warping, **56,** 123–125
resolution maximization, **56,** 119–120
temporal intensity changes, **56,** 133
transformations, **56,** 125–126
two-camera imaging, **56,** 125
Video microscopy, for centrosome reproduction assay, **67,** 275–276, 300
Video monitors
basic adjustments, **56,** 12–17
brightness, **56,** 13
contrast, **56,** 13–14
coordinate camera adjustments, **56,** 14–17
monitor response, **56,** 14
image display, digital-to-analog converter, **56,** 76
low-light microscopy, brightness, **56,** 47
Video printers, video-enhanced differential interference contrast light microscopy, **56,** 178–179
Video-rate confocal laser scanning microscopy
Bio-Rad laboratories, **70,** 43
NORAN instruments, **70,** 39–42
overview, **70,** 39
Video recorders, video-enhanced differential interference contrast light microscopy, **56,** 178–179
Vinblastine, *see* Drug role in analyzing mitosis
Vincristine, plus adriamycin, **64,** 373–374
Virakit purification, **71,** 402
VIP receptor, mapping on osteoclasts, **68,** 184
Viral concentration
retroviral transduction, **66,** 208–209
Viral infection
tumor necrosis factor-related ligands, **66,** 520–524
Viral plaque assay, for Bacmid DNA transfections, **67,** 15
Viral supernatant collection
MEFs, **66,** 206–207
Viral toxicity, **71,** 411
Viral vectors
and species specificity, **71,** 9
disadvantages compared with biolistics, **71,** 354
transfection by, **71,** 310–313
VirE2, mass measurement with scanning transmission electron microscopy, **53,** 146, 156
Virtual Embryo (http://www.ucalgary.ca), **71,** 85
Viruses, methods for purification of, **71,** 401
Viscosity, local, PFM studies, **68,** 28–29
Visible-light [Ca^{2+}] indicators, pCa_i
loading, measurement, calibration, **70,** 457–459
overview, **70,** 455–456
pitfalls, **70,** 459–460
probes, **70,** 456
Visual information, morphogenetic movements, analysis, **59,** 200
Visual scoring
plant senescence, **66,** 447
Visual systems, wild-type and mutant, **59,** 280–281
Visualization, labeled DNA, for PRINS, **64,** 65–66
Vital dye exclusion assays, **66,** 465
Vital dyes, **65,** 338
Vital staining, general principles, **59,** 181–192

Vitamins, media supplementation, **57,** 27
Vitronectin, **69,** 283
Vitronectin, in *Xenopus* cultures, **71,** 146
Voltage, as parameter of electroporation, **71,** 346
Voltage-gate ionic currents, **71,** 232
Voltage-sensitive ion channels,
AFM–path–clamp combination, **68,** 12–13
Voltage, converting current to, **63,** 123–124
Volume, particle, measurement, **63,** 329
Volume visualization, 3D
reconstruction, **70,** 65–66
Voxels
definition, **56,** 67–68
digital video imaging, **56,** 66–70
VSV-G retroviral vectors, **66,** 203

W

Washing, specimens, controversy, **63,** 232–234
Water bath, animal cell culture, **57,** 9
Water dipping, objective lens, **70,** 258
Water immersion objective lens, **70,** 264–265
Water purification, animal cell culture, **57,** 8, 26, 53
Water, SFM of DNA, **68,** 370–371
Wavelengths
higher, and background noise, **63,** 93
laser light, in flow cytometry, **63,** 21
selection, in flow cytometry, **63,** 115–118
WD, *see* Working distance
Web sites
Dictyostelium, **66,** 494
proteinase, **66,** 249
proteinase inhibitors, **66,** 249
Web sites, confocal microscopy, **70,** 77–80
Web sites, for spinal and motor neuron
cultures, **71,** 85
Website
image analysis software, **65,** 442
ORFs
yeast genome, **65,** 446
Pat Brown laboratory microarray
guide, **65,** 440
Western blot
activity, **66,** 294–296, 295
coimmunoprecipitation assay, **66,** 41–42
JNK activation
antiphospho-JNK antibody, **66,** 192
PARP-1, **66,** 292–294
electrophoresis, **66,** 292
immunoblotting, **66,** 292–293
low levels, **66,** 293–294
stripping, **66,** 294
TNF ligands, **66,** 507–508
vs. activity Western Blot, **66,** 295
Western blot analysis
cell-free nuclear assembly and
disassembly, **53,** 363, 402, 410–412
lamin aggregation and polymerization, **53,** 405, 440–441
nucleolar proteins, **53,** 43
Western blot assay, centrosome proteins,
67, 317
Western blot, probing antibody
specificity, **63,** 290–292
Western blotting, **62,** 463–464
cell–matrix adhesion assays, **69,** 372–374
instructive ECM, **69,** 47
Whole cell patch-clamp recording, **71,** 229–232
Whole-mount confocal microscopy, *Xenopus*
oocytes and eggs
antibody incubations, **70,** 391–392
bleaching, **70,** 391
borohydride reduction, **70,** 391
chromosome staining, **70,** 393–394
clearing, **70,** 394–395
handling procedures, **70,** 390–391
mounting procedures, **70,** 395–397
photobleaching, **70,** 397
secondary antibody choice, **70,** 392–393
Whole-mount embryos, embedding and
sectioning, **59,** 82–83
Whole-mount RNA, *in situ* hybridization, **59,** 263
Wide-field confocal microscopy, **70,** 249–250
WIK strain, gene mapping, **60,** 358
Wild Leitz Instrument, **71,** 96
Wild-type adenovirus infection issues, **71,** 405–406
Wild-type cybrids
ATP synthesis rates, **65,** 141
luminescence curve, **65,** 141
Wild-type strains
Dictyostelium discoideum, **66,** 478–480
Wild-type reference lines, **71,** 202
Windowing, power spectrum
interpretation, **55,** 133–134
WinMidi program, **66,** 349
WLC, *see* Worm-like chain model
Wolff, Kaspar Friedrich, **71,** 68
Working distance, objective lens, **70,** 259
World Precision Instruments
(www.wpiinc.com), **71,** 71
World Wide Web resources, zebrafish laboratory
addresses, **60,** 374–385

World Wide Web resources, zebrafish laboratory addresses (*cont.*)
Web servers, **60,** 373
ZFIN, *see* Zebrafish Information Network
Worm-like chain model, protein folding, **68,** 17
Wound healing assay, cell migration assay, **69,** 292–293
Wound healing assay, for cell migration, **63,** 551–552

X

X-biotin, **66,** 120
TGase glutamyl donor substrate proteins, **66,** 123–124
X-ray crystallography, green fluorescent protein, **58,** 20–21
X-rhodamine labeling, **61,** 159, 392
X/Y axis resolution, *see* Lateral resolution
Xanthylium dyes, probes for mitochondria, **64,** 119
Xenomitochondrial cybrid production, **65,** 406–407
Xenopus laevis, **66,** 366; **71,** 130
Xenopus
cell-free chromatin remodeling
decondensation and remodeling, incubation conditions, **53,** 507
extract preparation, **53,** 500–504
frog selection and preparation, **53,** 501, 503
nucleoplasmin role, **53,** 498–499
overview, **53,** 498
sperm chromatin preparation, **53,** 505–507
cell-free nuclear assembly
advantages, **53,** 358–359, 370
egg extract preparation
buffer preparation, **53,** 376
centrifugation, **53,** 376–377
dejellying, **53,** 375–376
laying of eggs, **53,** 374–375
stability, **53,** 378
immunofluorescence microscopy
antibodies, **53,** 382
nuclei recovery, **53,** 382–383
troubleshooting, **53,** 383–384
nuclear envelope assembly
electron microscopy, **53,** 381–382
incubation conditions, **53,** 378, 390
membrane staining, **53,** 381
model, **53,** 391–393
morphology and DNA staining, **53,** 378, 380
protein import assay, **53,** 380–381
nuclear envelope disassembly
cyclin response, **53,** 384–385
metaphase extract response, **53,** 385
nuclear envelope precursor fraction A, isolation, **53,** 386–389
nuclear envelope precursor fraction B
isolation, **53,** 386–387, 389–390
sucrose floatation of vesicles, 390, **53,**
S_{200} fraction
isolation, **53,** 386–387
properties, **53,** 387–388
sample preparation for electron microscopy, **53,** 134–137
sperm permeabilization
buffers, **53,** 371
lysolecithin treatment, **53,** 373
recovery from testes, **53,** 371–373
storage, **53,** 373–374
lamins, purification of recombinant proteins
cation-exchange chromatography, **53,** 597
chloroform/methanol precipitation, **53,** 597
expression vectors, **53,** 595–596
inclusion body solubilization, **53,** 596
Miller chromatin spreading of oocytes
braking pipettes, **53,** 480–481
comparison to other systems, **53,** 474
optimization, **53,** 490–491
solution preparation, **53,** 477–478
technique, **53,** 485–486
nuclear protein import *in vitro*
egg extract preparation, **53,** 528–529
oocyte extract preparation, **53,** 527–528
overview, **53,** 518
nucleus and nuclear envelope isolation from oocytes, **53,** 126–128, 292–293
RNA transport in oocyte system
advantages, **53,** 560, 584
enucleation, manual techniques, **53,** 572–573
immunoprecipitation analysis of RNA-associated proteins, **53,** 574–575
kinetic analysis
competition analysis, **53,** 581
initial rates, **53,** 579–580
rate-limiting step, **53,** 577
RNA cap effects, **53,** 577
transport time course, **53,** 579
microinjection
cytoplasmic injection technique, **53,** 572

incubator, **53,** 569
microinjector, **53,** 569
micromanipulator, **53,** 569
needle puller and beveler, **53,** 569
nuclear injection technique, **53,** 571
sample preparation, **53,** 570–571
stereomicroscope, **53,** 568
oocyte production
frog colony maintenance, **53,** 560, 562
isolation for microinjection, **53,** 564–565
ovariectomy, **53,** 562–564
RNA, analysis of extractions, **53,** 573–574
stability of injected RNA, **53,** 579
RNA substrate preparation
metabolic radiolabeling, **53,** 567
modified RNA synthesis, **53,** 566
purification, **53,** 566
template DNA, **53,** 565
transcription, **53,** 565–566
transcriptionally active gene visualization with transmission electron microscopy
grid
contrast enhancement, **53,** 488–489
deposition of dispersed nuclear contents, **53,** 486, 488
materials, **53,** 478–481
microinjection of plasmids, **53,** 483–485
microscopy parameters, **53,** 489–490
Miller chromatin spreading, **53,** 485–486, 490–491
oocyte removal, **53,** 481–482
preparation for microinjection, **53,** 482–483
principle, **53,** 474–476
promoter selection for plasmid genes, **53,** 491–492
regulatory sequence analysis, **53,** 493–495
ribonucleoprotein fibril mapping, **53,** 492–493
solution preparation, **53,** 477–478
trans-acting factor analysis, **53,** 493–494
Xenopus egg
centrosomal and noncentrosomal aster formation, **67,** 243
centrosome duplication assay logic, **67,** 289–290
centrosome-independent spindle assembly, **67,** 245–247
centrosome kinase assays, **67,** 315–316
centrosome reproduction, **67,** 270, 283
egg preparation, **67,** 241
extract storage, **67,** 325
mitotic extract preparation, **67,** 170
parthenogenesis *vs.* fertilization, **67,** 212–214
permeabilized sperm nuclei, **67,** 280–281
priming for extract preparation, **67,** 276
S-phase-arrested extracts, **67,** 276–279, 281
spindle assembly assays, **67,** 243–245
spindle assembly, associated proteins, **67,** 247–251
Xenopus mitotic spindle assembly and function *in vitro*
advantages of *Xenopus* extracts, **61,** 386, 401
anaphase *in vitro*
anaphase manipulation
nondegradable cyclin addition, **61,** 410
pharmacological agent addition, **61,** 410
preparation of anaphase-competent extracts
spin speed, **61,** 406
squeezed eggs, **61,** 405–406
real-time analysis
lengthening extract life span, **61,** 407
sample observation, **61,** 409–410
sample preparation, **61,** 408
slide and coverslip cleaning, **61,** 407–408
setup and monitoring anaphase reactions
extract age considerations, **61,** 407
procedure, **61,** 406
tight metaphase plate considerations, **61,** 406, 407
cytostatic factor (CSF)
calcium and CSF arrest, **61,** 386
CSF extracts, **61,** 386
EGTA chelation, **61,** 386–387
data analysis and interpretation
spindle assembly defect analysis
antibody addition effect, **61,** 404
combining antibody inhibition with dominant-negative approach, **61,** 404
genetic complementation equivalent, **61,** 405
protein depletion effect, **61,** 404
strength of observed phenotype, **61,** 405
manipulation of extracts
immunodepletion
antibody quality, **61,** 401–402
perturbation minimization, **61,** 403
preventing CSF arrest loss, **61,** 401
protocol, **61,** 402
random IgG control, **61,** 402
reagent addition
antibody amounts, **61,** 404
buffers, **61,** 403
dominant-negative protein concentration, **61,** 404
extract dilution, **61,** 403

Xenopus mitotic spindle assembly and function
in vitro (*cont.*)
uses, **61,** 403
monitoring spindle assembly reactions
pelleting spindles onto coverslips
advantages, **61,** 398
method, **61,** 1, 2, 399–400
requirements, **61,** 399
time-course experiments, **61,** 400–401
preparation of CSF extracts for spindle assembly
egg quality, **61,** 390
fluorescent tubulin preparation extract stability, **61,** 392
X-rhodamine label, **61,** 392
procedure, **61,** 391–392
protocol, **61,** 389–390
requirements
buffer and reagent stocks, **61,** 387–388
equipment, **61,** 388–389
primed frogs, **61,** 389
sperm nuclei preparation, **61,** 392
temperature requirements, **61,** 390
time requirements, **61,** 390–391
spindle assembly reactions
CSF spindle assembly procedure, **61,** 395
cycled spindle assembly procedure, **61,** 395–397
extract sensitivity consideration, **61,** 393
reagents, **61,** 395
set up
incubation temperature, **61,** 395
reaction volume, **61,** 395
types
CSF spindle assembly, **61,** 393
cycled spindle assembly, **61,** 393
DNA-coated bead assembly, **61,** 393
Xenopus motor neurons, **71,** 5
ambient laboratory temperatures for, **71,** 12
high rate of outgrowth in, **71,** 9
quantity of yield from, **71,** 11
size of, **71,** 9
suppliers of, **71,** 13
Xenopus neural tube, **71,** 3
Xenopus species, facilities for, **71,** 13
Xenopus spinal neurons
advantages of using, **71,** 130, 139–140
cocultures of, **71,** 145
culture substrata for, **71,** 145–148
culturing, **71,** 139–148
developmental properties of, **71,** 5
dissociated mixed cultures of, **71,** 141–143
easy culture of, **71,** 130
lipofection of, **71,** 136–138
live cell culture of, **71,** 129–154, 148–154
neural-enriched cultures of, **71,** 143–145
neuronal labeling of, **71,** 131–139
preparation of injection samples for, **71,** 132–133
protracted generation time of, **71,** 130
rapid rostral-to-caudal development of, **71,** 130
RNA/DNA injections of, **71,** 132–133
simple anatomy of, **71,** 130
single-cell injections into, **71,** 138–139
Xenopus, confocal microscopy
oocytes
brightness and contrast extremes, **70,** 402
digital image processing, **70,** 403–404
image brightness optimization, **70,** 400–402
image scale optimization, **70,** 399
objective choice, **70,** 398–399
optical sectioning, **70,** 399–400
presentation, **70,** 404–408
oocytes and eggs
antibody incubations, **70,** 391–392
bleaching, **70,** 391
borohydride reduction, **70,** 391
chromosome staining, **70,** 393–394
clearing, **70,** 394–395
handling procedures, **70,** 390–391
mounting procedures, **70,** 395–397
photobleaching, **70,** 397
secondary antibody choice, **70,** 392–393
oocytes and embryos
cytoskeletal and nuclear probes, **70,** 410–413
keratin filaments, **70,** 386–388
microfilaments, **70,** 388–390
microtubule fixation, **70,** 383–386
overview, **70,** 380–382
recipes and reagents, **70,** 409–410
Xenopus, maternal factors governing development, **60,** 2–3
Xenopus oocyte parthenogenetic assay, **61,** 28–29
XTT, apoptosis assay, **57,** 262

Y

Y66H green fluorescent protein, *see* Blue fluorescent protein
YAC, *see* Yeast artificial chromosome
YAC, *see* Yeast artificial chromosomes
YC, *see* Yellow cameleons
Yeast

actin purification
SDS-PAGE, **65,** 163
apoptosis protein expression, **66,** 458–462
caspase activity, **66,** 462
caspase activity measurement, **66,** 459–460
cell death assays, **66,** 462–465
extract preparation, **66,** 459–460
apoptosis proteins, **66,** 453–467
plasmids, **66,** 456
arrays
internally redundant genes, **65,** 447
Bax-induced cell death inhibition, **66,** 465–466
caspase-induced cell death inhibition, **66,** 465–466
cassette purification, **65,** 261
methods, **66,** 456–465
mitochondria
marker antigens, **65,** 265
PCR amplification, **65,** 261
PCR-based gene modification, **65,** 259
plasmids, **66,** 458
proapoptoic protein assays, **66,** 464
sedimentation assays, **65,** 164
strains and media, **66,** 456–457
transformation, **65,** 261–262
Yeast artificial chromosome
comparison with other artificial chromosome systems, **60,** 236–238, 272–273
positional cloning
candidate gene isolation
direct cDNA selection, **60,** 248
direct hybridization of clones to cDNA libraries, **60,** 249, 279, 281–282
exon trapping, **60,** 248
gene density in genome, **60,** 247
validation of genes, **60,** 250, 280–281
chromosomal walking
end-isolation of inserts, **60,** 244–246, 277–278
large-insert clone characterization, **60,** 246–247
overview, **60,** 244
genetic mapping, **60,** 243–244
overview of steps, **60,** 243
vector arms and elements, **60,** 237
zebrafish genomic libraries
availability, **60,** 239–241
hybridization-based screening, **60,** 241
polymerase chain reaction-based screening, **60,** 241
Yeast artificial chromosomes
clones
DNA prepared from, **64,** 42–43, 104–107
spanning break point, **64,** 109–111
in chromosome translocation study, **64,** 100
plugs containing, **64,** 39
Yeast cells
fixed
immunofluorescent staining, **65,** 269–274
walls
antibodies, **65,** 270–271
Yeast cells, AFM studies, **68,** 38
Yeast cells, electron microscopy
embedding protocols, **67,** 34
freezing techniques, **67,** 29–34
imaging, **67,** 34–35
immunolabeling for, **67,** 35–36
sectioning, **67,** 34–35
SPB 3D modeling, **67,** 37–40
Yeast centromeric plasmid, fusion protein construction, **58,** 89
Yeast genome
ORFs
Website, **65,** 446
Yeast homologous recombination, applications in zebrafish genetics, **60,** 250–251
Yeast-induced cell death
caspase, **66,** 10, 463
Yeast integrating plasmid, fusion protein construction, **58,** 89
Yeast mitochondria, **65,** 44–48
actin–binding assay, **65,** 150–172
immunostaining analysis, **65,** 44
isolation, **65,** 39–44
crude fraction preparation, **65,** 40
differenting centrifugation, **65,** 40–43
Nycodenz density gradient centrifugation, **65,** 43–45
motility, **65,** 333–349
growth chamber, **65,** 343–344
image acquisition, **65,** 344–346
quantitation, **65,** 346–349
yeast cell growth and synchronization, **65,** 335–337
yeast mitochondria vital staining, **65,** 337–342
outer membrane vesicle purification, **65,** 48–49
proteins, **65,** 430
purity determination, **65,** 44
quality, **65,** 44–46
RNA polymerase
core, **65,** 420
salt exraction, **65,** 47
sodium carbonate exraction, **65,** 47

Yeast mitochondria (*cont.*)
vital staining, **65,** 337–342
DNA-binding dyes, **65,** 338–339
membrane potential-sending dyes, **65,** 339–342
SYTO, **65,** 18, 339
Yeast mutants
protein import assaying, **65,** 206–207
Yeast nuclear-encoded mitochondrial proteins, **65,** 257–274
genes, **65,** 258
immunostaining, **65,** 265–274
modified gene product, **65,** 258–265
Yeast two-hybrid
Bcl-2 family, **66,** 35–36
Yeast, confocal objective lens selection, **70,** 275
Yeast, SPBs
activation, γ-tubulin role, **67,** 173
activator identification, **67,** 173–174
for two-hybrid system, **67,** 79–81
component interaction, **67,** 74
for identification, **67,** 73
interphase activation, **67,** 171–173
MT nucleation
mitotic, **67,** 171
solutions, **67,** 169–170
tubulin preparation, **67,** 170
permeabilization
cell culture, **67,** 167–168
cell wall digestion, **67,** 168–169
solutions, **67,** 167
with Triton X-100, **67,** 169
Yellow cameleons, pCa_i measurement
calibration, **70,** 468–469
expression, **70,** 464–465
measurements, **70,** 467–468
overview, **70,** 460–461
pitfalls, **70,** 469
targeting, **70,** 461–464
YFP calibration
targeted GFP mutants, **65,** 376
yobo, developmental role, **60,** 3
Yokogawa CSU-10
3D and 4D image viewing, **70,** 115–116
3D projection and stereo image generation, **70,** 113
arithmetic precision, **70,** 111–112
clam sperm, **70,** 97, 100
dandelion pollen grain, **70,** 92–93
design, **70,** 91–92
image stacks to projections, **70,** 115
image storage requirements, **70,** 116–117
low bleaching of fluorescence, **70,** 101–104
mechanical and optical performance, **70,** 117–118
neighborhood-based processing, **70,** 107–108
overview, **70,** 105–107
potentials, **70,** 118–122
Schizosaccharomycetes pombe, **70,** 97
sharpening, **70,** 108–110
slime mold, **70,** 96
Tetrahymena, **70,** 96–97
unsharp masking, **70,** 108–110
unsharp masking significance, **70,** 112–113
very-high-speed full-frame confocal imaging, **70,** 104–105
Yolk cell microtubles, zebrafish, **59,** 215–218
Yolk lipoprotein, preparation, **57,** 79
YO-PRO-1, *Xenopus* oocytes and egg chromosome staining, **70,** 393–394
you-too, mutation effects in development, **60,** 31–33, 38
Yta 10p-green fluorescent protein fluorescence patterns, **65,** 280

Z

Z-axis resolution, *see* Axial resolution
Z-series, confocal microscopy
3D reconstruction, **70,** 61–66
4D reconstruction, **70,** 66
Drosophila embryo, **70,** 366, 371
overview, **70,** 54, 56–57
z-range scanner, AFM, **68,** 177
Zebrafish
androgenic, haploid and diploid, **59,** 46–59
conclusions and perspectives, **59,** 58–59
equipment and materials, **59,** 48–50
salmonid ovarian fluid, **59,** 48–49
solutions, **59,** 50
water baths, **59,** 50
introduction, **59,** 46–48
methods, **59,** 50–54
diploid androgenotes production, **59,** 53–54
eggs, *in vitro* fertilization, **59,** 50–51
eggs, irradiation, **59,** 52
in vitro fertilization, **59,** 52–53
sperm, *in vitro* fertilization, **59,** 51–52
results and discussion, **59,** 54–58
embryos, androgenetic, **59,** 54–55
irradiation, alternative sources, **59,** 56–58
maternal gene transmission, **59,** 56

cell cultural, embryonal, **59,** 29–35
applications to other fish models, **59,** 34–35
conclusions, **59,** 35
general considerations, **59,** 30–31
introduction, **59,** 29–30
primary and multipassage, **59,** 31–33
transfection, **59,** 33–34
cell culture, primary fibroblast, **59,** 40
chromosomes, *see* Chromosomes, Zebrafish
development, analysis with explant culture assay, **59,** 127–155
assay induction with explants, **59,** 150–153
animal cap/shield induction assay, guide, **59,** 151–152
common procedures and considerations, **59,** 150–151
purified molecules, assaying, **59,** 153
explant isolation and culture, **59,** 140–150
common procedures, **59,** 140–145
guide to specific explants, **59,** 145–150
future directions, **59,** 155
general considerations, **59,** 131–132
assay method, **59,** 132
explant size, **59,** 132
questions, **59,** 132
reproducibility in isolating tissue, **59,** 132
induction and specification assays, **59,** 153–154
induction analysis, **59,** 154
specification analysis, **59,** 153–154
introduction, **59,** 128–131
application of analysis, **59,** 130–131
lineage commitment, analysis, **59,** 128–130
materials, **59,** 133–140
dissection tools, **59,** 136–139
embryos, **59,** 139–140
equipment, **59,** 133
solutions and culture media, **59,** 134–135
development, perturbation strategies, **59,** 87–108
gene product experiments, **59,** 95–104
gain-of-function studies, **59,** 96–98
loss-of-function studies, **59,** 98–104
introduction, **59,** 87–88
perspectives: inducible systems, **59,** 104–108
gene expression, regulation, **59,** 106–108
protein activity, regulation, **59,** 105–106
reagent administration, **59,** 88–89
microinjection, **59,** 88–89
other delivery methods, **59,** 89
via the medium, **59,** 88
reagents, **59,** 89–95
DNA, **59,** 93–94
oligonucleotides, **59,** 94–95
proteins and peptides, **59,** 95
RNA, **59,** 89–93
developmental screening, *see* Developmental mutant screening, zebrafish, Early pressure screens, zebrafish, Haploid screen, zebrafish, Maternal-effect genes, Gynogenesis-based screening in zebrafish, Retrovirus-mediated insertional mutagenesis, Zebrafish
ectopic gene expression, **59,** 117–125
introduction, **59,** 117–118
misexpression from DNA, **59,** 118
misexpression from RNA, **59,** 118–125
fluorescence dissection microscope, **59,** 124–125
injected mRNA, distribution, **59,** 118–120
microinjection procedures, **59,** 123–124
microinjection station, **59,** 124
mRNA synthesis, **59,** 120–121
transcription vectors, **59,** 121–123
embryo, cytoskeletal dynamics, **59,** 205–224
cleavage and blastula period, **59,** 212–215
vegetal array role, **59,** 212–215
cytoskeleton in zygote, organization, **59,** 209–212
cortical granule exocytosis, **59,** 211
ooplasmic segregation, **59,** 211–212
introduction, **59,** 205–206
methods, **59,** 219–224
tubulin dynamics, **59,** 218
unfertilized eggs, **59,** 206–209
yolk cell microtubles, **59,** 215–218
embryonic axis formation, **59,** 157–172
dorsal appearance, **59,** 157–158
dorsal–ventral patterning, **59,** 159–165
BMP signalling pathway, **59,** 163
dorsalized mutants, **59,** 162–163
mutants, **59,** 163–164
neural crest specification, **59,** 164–165
ventralized mutants, **59,** 160–162
dorsal–ventral polarity, **59,** 158–159
midline mesodermal tissues, development, **59,** 165–172
chordamesoderm mutants, **59,** 166–168
defects, **59,** 168–169
deficits in *spadetail* mutant, **59,** 169–170
double mutants, **59,** 170–171
genetic model, **59,** 171–172
prechordal plate mutants, **59,** 166

Zebrafish (*cont.*)
shield formation, **59,** 165–166
embryonic, cell cycles and development, **59,** 11–25
cleavage period, **59,** 14–16
introduction, **59,** 11–12
segmentation period, **59,** 24–25
zygote period, **59,** 12–14
embryos, **59,** 367–383
bone morphogenetic protein, reactions to antagonists, **59,** 371–372
cell ablation in anterior ectoderm, **59,** 373
dorsal neuron induction, **59,** 375
fibroblast growth factor-8, **59,** 372
germ ring interactions, **59,** 373
motorneuron subtype determination, **59,** 378–379
nervous system, simple, **59,** 368–370
neural plate patterning, **59,** 373–375
neurogenesis patterning, **59,** 380–381
neuron number control, **59,** 375–378
origins of nervous system, **59,** 370–371
perspectives, **59,** 382–383
protosegment formation, role in rhombomere organization, **59,** 381–382
system, **59,** 367–368
gene nomenclature, **60,** 159–160
genome
duplication of genes and evolution, **60,** 153, 156–159
mapping, *see also* Amplified fragment length polymorphisms, Random amplified polymorphic DNAs, Simple sequence length polymorphisms, Single strand conformation polymorphisms
comparative mapping, **60,** 149–150, 152–153, 156–157
high-resolution mapping, **60,** 268–270, 274
meiotic recombination rate mapping, **60,** 151
mouse/zebrafish somatic cell hybrids, **60,** 304, 319–320
overview of techniques, **60,** 150–151, 170–171
radiation hybrids, **60,** 298–299
strains, **60,** 358–359
Zebrafish Information Network, **60,** 344–348
orthologous chromosomes, **60,** 149, 152
size, **60,** 99–100, 150
genomic DNA preparation
cell lysis, **60,** 198
digestion, **60,** 196, 199
extraction, **60,** 198–199
rapid isolation from embryos, **60,** 270–271
hemostasis, analysis, **59,** 337–355
assay development and applications, **59,** 352–355
hemostasis applications, **59,** 353–355
introduction, **59,** 337–343
fish hemostasis, **59,** 343
hemostasis and thrombosis, disorders, **59,** 341–342
mammalian hemostasis, **59,** 338–341
model, **59,** 342–343
methods, **59,** 343–352
adult hemostasis, **59,** 349–352
coagulation factor genes, **59,** 352
embryonic and larval hemostasis, **59,** 343–349
hybrid cells, *see* Hamster/zebrafish radiation hybrid, Mouse/zebrafish somatic cell hybrid
Internet resources, *see* World Wide Web resources, Zebrafish
mutations, **59,** 259–262
system, overview, **59,** 3–8
cell and developmental biology, **59,** 4–6
conclusions, **59,** 8
future prospects, **59,** 8
genetics and genomics, **59,** 7–8
history and advantages/disadvantages, **59,** 4–6
human disease, **59,** 6–7
introduction, **59,** 3–4
organogenesis, **59,** 6–7
transgenesis, *see* Transgenic zebrafish
Zebrafish, **71,** 3
as forward genetic system, **71,** 228
compared with *Xenopus* culture, **71,** 130
Zebrafish fin, growth control, **59,** 301–333
cellular basis of growth, **59,** 308–310
checkpoints in growth, **59,** 307–308
conclusions, **59,** 310
introduction, **59,** 301–303
Zebrafish Information Network
address, **60,** 354
browser requirements and settings, **60,** 351–352
Community Resources data classes, **60,** 342, 344
data model description, **60,** 342, 344–351
data submission
authorized users, **60,** 354
staff entries, **60,** 353

updating data, **60,** 354
database design process, **60,** 340–342
Experimental Data classes, **60,** 344–348
origins, **60,** 340
passwords and security, **60,** 352–354
queries, **60,** 353
Reference Resource, **60,** 348–350
relational database searches, **60,** 350–351
Zeiss objective lenses, **70,** 284
Zeugmatin–green fluorescent protein fusion protein
antibody competition studies, **58,** 257
construction
primer design, **58,** 242–243
terminus selection for protein fusion, **58,** 241–242
fluorescence microscopy
charge-coupled device camera, **58,** 256
culture chambers, **58,** 255–256
myofibril stability analysis, **58,** 252–254
time-lapse acquisition and image processing, **58,** 256
microinjection, **58,** 247
photobleaching, **58,** 258
transfection of cells
calcium phosphate, **58,** 246–247
efficiency, **58,** 246
embryonic avian cardiomyocyte preparation, **58,** 243–244
embryonic avian myoblast preparation, **58,** 243, 245–246
lipofection, **58,** 246
troubleshooting, **58,** 257–258
ZFIN, *see* Zebrafish Information Network
Zona pellucida, laser drilling, **55,** 88–92
Zygote (http://zygote.swarthmore.edu/), **71,** 85
Zygote period, embryonic zebrafish, **59,** 12–14
Zymogen granules, AFM studies, **68,** 39
Zymogen procaspases, **66,** 94
Zymogens, **66,** 3

CONTRIBUTOR INDEX

Boldface numerals indicate volume number.

A

Adam, Myriam, **66**:469
Agard, David A., **67**:141
Ahmad, Manzoor, **66**:1
Ahmed, Zimran, **58**:75
Ainscow, Edward K., **65**:353
Akiyama, Steven K., **69**:281
Alam, Shahabuddin, **57**:69
Alberts, Bruce M., **61**:1; **67**:141
Allen, Sally Lyman, **62**:241, 561
Allen, T. D., **53**:125,
Allis, C. David, **62**:379
Alm, Eric J., **56**:91
Alnemri, Emad S., **66**:1
Altmann, S. M., **68**:311
Amemiya, Chris T., **60**:235
Amores, Angel, **60**:149, 323
Amsterdam, Adam, **60**:87
Annis, Douglas S., **69**:69
Aris, J. P., **53**:33,
Arnaudeau, Serge, **70**:453
Asai, David J., **62**:433
Ashkin, A., **55**:1
Ashwell, Jonathan D., **66**:417
Asmussen, Hannelore, **69**:341
Aten, Jacob A., **53**:455
Attardi, Giuseppe, **65**:114
Aufderheide, K., **70**:337
Autexier, Chantal, **62**:395
Ayoob, Joseph C., **58**:239

B

Balan, Irina, **71**:353
Balcar, Vladimir J., **71**:89
Balczon, Ron, **67**:257
Bamburg, James R., **71**:1, 387
Banáth, Judit P., **64**:235
Bangs, Peter, **66**:321
Barnes, David, **57**:3; **59**:29
Barrientos, Antoni, **65**:397
Bass, Peter W., **71**:17
Bastianutto, Carlo, **58**:75
Battaglia, Alessandra, **63**:433
Bauer, Kenneth D., **57**:265
Bayne, Christopher J., **57**:187
Beach, Dale L., **61**:369
Beattie, Christine E., **60**:71
Becker, Nathan, **68**:213
Bedner, Elżbieta, **63**:527; **64**:421
Bedner, Elzbieta, **66**:69
Beier, David R., **60**:185
Belmont, Andrew S., **53**:99; **58**:203
Benavente Ricardo, **53**:591
Benedict, Chris A., **66**:499
Beningo, Karen A., **69**:325
Bennett, Paul, **59**:301
Bennion, G. R., **53**:125
Benoit, Martin, **68**:91
Berditchevski, Fedor, **69**:223
Berland, Keith M., **56**:277
Berland, Keith, **56**:19
Berns, Michael W., **55**:71
Bernstein, Barbara W., **71**:37, 387
Berrios, Miguel, **53**:3, 397
Beyer, Ann L., **53**:471
Bhattacharyya, A., **62**:417
Birch-Machin, Mark A., **65**:97
Blackburn, Elizabeth, **62**:417
Blader, Patrick, **59**:87
Blomberg-Wirschell, Maureen, **58**:223
Bloom, Kerry S., **61**:369
Bloom, Kerry, **56**:185
Böhmer, Ralph M., **64**:139
Boise, Lawrence H., **66**:29
Boldogh, Istvan R., **65**:159
Bolton, Wade E., **63**:253, 487
Bolund, Lars, **64**:55
Bonilla, Eduardo, **65**:311
Bonnefoy, Nathalie, **65**:381
Bonner, Jennifer, **71**:171
Borisy, Gary, **67**:43
Bornens, Michael, **61**:13
Bornens, Michel, **67**:213, 289

Bornstein, Paul, **69**:7
Bortner, Carl D., **66**:49
Boucher, Claudia, **66**:289
Boukhelifa, Malika, **71**:287
Bowden, Emma T., **63**:613
Bräutigam, Edeltraut, **64**:487
Brand, Andrea, **58**:165
Brandhorst, Bruce P., **59**:45
Braunfeld, Michael B., **67**:141
Brelje, Clark T., **70**:165
Bridgman, Paul C., **71**:353
Brini, Marisa, **58**:75
Brinkley, William R., **61**:57
Broadie, Kendal, **71**:195
Brockenbrough, J. S., **53**:33
Brown, Anthony, **71**:305
Brown, Michael E., **71**:353
Bruns, Peter J., **62**:213, 229; 501
Burke, Brian, **53**:357
Butow, Ronald A., **65**:277, 439

C

Campagnola, Paul, **70**:429
Cande, W. Zacheus, **67**:167
Cardullo, Richard A., **56**:91
Carminati, Janet L., **58**:87
Carter, William G., **69**:27
Cassels, Grisel, **61**:297
Cassidy-Hanley, Donna, **62**:213, 229, 501
Centonze, Victoria E., **70**:129
Cepero, Enrique, **66**:29
Chalker, Douglas L., **62**:469
Charras, Guillaume, **68**:171
Checcarelli, Nicoletta, **65**:133
Chen, Aileen, **68**:301
Chen, Ming-Chyuan, **59**:227
Chevrette, Mario, **60**:303
Chiesa, Anna, **65**:353
Chilcoat, N. Doane, **62**:347
Chitnis, Ajay B., **59**:367
Cho, Sang-Joon, **68**:33
Chouaib, Salem, **66**:417
Cidlowski, John A., **66**:49
Clatch, Richard J., **64**:313
Clegg, Dennis O., **69**:163
Cohan, Christopher S., **71**:157
Coimbra, Silvia, **66**:437
Cole, Eric S., **62**:291, 313; **70**:337
Cole, Richard W., **56**:253
Collas, Philippe, **53**:417
Compton, Duane A., **67**:225
Cook, Amy, **66**:187
Cooper, Mark, **59**:179
Coopman, Peter J., **63**:613
Corley-Smith, Graham E., **59**:45
Cornillon, Sophie, **66**:469
Cossarizza, Andrea, **63**:467
Costa, Aurora, **64**:359
Crissman, Harry A., **63**:131
Critcher, Ricky, **60**:287
Cubitt, Andrew B., **58**:19
Czajkowsky, Daniel M., **68**:231

D

D'Amico, Leonard A., **59**:179
Dabiri, Guissou A., **58**:239
Dai, Jianwu, **55**:157
Daidone, Maria Grazia, **64**:359
Darzynkiewicz, Zbigniew, **63**:343, 527; **64**:421; **66**:69
Davis, Trisha N., **67**:95
Dawid, Igor B., **59**:367
Day, Richard N., **58**:293
De Giorgi, Francesca, **58**:75
de Jong, Luitzen, **53**:455
de Kroon, Anton I. P. M., **65**:37
Debec, Alain, **67**:113
deBelle, I., **53**:323
Degli Esposti, Mauro, **65**:75
DeMaggio, Susan, **70**:475
Demaurex, Nicolas, **70**:453
den Eijnde, Stefan van, **66**:339
Dentler, William, **62**:323
Deptala, Andrzej, **64**:385, 421
Dernburg, Abby F., **53**:187,
Desai, Arshad, **61**:155, 385
Detrich, III, William H., **59**:3, 227; **60**:361
Dey, Runu, **65**:397
Dickinson, L. A., **53**:323
Diekert, Kerstin, **65**:37
Dingwall, Colin, **53**:517
Dobrowsky, Rick T., **66**:135
Dodge Zantek, Nicole, **63**:549
Doerry, Eckehard, **60**:339
Doležel, Jaroslav, **64**:3
Donahue, Christopher J., **57**:265
Dong, Chen Y., **56**:277
Douglas, Sarah A., **60**:339
Dove, J. E., **53**:33

Doxsey, Stephen J., **58**:223; **67**:53
Dunn, Kenneth, **56**:217
Durand, Ralph E., **64**:211, 235
Dykens, James A., **65**:285

E

Earnshaw, William C., **66**:289
Eble, Johannes A., **69**:223
Edwards, Pat, **60**:373
Eisen, Judith S., **60**:71
Ekker, Marc, **60**:303
Ekker, Stephen C., **59**:117
Ellenberg J., **58**:261
Emanuelsson, Olof, **65**:175
Endl, Elmar, **63**:399
Endow, Sharyn A., **58**:153
Engebrecht, Jo Anne, **53**:257
Engel, Andreas, **68**:257
Ensrud, Kathy, **61**:233
Epstein, Charles B., **65**:439

F

Fésüs, László, **66**:111
Fan, Qichang, **62**:533
Faragher, Alison J., **67**:305
Fath, Angelika, **66**:437
Fattorossi, Andrea, **63**:433; **64**:153
Fearnhead, Howard O., **66**:167
Fehrenbacher, Kammy, **65**:453
Fekete, Donna M., **71**:369
Felgner, Harald, **55**:195
Ferlini, Cristiano, **63**:433
Fernandes-Alnemri, Teresa, **66**:1
Ferrigno, Paul, **58**:107
Finer, Jeffrey T., **55**:47
Fisher, Paul A., **53**:23, 397
Fishman, Mark C., **59**:359; **60**:235
Florin, Ernst-Ludwig, **68**:193
Foernzler, Dorothee, **60**:185
Force, Allan, **60**:149
Forsgren, Per-Ola, **70**:149
Fox, Thomas D., **65**:381
Franc, Nathalie, **66**:321
Frankel, Joseph, **62**:27
French, Todd, **56**:19, 277
Fritz, Andreas, **59**:253
Fry, Andrew M., **67**:305
Fujiwara, Takahiro, **55**:173
Funatsu, Takashi, **55**:117
Fung, Jingly, **64**:97

G

Gabridge, Michael G., **57**:49
Gaertig, Jacek, **62**:485, 513
Galande, S., **53**:323
Galbraith, David W., **58**:315
Gallo, Gianluca, **71**:325
Garcés, Jorge A., **62**:333
Gard, David L., **70**:379
Gates, Michael A., **60**:165
Gavin, Ray H., **62**:333
Gerdes, Johannes, **63**:399
Gerrow, Kimberly A., **71**:171
Gibson, Tracie M., **62**:433
Giddings Jr., Thomas H., **67**:27
Gil, Susana G., **69**:27
Gilley, Dave, **62**:417
Ginsberg, Mark, **69**:209
Giocondi, Marie-Cécile, **68**:51
Gittes, Frederick, **55**:129
Givan, Alice L., **63**:19
Gliksman, Neal, **56**:63
Gobeil, Stéphane, **66**:289
Goldberg, M. W., **53**:125
Goldfarb, David S., **53**:545, 559
Goldstein, Joshua C., **66**:365
Golstein, Pierre, **66**:469
Gómez, Timothy M., **71**:129
Gong, Zhiyuan, **60**:213
Gonzalez, J. A., **71**:387
Goodwin, Paul C., **58**:343
Gorczyca, Wojciech, **64**:421
Gorovsky, Martin A., **62**:513, 549
Gräf, Ralph, **67**:337
Granato, Michael, **60**:21
Gratton, Enrico, **56**:277
Green, Douglas R., **66**:365
Greene, Lloyd, **66**:417
Grimellec, Christian Le, **68**:51
Grinblat, Yevgenya, **59**:127
Grishanin, R. N., **71**:267
Grolig, Franz, **55**:195
Grote, Monika, **53**:287
Gunawardane, Ruwanthi N., **67**:1

H

Hörber, Heinrich J. K., **68**:1
Hackett, Perry B., **60**:99
Haddad, Alex, **62**:347
Hadjantonakis, Anna-Katerina, **57**:279
Haffter, Pascal, **60**:21

Hai, Bing, **62**:513
Hainfield, James F., **53**:139
Hale IV, Walker, **65**:439
Hale, Irene L., **70**:301
Hamilton, Eileen P., **62**:189, 219, 253, 265
Hammerschmidt, Matthias, **59**:87
Hansma, Helen G., **68**:213; **69**:163
Hanyu, Kazuko, **62**:455
Harada, Yoshie, **55**:117
Harrigan, Dan, **71**:129
Hasbold, Jhagvaral, **63**:375
Haseloff, James, **58**:139
Hay, Robert J., **57**:31
He, Yan, **71**:17
Heald, Rebecca, **67**:241
Heck, Margarete M. S., **61**:317
Heidemann, Steven R., **71**:51
Heim, Roger, **58**:19
Helmrich, Angela, **57**:3; **59**:29
Henion, Paul D., **60**:71
Henley, John, **71**:129
Hennessey, Todd M., **62**:363
Henry, Clarissa A., **59**:179
Hepler, Peter K., **61**:413
Herrmann, Johannes M., **65**:217, 429
Higashikubo, Ryuji, **64**:269
Higuchi, Hideo, **55**:117
Hinchcliffe, Edward H., **56**:1; **61**:439; **67**:269
Hindkjaer, Johnny, **64**:55
Hinsch, Jan, **56**:147
Hinterdorfer, Peter, **68**:115
Hiraoka, Yasushi, **67**:167
Hirschberg, K., **58**:261
Hodgkin, Philip D., **63**:375
Hoey, John G., **62**:333
Hoffman, Robert A., **63**:299
Hofmann, Marie-Claude C., **57**:93
Hollenbeck, Peter J., **71**:1,339,369
Hollmann, Christiane, **63**:399
Holmes, Kevin L., **63**:185
Hopkins, Nancy, **60**:87
Horowitz, R. A., **53**:167
Horton, Mike, **68**:171
Horwitz, Alan F., **69**:341
Hu, Yanhui, **66**:417
Hukku, Bharati, **57**:203
Humphris, A. D. L., **68**:337
Huwiler, Kristin G., **69**:69
Hyatt, Tana M., **59**:117
Hyman, Tony, **61**:137, 145

I

Ingber, Donald, **69**:385
Inoué, Shinya, **70**:87
Inoué, Ted, **56**:63; **70**:87
Iozzo, Renato V., **69**:53
Ishii, Yoshiharu, **55**:117
Ivics, Zoltan, **60**:99
Izsvak, Zsuzsanna, **60**:99

J

Jacobberger, James W., **63**:271
Jacobson, Ken, **56**:19
Jagadeeswaran, Pudur, **59**:337
Janke, C., **67**:71
Jeffery, Peter L., **71**:89
Jena, Bhanu P., **68**:33
Jensen, Peter Østrup, **64**:129
Jessen, Jason R., **60**:133
Johnson, Bryan W., **66**:29
Johnson, Stephen L., **59**:301; **60**:357
Johnston, Peter J., **64**:235
Joly, Lucille, **60**:303
Jones, Alan M., **66**:437
Jordan, Mary Ann, **61**:267
Joshi, Harish C., **67**:179
Jouaville, Laurence Sophie, **58**:75
Jowett, Trevor, **59**:63
Juan, Gloria, **63**:343

K

Kahle, Hanna, **64**:487
Kamentsky, Louis A., **63**:51
Kane, Donald A., **59**:11; **60**:361
Kaplan, Joseph, **57**:203
Kaplan, Ken, **61**:145
Kapler, Geoffrey, **62**:485
Karnes, James L., **71**:157
Karrer, Kathleen M., **62**:127, 565
Katakura, Yoshinori, **57**:69, 111
Kay, Steve A., **58**:283
Keller, H. Ernst, **56**:135
Kestens, Luc, **64**:487
Kinch, Michael S., **63**:549, 599
Kindt, Johannes H., **68**:213
Kirk, K., **62**:417
Kirkpatrick, Arthur E., **60**:339
Kiseleva, E., **53**:125
Kispal, Gyula, **65**:37
Klein, Franz, **53**:257

Kluck, Ruth M., **66**:365
Knobel, Susan M., **58**:31
Koester, Steven K., **63**:253, 487
Kohwi, Y., **53**:323
Kohwi-Shigematsu, T., **53**:323
Koitschev, Assen, **68**:141
Kokkoli, Efrosini, **69**:163
Kolesnick, Richard N., **66**:135
Kolvraa, Steen, **64**:55
Koyama, Carol, **67**:113
Kreidberg, Jordan A., **69**:309
Krishan, Awtar, **64**:193
Kroemer, Guido, **65**:147
Krohne, Georg, **53**:591
Kubaláková, Marie, **64**:3
Kubitscheck, Ulrich, **53**:79
Kuhn, Thomas B., **71**:67
Kuo, Scot C., **55**:43
Kurada, Phani, **66**:321
Kuriyama, Ryoko, **61**:233; **67**:125
Kurtz, Ira, **70**:417
Kuruvilla, Heather G., **62**:363
Kusumi, Akihiro, **55**:173
Kwok, Cheni, **60**:287

L

Lacombe, Francis, **64**:343
Lagroye, Isabelle, **64**:251
Lamoureux, Phillip, **71**:51
Landberg, Göran, **63**:419
Lane, Mary Ellen, **59**:127
Langer, Matthias G., **68**:141
Langford, Kevin, J., **69**:297
Lansdorp, Peter M., **64**:69
Lantz, Larry M., **63**:185
Larsen, Jørgen K., **63**:419; **64**:129
Larsen, Jacob, **64**:129
Leaver, C. J., **65**:53
LeDuc, Philip, **69**:385
Lehenkari, Petri, **68**:171
Lenaz, Giorgio, **65**:1
Lenne, P.-F., **68**:311
Leno, Gregory H., **53**:497
Lesniewska, Eric, **68**:51
Levraud, Jean-Pierre, **66**:469
Lewis, Aaron, **70**:429
Lewis, Joseph, **66**:187
Li, Gang, **58**:203
Li, Ronghao, **57**:167
Li, Xun, **64**:421; **66**:69
Liang, Hong, **55**:71
Liao, Eric C., **60**:181
Liddell, A., **65**:53
Lill, Roland, **65**:37
Lin, Shuo, **60**:133
Lincoln, Carolyn Kay, **57**:49
Lingle, Wilma L., **67**:325
Lippincott-Schwartz, J., **58**:261
Lisi, Simonetta, **66**:321
Liu, Yuan C., **59**:337
Liu, Zheng-gang, **66**:187
Lo, Chun-Min, **69**:325
Lobb, Roy R., **69**:17
Loew, Leslie M., **70**:429
Lohka, M. J., **53**:367
Loidl, Josef, **53**:257
Loo, Deryk T., **57**:251
Lowe, Scott W., **66**:197
Lucretti, Sergio, **64**:3
Lyons, Bruce A., **63**:375
Lysák, Martin A., **64**:3

M

Mádi, András, **66**:111
Métivier, Didier, **65**:147
Müller, Daniel J., **68**:257
Müller, Otto, **55**:195
Macas, Jiří, **64**:3
Maddox, Paul S., **56**:185
Maddox, Paul, **61**:369
Maisse, Carine, **65**:147
Majlof, Lars, **70**:149
Malicki, Jarema, **59**:273
Maloney, M. T., **71**:387
Malyapa, Robert S., **64**:251
Maness, Patricia, **71**:287
Manders, Erik M. M., **53**:455
Manfredi, Giovanni, **65**:133
Mannella, Carmen A., **65**:245
Marekov, Lyuben N., **66**:111
Marko, Michael, **61**:81
Marsault, Robert, **58**:75
Marsh, T. C., **70**:337
Martin, T. F. J., **71**:267
Martinez, Cecilia Y., **71**:339
Mastronarde, David N., **67**:27
Masuda, Hirohisa, **67**:167
Mather, Jennie P., **57**:19, 219, 265
Mathus, Todd, **69**:111
Matsumoto, Brian, **70**:301
Matthies, Heinrich J. G., **71**:195

Maxfield, Frederick R., **56**:217
Mayer, Sharon P., **64**:385
Mc Closkey, Thomas W., **64**:567
McConkey, David J., **66**:229
McCoy Jr., Philip J., **63**:207
McCurrach, Mila E., **66**:197
McEwen, Bruce F., **61**:81
McIntosh, J. Richard, **67**:27
Meany, Jennifer A., **71**:87
Meberg, Peter J., **71**:111
Mehta, Amit D., **55**:47
Meissner, Hildegard, **57**:147
Melamed, Myron R., **63**:3; **64**:421
Meng, Anming, **60**:133
Merrill, Gary F., **57**:229
Midwood, Kim S., **69**:145
Miles, M. J., **68**:337
Milhiet, Pierre Emmanuel, **68**:51
Millan, Jose Luis, **57**:93
Millar, A. H., **65**:53
Miller, C. D., **58**:261
Miller, Frederick J., **61**:439
Miller, Matthew W., **71**:111
Minamide, L. S., **71**:387
Miranti, Cindy, **69**:359
Misenheimer, Tina M., **69**:69
Mitchison, Timothy J., **61**:385
Mizzen, Craig A., **62**:379
Moens, Cecilia B., **59**:253
Molloy, Justin E., **55**:205
Moore, Alison, **57**:265
Moraes, Carlos T., **65**:397
Moritz, Michelle, **61**:1; **67**:141
Morphew, Mary, **67**:27
Mosher, Deane F., **69**:69
Moudjou, Mohammed, **61**:13
Moy, Vincent T., **68**:301
Mueller, Susette C., **63**:613
Mullins, Mary C., **59**:157
Munné, Santiago, **64**:97
Munshi, H. G., **69**:195
Murase, Shin-ichi, **69**:341
Murgia, Marta, **58**:75
Murray, Andrew, **61**:385
Mykles, Donald L., **66**:247

N

Naber, Stephen, **66**:393
Nadezhdina, Elena, **67**:43
Nagy, András, **57**:279
Naini, Ali, **65**:133
Nanney, David L., **62**:3
Nemes, Zoltán, **66**:111
Neupert, Walter, **65**:217, 429
Newmeyer, Don D., **66**:365
Ng, Tony, **69**:261
Ngo, Kha, **71**:51
Nicklas, Bruce R., **61**:209
Norris, Paula S., **66**:499
Nowakowski, Dan W., **65**:257
Numata, Osamu, **62**:455
Nutt, Leta K., **66**:229

O

Oakley, Berl R., **61**:317; **67**:195
Ochs, Robert L., **53**:303
O'Connor, Timothy P., **71**:171
Ödenthal, Jorg, **60**:21
Oegema, Karen, **67**:1
Ohta, Toshiro, **67**:125
Okamoto, Koji, **65**:277
Oldenbourg, Rudolf, **61**:175
Olive, Peggy L., **64**:211, 235
Opas, Michal, **70**:453
Orias, Eduardo, **62**:189, 219, 253, 265
Orias, Judith D., **62**:189
Orme-Johnson, Nannette, **65**:483
Oroudjev, Emin, **68**:213; **69**:163
Osborne, Barbara A., **66**:417
Osheim, Yvonne N., **53**:471
Oshiro, Masafumi, **56**:45
Ostuni, Emanuele, **69**:385
Otey, Carol A., **71**:287
O'Toole, Eileen T., **67**:27
Otto, Friedrich, **64**:487
Ovechkina, Yulia, **67**:195

P

Paddock, Stephen W., **70**:361
Paddy, Michael R., **53**:49
Palacios, Isabel, **53**:517
Palazzo, Robert E., **61**:35; **67**:149
Pallotti, Francesco, **65**:1
Parker, Leon H., **59**:313
Parsons, Maddy, **69**:261
Patterson, George H., **58**:31
Paw, Barry H., **59**:39
Pelegri, Francisco, **60**:1
Peloquin, John, **67**:43
Peng, Gang, **67**:125

Pennock, David G., **62**:281
Periasamy, Ammasi, **58**:293
Perlman, Philip S., **65**:277
Peters, Reiner, **53**:79, 287
Pfanner, Nikolaus, **65**:189
Phillips, David M., **57**:297
Piacentini, Mauro, **66**:111
Piel, Matthieu, **67**:289
Pierce, Daniel W., **58**:49
Pierce, Robert H., **64**:117
Pierelli, Luca, **64**:153
Pietrasanta, Lia I., **68**:213
Pihan, German A., **67**:53
Pinton, Paolo, **58**:75; **65**:353
Pirity, Melinda, **57**:279
Piston, David W., **58**:31
Plautz, Jeffrey D., **58**:283
Poccia, Dominic, **53**:417
Poirier, Guy G., **66**:289
Pon, Liza A., **65**:159, 257, 333
Poon, Steven S. S., **64**:69
Poot, Martin, **64**:117
Porcelli, Anna M., **65**:353
Postlethwait, John H., **59**:45; **60**:149, 165, 323
Pozzan, Tullio, **58**:75
Pralle, Arnd, **68**:193
Prendergast, F. G., **58**:1
Presley, J. F., **58**:261
Purohit, Aruna, **67**:53

R

Radmacher, Manfred, **68**:67
Raible, David W., **60**:71
Rajwa, Bartlomiej, **63**:583
Ramos, Joe W., **69**:209
Randolph, Caterina, **62**:469
Ransom, David G., **60**:195, 365
Rapraeger, Alan C., **69**:83
Reed, John C., **66**:453
Rew, David A., **64**:445
Reynolds, Matthew, **71**:51
Rieder, Conly L., **56**:253; **61**:297
Riepert, S., **53**:125
Riggs, Blake, **67**:113
Rillema, Jill R., **57**:251
Ringlien, W., **70**:337
Rizzuto, Rosario, **58**:75; **65**:353
Roberts, Paul M., **53**:545
Robinett, Carmen, **58**:203
Robinson, J. Paul, **63**:89, 583; **64**:511
Robles, Estuardo, **71**:129
Rodionov, Vladimir, **67**:43
Rooney, Isabelle A., **66**:499
Roos, Göran, **63**:419
Roti Roti, Joseph L., **64**:251, 269
Rugolo, Michela, **65**:353
Rutherford, S. A., **53**:125
Rutter, Guy A., **65**:353
Ryan, Michael T., **65**:189

S

Sagerström, Charles, **59**:127
Saito, Kiwamu, **55**:117
Sako, Yasushi, **55**:173
Saleh, Ayman, **66**:1
Salisbury, Jeffrey L., **67**:325
Salmon, E. D., **56**:153,185
Salmon, Edward D., **61**:155, 369
Salvioli, Stefano, **63**:467
Samejima, Kumiko, **66**:289
Samorì, Bruno, **68**:357
Sanderson, Ralph D., **69**:297
Sanger, Jean M., **58**:239
Sanger, Joseph W., **58**:239
Sarmiere, P. D., **71**:387
Sawin, Kenneth E., **58**:123
Scambia, Giovanni, **64**:153
Schägger, Hermann, **65**:231
Schiebel, E., **67**:71
Schier, Alexander F., **60**:259
Schlegel, Robert A., **66**:339
Schliwa, Manfred, **55**:195
Schmidt, Christoph F., **55**:129
Schmitt, Karin, **60**:287
Schnackenberg, Bradley J., **67**:149
Schneider, Gisbert, **65**:175
Schon, Eric A., **65**:457, 461, 463
Schramm, C., **67**:71
Schulte-Merker, Stefan, **60**:1
Schwartz, Lawrence M., **66**:393, 417
Schwartz, Martin A., **69**:13
Schwarzbauer, Jean E., **69**:145
Seamer, Larry, **63**:169
Sedat, John W., **53**:187
Seidel-Rogol, Bonnie L., **65**:413
Serluca, Fabrizio C., **59**:359
Severin, Fedor, **61**:145
Shadel, Gerald S., **65**:413
Shao, Zhifeng, **68**:231, 243
Shapiro, Howard M., **63**:107, 149
Shaw, A. E., **71**:387
Shaw, Sidney L., **56**:185; **61**:369

Sheehan, John P., **59**:337
Sheetz, Michael P., **55**:29, 157
Shelby, Richard D., **58**:183
Sheng, Sitong, **68**:243
Shibata, Takehiko, **67**:167
Shirahata, Sanetaka, **57**:69, 111
Sigle, Randy O., **69**:27
Silver, Pamela A., **58**:107
Silver, Randi B., **56**:237
Silverman, Gary A., **60**:235
Silvestrini, Rosella, **64**:359
Šimková, Hana, **64**:3
Simon, Ellen M., **62**:3
Simon, Martha N., **53**:139
Simon, Viviana, **65**:333
Sincock, S. A., **64**:511
Sitko, John C., **68**:213
Sive, Hazel, **59**:127
Sklar, Larry A., **63**:169
Sluder, Greenfield, **56**:1; **61**:439; **67**:269
Smirnov, Sergei, **69**:111
Smith, Helen R., **62**:213
Smith, Jeffrey W., **69**:249
Smith, Paul J., **64**:173
Sneider, J. M., **71**:387
So, Peter T. C., **56**:277
Solari, Alberto J., **53**:235
Solnica-Krezel, Lilianna, **59**:205
Sorenson, Robert L., **70**:165
Sorger, Peter, **61**:145
Sottomayor, Mariana, **66**:437
Spinazzola, Antonella, **65**:133
Spitz, Douglas R., **64**:269
Spudich, James A., **55**:47
Srinivasula, Srinivasa M., **66**:1
Stack, Sharon M., **69**:195
Stainier, Didier Y. R., **59**:313
Stamenkovic, Ivan, **66**:307
Stap, Jan, **53**:455
Stearns, Tim, **58**:87
Steen, Harald B., **64**:539, 553
Steinert, Peter M., **66**:111
Steinkamp, John A., **63**:131
Sterba, Ronald E., **55**:29
Stewart, Carleton C., **63**:217; **64**:289
Stewart, Sigrid J., **63**:217; **64**:289
Stout, Amy K., **65**:285
Stout, Andrea L., **55**:99
Strähle, Uwe, **59**:87
Streuli, Charles H., **69**:403
Stuart, K. R., **70**:337
Stuart, Kathleen R., **62**:291, 313
Sudlow, Gail, **58**:203
Sullivan, Kevin F., **58**:183; **61**:113
Sullivan, William, **67**:113
Swayne, Theresa C., **65**:257, 333
Sweeney, Rosemary, **62**:533

T

Tadir, Yona, **55**:71
Takada, Saeko, **67**:167
Takeda, Tetsuya, **62**:455
Talbot, William S., **60**:259
Tanji, Kurenai, **65**:311
Tellis, Patricia, **60**:303
Terns, Michael P., **53**:559
Terry, Nicholas H. A., **63**:355
Teruya, Kiichiro, **57**:111
Theurkauf, William, **61**:317
Thomas, Howard, **66**:437
Tirrell, Matthew, **69**:163
Tokunaga, Makio, **55**:117
Tolhurst, Ornella, **71**:89
Tomishige, Michio, **55**:173
Topczewski, Jacek, **59**:205
Tournier, Frédéric, **67**:213
Traganos, Frank, **63**:343, 527
Tram, Uyen, **67**:113
Tran, Phong, **56**:153
Tromberg, Bruce, **55**:71
Turkewitz, Aaron P., **62**:347
Turnacioglu, Kenan K., **58**:239
Turnbull, Douglass M., **65**:97

V

Valavanis, Christos, **66**:393, 417
Vale, Ronald D., **58**:49
Valet, Günter, **64**:487
Valster, Aline H., **61**:413
van Driel, Roel, **53**:455
van Eeden, Fredericus J. M., **60**:21
Van Hooser, A., **61**:57
VanderWaal, Robert P., **64**:269
Verbsky, John W., **62**:347
Vial, J-P., **64**:343
Viani, Mario B., **68**:213
Villani, Gaetano, **65**:119
Vogel, Jacalyn M., **61**:35
Vogel, Jacalyn, **67**:125
von Heijne, Gunnar, **65**:175
Voos, Wolfgang, **65**:189

Vos, Jan W., **61**:413
Voytik-Harbin, Sherry L., **63**:561, 583

W

Wadsworth, Patricia, **61**:219
Wagenbach, Mike, **61**:245
Walberg, Mette, **64**:553
Walczak, Claire E., **61**:385
Walker, Charline, **59**:45; **60**:43
Wall, Joseph S., **53**:139
Wang, H., **62**:417
Wang, Tzu-Hao, **66**:187
Wang, Yu-li, **56**:305; **57**:313
Wang, Yu-Li, **69**:325
Ward, John G., **62**:469
Ware, Carl F., **66**:499
Ware, Tracy, **62**:417
Watanabe, Yoshio, **62**:455
Waterhouse, Nigel J., **66**:365
Waterman-Storer, Clare M., **56**:185
Waterman-Storer, Clare, **61**:155
Waters, Jennifer, **56**:185
Watkin, Harriet, **69**:403
Webb, Donna J., **69**:341
Webb, Watt W., **55**:99
Weier, Heinz-Ulli G., **64**:33, 97
Weinberg, Ronald P., **71**:89
Weir, Iona E., **63**:505
Wessendorf, Martin W., **70**:165
Westerfield, Monte, **59**:3; **60**:339
Westermann, Benedikt, **65**:217, 429
Westphal, Manfred, **57**:147
Wheatley, Sally P., **57**:313
White, R. Allen, **63**:355
White, Kristin, **66**:321
Whitelock, John M., **69**:53
Whitesides, George, **69**:385
Wierzbicka-Patynowski, Iwona, **69**:145
Wiese, C., **53**:125
Wiese, Christiane, **67**:1
Wiggan, O. P., **71**:387
Wignall, Sarah M., **67**:241
Wiley, Emily A., **62**:379
Williams, Norman E., **62**:441, 449
Williamson, Patrick, **66**:339
Wilson, Leslie, **61**:267
Wiltshire, Marie, **64**:173
Winey, Mark, **67**:27
Wittmann, Torsten, **61**:137
Wolf, David E., **56**:117
Woodcock, C. L., **53**:167
Woodruff, Teresa K., **57**:333
Woollenweber, Leslie A., **58**:19
Wordeman, Linda, **61**:245
Wright, David J., **70**:1
Wright, Shirley J., **70**:1
Wright, William D., **64**:251, 269
Wuitschick, Jeffrey D., **62**:565
Wuskell, Joseph P., **70**:429

X

Xu, Mai, **64**:269

Y

Yan, Yi-Lin, **60**:149, 165
Yanagida, Toshio, **55**:117
Yang, Hyeong-Cheol, **65**:333
Yang, Yili, **66**:417
Yao, Meng-Chao, **62**:469, 533
Ye, Fengchun, **60**:303
Yeh, Elaine, **56**:185; **61**:369
Yip, Kay-Pong, **70**:417
Yu, Lanlan, **62**:549
Yurchenco, Peter D., **69**:111

Z

Zaal, K. J. M., **58**:261
Zamzami, Naoufal, **65**:147
Zhang, Dahong, **61**:209
Zhang, Hong, **66**:453
Zhang, Peng, **64**:251
Zheng, Yixian, **67**:1
Zhong, Tao P., **60**:235
Zhou, Feng-Quan, **71**:157
Zhou, Jun, **67**:179
Zon, Leonard I., **59**:3, 39, 313; **60**:181, 195, 235, 357, 361, 365
Zuccheri, Giampaolo, **68**:357

Printed in Dunstable, United Kingdom

91897607R00147